TEXTILE SCIENCE AND TECHNOLOGY 11

TEXTILE PROCESSING AND PROPERTIES

Preparation, Dyeing, Finishing and Performance

TEXTILE SCIENCE AND TECHNOLOGY

TEXTILE SCIENCE AND TECHNOLOGY 11

TEXTILE PROCESSING AND PROPERTIES

Preparation, Dyeing, Finishing and Performance

BY

TYRONE L. VIGO

*U.S. Department of Agriculture, Agricultural Research Service,
Southern Regional Research Center, 1100 R.E. Lee Boulevard,
New Orleans, LA, U.S.A.*

ELSEVIER
Amsterdam – Boston – London – New York – Oxford – Paris – San Diego
San Francisco – Singapore – Sydney – Tokyo

ELSEVIER SCIENCE B.V.
Sara Burgerhartstraat 25
P.O. Box 211, 1000 AE Amsterdam, The Netherlands

First edition 1994
Second impression: 1997
Third impression: 2002

Library of Congress Cataloging in Publication Data
A catalog record from the Library of Congress has been applied for.

British Library Cataloguing in Publication Data
A catalogue record from the British Library has been applied for.

ISBN: 0 444 88224 3 (hardbound)
ISBN: 0 444 82623 8 (paperback)

Transferred to digital printing 2005

TO MY SPOUSE AND BEST FRIEND *EILEEN* AND
IN MEMORY OF HER PARENTS FRANK AND IRMA
CASTEL WHO NEVER DOUBTED ITS COMPLETION

PREFACE

Textile science and technology encompass diverse disciplines and areas of expertise. Moreover, advances in the basic and applied sciences and engineering have been used to produce textile materials and substrates that did not exist several years ago. There are several useful texts and monographs that separately cover preparatory processes, dyeing, finishing and textile end use performance. However, there is a need to have one book that critically evaluates all of these topics and that also provides salient references for further information. Thus, this book was written to integrate all aspects of textile wet processing and modification and the subsequent performance of textile products. Such integration is achieved by the careful selection of fundamental concepts and practical guidelines in each of these areas and citation of critical reviews, books and specific references.

Organization

The book is organized into six chapters covering four major textile topics: preparatory processes, dyeing and printing, improvement of functional and aesthetic properties and tests for evaluation of end use performance. Preparation and purification of textiles are differentiated by fiber type and discussed in terms of fundamental and practical aspects (**Chapter 1**). Dyeing and printing are also discussed from fundamental aspects such as theories of dyeing (**Chapter 2**) and from practical aspects such as classification of dyes by structure and method of application (**Chapter 3**). Improvement of textile properties is presented in a much broader and unified context than just chemical finishing processes. Thus, property improvement (**Chapter 4**) is discussed from physical, chemical and physicochemical concepts and processes. Evaluation of textiles in use is discussed in the context of wet and instrumental methods (**Chapter 5**) and in the domain of standardized and related tests (**Chapter 6**). Each chapter has general and specific references that include comprehensive and informative critical reviews of important topics.

Coverage

The text has several unique features that should be useful to educators and to those working in the textile and allied industries. It contains a detailed and logical progression of the initial purification of textiles to their performance and care under all end use conditions. Highlights include: (a) effective retrieval of a variety of information from the *Colour Index* and classification of relevant organic reactions for synthesis of classes of dyes (**Chapter 2**); (b) comprehensive classification of textile property improvements by method of application (**Chapter 4**); (c) detailed classification of wet and instrumental methods for textile characterization and evaluation (**Chapter 5**); (d) discussion of textile performance in terms of physical, chemical and biological influences and the interactivity of multiple agents on their performance (**Chapter 6**); and (e) relevant examples and illustrations of textile machines and apparatuses for processing, property improvements and evaluation.

Instructional Aspects

Each topic and subtopic discussed (e.g. carbonizing of wool in **Chapter 1**, thermal analysis in **Chapter 5** and comfort in **Chapter 6**) are succinctly discussed and supported by appropriate and current general and specific references. Although many of the references are derived from American, British and German sources, the references cited are necessarily global in scope to provide the best and most current information on all topics.

References provided (particularly critical reviews) may be retrieved to augment information in the text for evaluation of students in advanced undergraduate courses and/or for seminar/course topics in graduate studies. The book is designed to be used in academic environments and as a comprehensive source of information for textile scientists, engineers and others who need information in the areas of textile wet processing and end use performance.

Acknowledgments

The author gratefully acknowledges the editorial and critical review by several of his colleagues and peers to improve the content and clarity of this text: Dr. Al Turbak, Consultant, Marietta, GA; Drs. Howard L. Needles and S. Haig Zeronian, Division of Textiles and Clothing, University of California, Davis; Dr. Roger Barker,

School of Textiles, North Carolina State University, Raleigh, N. C.; and Drs. J. Nolan Etters and Charles Yang, Dept. of Textiles, Merchandising and Interiors, University of Georgia, Athens, GA.

The author is also indebted to Dr. Menachem Lewin, Dept. of Chemistry, Polytechnic Institute, Brooklyn, N. Y.; Dr. Howard L. Needles, Division of Textiles and Clothing, University of California, Davis; and Emeritus Professor Richard Gilbert, School of Textiles, North Carolina State University, Raleigh, N. C. for their encouragement and support to undertake this project. The author is especially indebted to his spouse Eileen (to whom this book is dedicated) for her kindness and dedication for providing the necessary time and motivation to initiate and complete this endeavor.

CONTENTS

Chapter 1

PREPARATORY PROCESSES

1.1 INTRODUCTION

Preparatory processes are necessary for removing impurities from fibers and for improving their aesthetic appearance and processability as fabrics prior to dyeing, printing, and/or mechanical and functional finishing. **Singeing** may be necessary to produce a smooth and uniform fabric surface, while **sizing** is necessary to prevent breakage and lower processing speeds of a variety of natural and synthetic fiber yarns during their weaving. **Scouring** is practiced to remove impurities from all types of natural and synthetic fibers; however, special scouring processes and **carbonization** methods are required to remove a variety of impurities and waxes from wool. **Bleaching agents and optical brighteners** are utilized on all types of fibers to improve their appearance and to render them more uniform for subsequent dyeing and finishing processes. **Mercerization** with alkali or **treatment with liquid ammonia** (for cellulosics and in some instances for cellulose/synthetic fiber blends) improves the moisture sorption, dye uptake and functional fabric properties. Although purification and pretreatments are generally conducted in certain sequences, they have also been employed at different stages of dyeing and finishing to obtain the desired fabric properties.

1.2 SINGEING

Singeing is the process of passing a yarn or fabric rapidly through a heat source to remove protruding fibers, then rapidly lowering the temperature of the fabric. The latter can be accomplished by quenching the fabric in a water trough, by passage of steam or by other comparable techniques. Most present singeing techniques are continuous, in contrast to earlier batch processes. Techniques for singeing may be

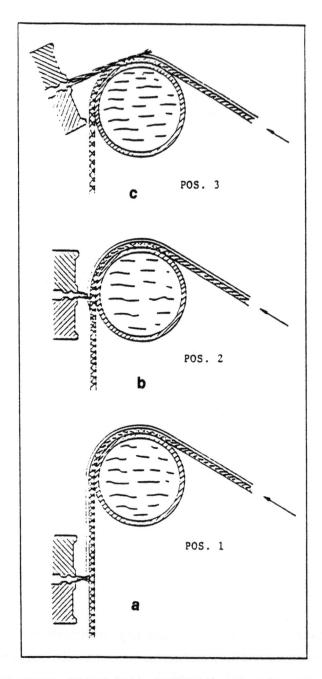

Figure 1.1. Direct (a) controlled singeing, (b) singeing on water-cooled roller, (c) tangential or open singeing (ref. 3). Courtesy of ITS Publishing.

direct (controlled flame, water-cooled roller or open methods) or indirect (generation of heat by infrared reflectors) (refs. 1, 2, 3). In the direct, controlled-flame method, the distances between the burner and the fabric are very short (usually 6-8 mm), but the actual exposure time to temperature as high as 1,300°C is less than 0.1 seconds [see Fig. 1.1 (a)]. In the direct method, the flame contacts the free-guided fabric at a right angle. The best singeing is obtained in this position; however, loosely woven fabrics may be readily oversinged. The water-cooled roller method [Fig. 1.1 (b)] is usually used for blended and synthetic fabrics as well as for open weave constructions that tend to be oversinged by the controlled flame method. The fabric also makes contact with the flame at a right angle, but lies flat on the off-side on a water-cooled roller. However, the water-cooled method has the disadvantages of poorer singeing intensity and maintenance of high water cooling temperatures to prevent spotting of fabrics due to water condensation. The open or tangential singeing method [Fig. 1.1 (c)] burns off only protruding fibers and the fabric is barely in contact with the flame. This position is particularly suitable for singeing lightweight and sensitive fabrics. Although this method is relatively inexpensive, it can lead to uneven singeing due to insufficient thermal energy in all parts of the active zone.

The indirect method differs markedly from the three direct methods. The fabrics are singed by heat emitted from infrared reflectors in a uniform radiation zone in which the temperature reaches a maximum of 1,000°C. This allows for singeing of a variety of fabric constructions by adjustment of a concave configuration of the reflectors to produce radiant energy that affects only fiber surfaces. Modern singeing technology has combined the best features in earlier machinery to process fabrics varying in fiber type and construction. Parameters for singeing fabrics comprised of diverse fiber types and weights and fabric constructions have been tabulated and critically evaluated (refs. 2, 3). These parameters include: flame intensity, singeing speed, distance between the flame and the fabric, method of singeing, and the special precautions and features that should be taken into account for each type of fabric processed.

Although the singeing process has been traditionally used to improve the smoothness and uniformity of fabrics derived from natural and synthetic fibers as well as their blends, it has more recently also been applied to cotton/polyester fabrics and synthetic fabrics to improve their resistance to pilling. Proper singeing thus produces a uniform fabric surface, but the sequence in which it is conducted is important. For example, cotton/polyester blends that are dyed in dyebaths containing dye carriers must be singed after dyeing rather than before dyeing to prevent dye specks (from the melted synthetic fibers) occurring in the finished and dyed fabrics.

1.3 SIZING AND DESIZING

1.3.1 Sizing

Most cotton and synthetic fiber staple yarns are sized prior to weaving with various types of natural and/or synthetic polymers. If the yarns are of high twist and/or plied, they will normally not be damaged when they are woven into fabrics, and thus application of sizes may not be necessary. However, if the yarns are of low twist and unplied, they are usually abraded, broken, or otherwise damaged, and thus need to be sized to prevent such damage. Thus, sizing is desirable for maintaining high processing speeds that are not reduced by stopping the units because of damaged fibers and yarns. Equally important, sizing can prevent processing damage during weaving that could subsequently lead to a nonuniform fabric for dyeing and finishing operations.

Earlier sizes utilized were natural polymers and their derivatives. These were primarily starches, modified starches and cellulose derivatives. However, with the advent of polyester, cotton/polyester blend fabrics and other blend fabrics, poly(vinyl alcohol), partially hydrolyzed poly(vinyl acetate) and several other synthetic polymers have been employed as sizing agents for yarns. Representative sizing agents for yarns and fabrics are listed in Table 1.1. The various natural and synthetic polymers that have been used in textile warp sizing have been summarized and critically reviewed with regard to their desirable features and chemical structures (ref. 4).

Starch and starch derivatives used for sizing include most types of native starches, the amylose and amylopectin fractions of starch, and starch derivatives such as dialdehyde starch, hydroxyethylated and acetylated starches. Carboxymethylcellulose is the most frequently used cellulose derivative for sizing. Of the synthetic polymers, poly(vinyl alcohol) and a variety of acrylic homopolymers and copolymers are frequently used as sizing agents. To a lesser extent partially hydrolyzed poly(vinyl

TABLE 1.1

Polymeric sizing agents used in textile processing

Natural Polymers	Synthetic Polymers
Starches	Poly(vinyl alcohol)
Modified or refined starches	Poly(vinyl acetate)
Starch derivatives	Acrylics
Cellulose derivatives	Polyesters
	Polyurethanes
	Styrene copolymers

acetate), and polyester dispersions, polyurethanes and styrene copolymers are utilized. Copolymers have also been synthesized specifically for use in solvent and hot melt sizing.

Although requirements for polymers used for sizing vary to some extent from one type of fiber to another, the polymer must ultimately be a good film-former with appropriate hardness properties, have good adhesion, tensile strength, abrasion resistance and flexibility, be compatible with other ingredients in the formulation, not corrosive to mill equipment, have low foaming properties, be easily removed from the yarns, and have a relatively low viscosity that allows it to penetrate the fibrous surface. Typical formulations have to maintain their stability for several hours at elevated temperatures and contain % solids from as little as 3% to over 20% that impart comparable weight gains to sized yarns after drying; a variety of additives such as antifoam agents, waxes and lubricants may also be present in a representative formulation (ref. 4).

Conventional processes for sizing warp yarns are conducted in machines called slashers (see schematic of representative slasher in Fig. 1.2). Warp sheets of yarn move from a battery of beam creels (A) through a container (size box B) that contains the sizing agent. The wetted yarns are subsequently squeezed of excess liquid polymer (wet split C), then passed through a series of heated cylinders (D) to dry the warp sheets (E) that are wound up on a beam (F) for weaving.

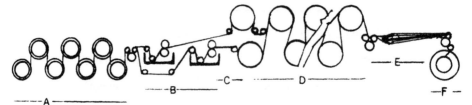

Figure 1.2. Major tension producing zones on the slasher: A. creel section, B. size box, C. wet split, D. drying cylinders, E. dry split, F. beaming. (ref. 5). Courtesy of Dr. David Hall, Auburn University.

Most of the newer and innovative sizing processes have been developed to save energy and reduce effluents from textile mills. Essentially, this is accomplished by (a) low wet pickup methods that reduce the amount of water to be evaporated from the sized yarns (high pressure and foam sizing), (b) application of polymer sizes without solvents (hot melt sizing), (c) application of polymer sizes from nonaqueous solvents such as perchloroethylene (solvent sizing), and (d) application of durable sizes. Methods (a) and (b) have been critically reviewed, with concepts and numerous schematics presented to illustrate how sizes are applied by these techniques (ref. 6).

In high pressure sizing, the wet pickup of the sized yarns is approximately half that normally obtained (50% rather than 100%) from a conventional sizing process after the yarns are passed through pressure rolls; care must be taken not to distort the yarn surface or cause premature breakage due to the high pressure (in some instances twenty times greater than the normal roll pressure) to which the warp sheets are exposed. Uniform distribution of pressure on the entire width of the warp sheets and a constant squeeze effect are also critical for the successful use of high pressure warp sizing.

The other low wet pickup method that has been increasingly used is foam sizing. Techniques for foam finishing (discussed later in section on **special finishing techniques**) have been adapted for use in foam sizing. One method of foam sizing that has been investigated is the horizontal pad system shown in Figure 1.3 (a). In this method, the foam is applied to both sides of the warp sheet and held in a reservoir above the two rubber-covered horizontal rolls. As the warp sheet passes through the center of the foam bank into the rollers, the nip pressure causes the foam to collapse and penetration of the yarns by the size occurs. Again, foam density and concentration of foam solution are important as well as constant nip pressure to produce uniform collapse of the foam. Another method that has received some attention is the knife-over-roll system shown in Figure 1.3 (b). The foam is deposited from a foam bank on one side of the warp sheet, and its thickness is regulated by adjustment of the doctor knife or blade; as the warp sheets pass under the knife and through the pressure rolls, the foam collapses; variables such as foam density, setting of the doctor knife, and uniform speed of the warp sheets through the apparatus must be carefully controlled to obtain reproducible and uniform sizing. Although neither of the foam sizing systems are extensively used on a commercial basis, several trials indicate that this technique has promise for mill use.

In hot melt sizing, copolymers (usually polyester or acrylic) that are solids at ambient temperature are melted and applied in the liquid state to a heated, coated roller in such a manner that the amount of size adhering to the roller is controlled, and the warp sheet makes contact with the heated roller. Schematics in Figures 1.3 (c) and 1.3 (d) show, respectively, a side view and a cross section of a commercially used applicator developed by Burlington and Westpoint Foundry and Machinery for this purpose. The roller is a grooved applicator that makes contact with the warp yarns in such a manner that the warp sheets move about 120 times faster than the rollers. This generates a differential friction between the yarns and the roller that causes protruding fibers to be laid down and effectively coats the entire yarn surface rapidly as the size solidifies when it is exposed to cooler temperatures. In addition to saving energy costs by not having to evaporate water or recover solvents, hot melt sizing also allows for greater processing speeds than conventional sizing, reduced shedding of sizes, and improved yarn elasticity due to optimum bonding of polymer on the fiber surface.

Solvent sizing has gained some commercial acceptance because less energy is required to heat the organic solvents containing the sizes and evaporate the solvents from the yarns. Most solvents utilized are chlorocarbons or fluorocarbons, particularly perchloroethylene and trichloroethylene. Changes in the composition and structure

8

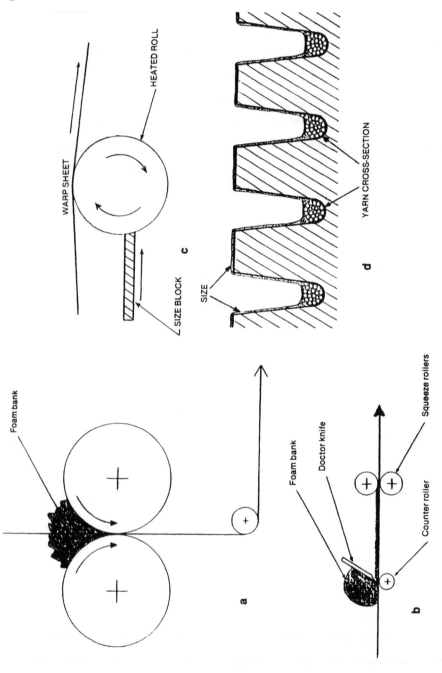

Figure 1.3. (a) Diagram of horizontal pad system, (b) knife-over-roll system, (c) hot-melt size applicator (side view), and (d) cross-section of applicator roll (ref. 6). Courtesy of South African Wool & Textile Research Institute for (a) and (b) and of West Point Foundry and Machine Co. for (c) and (d).

of the polymers used for sizes must be made, because polymers that are soluble in water may not be soluble in chlorinated hydrocarbons. Hydrophobic polymers or copolymers or chlorinated polymers, such as methyl methacrylate/butyl methacrylate or chlorinated ethylene-propylene copolymers are frequently employed for solvent sizing (ref. 4). In addition to reduced energy consumption, solvent sizing generally affords better fabric hand and aesthetics than that of aqueous conventional sizing.

It is possible to durably affix polymers as sizes on warp yarns by controlled crosslinking of appropriate polymers and copolymers; however, this technique is not frequently employed on a commercial scale. Both natural polymers (such as starch) and synthetic polymers (such as acrylic acid polymers and copolymers) have been applied as durable sizes by crosslinking with a variety of polyfunctional agents (e.g., acids, aldehydes, and urea-formaldehyde systems).

1.3.2 Desizing

After the sized fabric is woven, it is necessary to desize it prior to subsequent processes such as bleaching, dyeing and finishing. Most earlier desizing utilized enzymes such as α- and β-amylases at temperatures around 40-70°C and pH values of 4.6-7.0 to remove starch sizes from cellulosics, or by steeping the yarns in dilute solutions of acid. Oxidative desizing, i.e., use of one of several oxidizing agents to remove size from yarns, has become an increasingly popular alternative technique to enzymatic desizing. Alkaline scouring, particularly to remove size from yarns sized with poly(vinyl alcohol), is frequently used (see section on **scouring of cotton, silk and synthetic fibers**). Newer techniques for sizing include solvent sizing and low temperature plasma sizing. Commonly used chemical and biological desizing agents are listed in Table 1.2.

Although enzymatic desizing is effective, it is usually limited to the removal of starch sizes from yarns because an enzyme will only be active or affect a particular substrate. Each of the three different starch-specific enzymes depolymerizes the 1,4-glucosidic linkage by different pathways; thus, different enzyme concentrations and different desizing temperatures and pH conditions are required. Alkaline scouring will be discussed more in the section under **scouring of cotton, silk and synthetic fibers**. As a desizing technique, it essentially involves wetting sized yarns in a 1-3%

TABLE 1.2

Representative chemical and biological desizing agents

Biological agents	Oxidizing agents
α-amylase	NaOCl (sodium hypochlorite
β-amylase	NaClO$_2$ (sodium chlorite)
amyloglucosidase	NaBrO$_2$ (sodium bromite)
	H$_2$SO$_5$ (peroxymonosulfuric acid)
Acids and bases	K$_4$P$_2$O$_8$ (potassium peroxydiphosphate)
HCl (hydrochloric acid)	H$_2$O$_2$ (hydrogen peroxide)
H$_2$SO$_4$ (sulfuric acid)	NaBO$_3$.4H$_2$O (sodium perborate)
Na$_2$CO$_3$ (sodium carbonate)	2Na$_2$CO$_3$.3H$_2$O$_2$ (sodium carbonate-hydrogen peroxide)
NaOH (sodium hydroxide)	CH$_3$COOOH (peracetic acid)

soap-soda or soap-caustic solution at temperatures below the boiling point of water, steeping and thoroughly rinsing the warp sheets to remove all size.

Solvent desizing utilizes the same solvents that are used in solvent sizing, namely perchloroethylene and trichloroethylene. Although there are several advantages (e.g., low effluents, uniform desizing and improved aesthetics or hand for certain types of fabrics), solvent desizing has not yet been commercially adopted due to its high initial capital cost and its lack of versatility for removing different types of polymeric sizes (ref. 4).

While enzymatic desizing is suitable only for the removal of natural polymers, such as starches and cellulose derivatives, oxidative desizing may be employed to remove both natural and synthetic polymers. Although chlorine dioxide, sodium chlorite and sodium bromite were frequently used in earlier oxidative desizing processes, more recent emphasis has been placed on employing hydrogen peroxide or various inorganic peroxy compounds and peracids. Dickinson has critically evaluated oxidative desizing and lists all oxidizing agents separately, with additional comments about acceptable formulations and processing conditions; disadvantages of oxidative desizing are degradation of size that cannot be reused, adequate control of equipment and processing conditions, and limiting adverse effects of metals derived from fabrics or equipment that are incompatible and contaminate the oxidant. A representative, general purpose formulation that is suitable for continuous pad-steam desizing operations contains 5.0-7.5% H_2O_2, 1.5-2.5% NaOH, 1.5-2.5% sodium silicate (stabilizer), 0.3-0.5% emulsifier and 0.1-0.2% wetting agent (ref. 7). Desizing poly(vinyl alcohol) or poly(vinyl alcohol/carboxymethyl cellulose (50/50) sizes from polyester/cotton warps (67/33) may be optimized with various non-oxidizing additives (glycerol, urea or a nonionic surfactant) as well as with oxidative desizing agents such as sodium persulfate. However, overdrying of the sized warps prior to desizing led to a marked reduction in the amount of sizing removed from the fibers irrespective of the types of additives used (ref. 8).

1.4 SCOURING AND CARBONIZING

1.4.1 Wool scouring

In contrast to natural cellulosic and synthetic fibers, purification of wool prior to subsequent bleaching, dyeing, and finishing is absolutely essential, since raw wool contains 40% or more by weight of impurities in the form of waxes, suint, cellulosic material such as straw and dried grass, dirt, and proteinaceous material. Various scouring processes are employed to remove most of the wool waxes, suint and non-cellulosic components; the wool is then carbonized with sulfuric or other appropriate acids to free it from unwanted vegetable matter. The scouring of wool, particularly with regard to machinery and processes that effectively clean it, and remove and recover grease, suint, dirt, and desirable chemical by-products, has been comprehensively and critically reviewed by Gibson (ref. 9), Wood (ref. 10) and Christoe (ref. 11).

Wool waxes are recovered from the grease during scouring. These waxes are comprised of a variety of monocarboxylic, dicarboxylic and hydroxycarboxylic acids as well as steroidal alcohols (e.g., cholesterol and lanosterol). It has been determined that unscoured wool contains an unoxidized fraction of wool grease and other contaminants that is easily removable and readily recoverable and an oxidized fraction at the tip of the hair that is difficult to remove and separate from other oxidized contaminants (ref. 11). Suint is usually considered to be a variable composition of water-soluble materials (inorganic cations and anions plus various fatty acids) that is readily removed by scouring. The dirt that is removed from the scoured wool consists of both inorganic and organic materials. The proteinaceous material has recently been discovered to consist of skin flakes from the sheep and soluble peptides. The dirt and the proteinaceous matter, as well as the wax or grease, exist as unoxidized and oxidized fractions in varying proportions, and thus contribute to the complexity of effectively scouring the wool. Thus, the numerous processes and techniques that have been devised and that are currently used for wool scouring must by necessity address not only removal of all contaminants (excluding vegetable matter) but also effectively and economically recover useful chemical by-products and discharge minimum effluents during the processing.

In a conventional scouring operation (Fig. 1.4), the wool enters and is submerged by a device called a dunker (to prevent it from floating to the surface). It is then moved or propelled through a series of bowls of liquors with long-fingered rakes; this technique is used to prevent it from undergoing excessive entanglement or felting. Transfer of the wet wool to the squeeze rollers is most commonly accomplished by a washplate bowl; these bowls have false bottoms with a perforated plate to allow dirt to fall through and prevent the wool from entering circulating streams of liquor drawn into the bowl during scouring. In an ideal system, there are four bowls, with detergent being added to one or more of the first three bowls, and water added to the last. Countercurrent flow rate from bowl 4 to bowl 1 through bowls 2 and 3 is effected, and contaminants are discharged at bowl 1 as part of the recovery loop (ref.10).

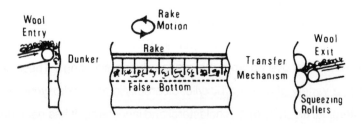

Figure 1.4. The elements of a conventional scouring bowl (ref. 10). Courtesy of Marcel Dekker Publishers.

A typical soap-alkali wool scouring process utilizes 0.7-0.8% soap and 0.2-0.3% sodium carbonate in the first bowl at approximately 50°C; the concentration of the soap and alkali and the temperature are progressively decreased until the last bowl contains only water at 40-43°C. Nonionic detergents such as condensates of nonyl phenol and ethylene oxide are utilized in a similar manner to scour wool, but scouring temperatures are usually higher than those employed in soap-alkali scouring. The use of organic solvents (such as trichloroethylene or petroleum distillates) produces cleaner wool when this technique is used for scouring; however, there remain problems with regard to solvent disposal, flammability and toxicity. As previously mentioned, the variety of processes and equipment that have been developed to effectively scour wool also have incorporated elaborate and useful devices to recover the contaminants removed: grease (the wool wax fraction that is heaviest and falls to the bottom during centrifugation or comparable separation processes); cream (the wool wax fraction that is skimmed from the top after separation); and sludge (wet dirt separated from grease and cream). Representative systems that achieve good decontamination of the fiber and separation of the contaminants into different components with recovery loops include: WRONZ Comprehensive Scouring System (first to incorporate both grease and dirt recovery devices), Lo-flo Scouring Process (one that allows for substantial recovery of grease by destabilization of the emulsion in the first scouring bowl), and SIROSCOUR System (one in which the recovery loops for dirt and grease are separate and thus allow for maximum recovery) (ref. 11).

In the Lo-Flo process (see schematic in Figure 1.5), three mini bowls are used in place of the first large bowl to effectively reduce water input and consumption. Destabilization in the first bowl is achieved by allowing naturally occurring contaminants to accumulate in the bowl liquor. It is believed that the suint in the fiber is responsible for causing such destabilization. This process (and other representative contaminant recovery processes) utilizes countercurrent flow of liquor from one of the rinse bowls to the first bowl that is subsequently discharged into a recovery loop of loop for concentrated liquors consisting of a decanter centrifuge connected in series with the primary grease centrifuge. This feature allows separation of the contaminants into spadeable sludge and the waxes into cream and heavy solid phases.

14

1.4.2 Wool carbonizing

After the wool is properly scoured, it is then carbonized. The purpose of carbonization is to remove vegetable matter (primarily cellulosic in nature) from the wool fibers. This matter is usually present in the form of seeds and burrs. The carbonization process consists of first immersing the scoured wool in dilute solutions

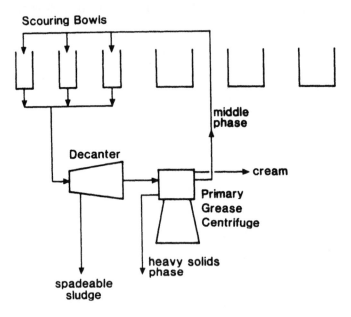

Figure 1.5. Contaminant recovery loop for Lo-Flo scouring system (ref. 11). Courtesy of International Wool Secretariat.

of strong acids (usually sulfuric or occasionally hydrochloric), followed by predrying, baking, crushing/dedusting the carbonized vegetable matter, and neutralizing the acid in cold water and soda solution to a slightly acidic pH prior to final drying. The carbonizing process has been reviewed with regard to conventional and rapid carbonizing processes, each of the five processing steps, and fundamental chemical changes occurring in the wool fiber after carbonization (refs. 12, 13).

A typical carbonizing train in Figure 1.6 shows the wool fabric entering the chamber containing dilute acid, removal of excess liquor through squeeze rollers, then entry into a drying and baking zone. The baked fabric is then put through pressure rolls that crush carbonized matter subsequently removed by dusting methods, neutralized in dilute alkaline solutions, and finally washed and dried.

The use of nonionic detergents is generally advocated to minimize fiber damage of the wool in the acid bath. Rapid carbonization of wool is claimed to be an effective method of minimizing fiber entanglement as well as acid damage. In this process, the wool is conveyed between two mesh belts in the acidifier while being jetted with a relatively strong acid solution containing large quantities of surfactants (ref. 15).

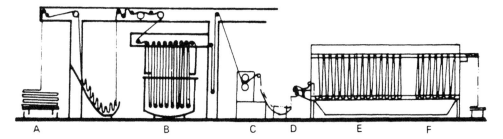

Figure 1.6. Typical sulfuric acid piece carbonizing train. A. Fabric entry. B. Acid impregnation chamber. C. Mangle. D. Looping scray and dryer feed. E. Drying zone. F. Baking zone. (ref. 14). Courtesy of Marcel Dekker Publishers.

Although a sizeable amount of wool is scoured and carbonized in fiber form, this type of process has also been utilized on wool yarns and fabrics.

There have been several important new developments and studies relating to the carbonization of wool. Thionyl chloride ($SOCl_2$), in vapor form or in perchloroethylene, has been proposed as an alternative carbonizing agent to sulfuric acid. It is claimed to produce comparable wool fiber properties and in some instances, eliminate the need for a baking step. Thionyl chloride and/or its hydrolysis products (sulfur dioxide and hydrochloric acid) were proposed to be the active carbonizing species (ref. 16). Pailthorpe and his colleagues (ref. 13) have conducted a series of studies that address: (a) causes and prevention of localized acid damage in carbonized wool, (b) critical concentration of sulfuric acid in wool carbonizing, (c) effects of time delay in rapid carbonizing, and (d) surface barrier effects in rapid carbonizing methods. Localized damage is shown to be due to big droplets of unevenly distributed acid that are trapped during drying and baking steps. Above temperatures of 70°C and sulfuric acid concentrations of 43%, damage in wool occurs. A time delay of 10 minutes in rapid carbonizing allows surface free acid to penetrate into the fibers and form bonds to the wool, thus minimizing the amount of acid during drying and baking. This also produces dye-resist behavior on the fabric surface and contributes to improved levelling during subsequent dyeing by increasing dye sites from bonding of the free acid to the fibers. It has also been determined that about half of the damage to the wool fiber occurs due to the cumulative processes of acidification, baking and the time interval between baking and neutralizing. Recommendations are made for minimizing fiber damage at each step of the carbonizing process (ref. 17).

1.4.3 Scouring of cotton, silk and synthetic fibers

Although other natural fibers such as cotton and silk contain impurities that are easier to remove than those that occur in wool, it is still necessary to scour them to insure uniform bleaching, dyeing and finishing as well as to enhance their wettability and absorbency. The composition of impurities in cotton, silk and various synthetic fibers, scouring agents used, and techniques and processing equipment available for effective scouring have been discussed in detail in monographs (refs. 18 and 19).

Cotton may contain from 4-12% by weight impurities in the form of waxes, proteins, pectins, ash, and miscellaneous substances such as pigments, hemicelluloses and reducing sugars. The hydrophobic nature of the waxes make their removal difficult relative to the removal of other impurities. The composition of cotton wax consists primarily of a variety of long chain (C_{15} to C_{33}) alcohols, acids, and hydrocarbons as well as some sterols and polyterpenes. Examples include gossypol ($C_{30}H_{61}OH$), stearic acid ($C_{17}H_{35}COOH$), and glycerol. Little is known about the structure of the proteins, and the pectins are essentially present as the methyl ester of poly-D-galacturonic acid. Ash is a mixture of inorganic compounds (particularly sodium, potassium, magnesium and calcium salts), while other impurities vary in composition but are readily hydrolyzed and removed under practical scouring conditions.

Effective removal of impurities in cotton, particularly waxes, is achieved by boiling in 3-6% sodium hydroxide or less frequently in dilute solutions of calcium hydroxide (lime) or sodium carbonate (soda ash). The proper choice of textile auxiliaries in the alkaline bath is essential for good scouring. These include sequestering or chelating agents such as ethylenediaminetetraacetic acid (EDTA) to solubilize insoluble inorganic substances present in hard water and surfactants such as the anionic sodium lauryl sulfate that serves as a detergent, dispersing agent, and emulsifying agent to remove unsaponifiable waxes. Synthetic fibers are scoured with milder formulations such as soap or detergents containing comparatively small amounts of alkali (e.g., 0.1-0.2% sodium carbonate). Cotton/synthetic fiber blends (such as cotton/polyester) require alkaline concentrations and conditions intermediate between those of all cotton and all synthetics for effective scouring.

Scouring of the silk fiber is also known as degumming. Silk scouring has been critically reviewed with respect to degumming processes and machinery and iden-tification of material removed from the fiber (ref. 20). The main contaminant to be removed from silk is the protein sericin, also known as gum, which may range from 17% to 38% by weight of the unscoured silk fiber. The sericin removed from the silk fiber has been separated into four fractions that differ both in their amino acid composition and their physical properties. There are five methods for degumming silk fibers: (a) extraction with water, (b) boiling-off in soap, (c) degumming with alkalis, (d) enzymatic degumming and (e) degumming in acidic solutions. Boiling-off in soap solutions remains the most popular degumming method. A variety of soaps and processing modifications give varying degrees of purification of the silk fiber. Although there are many qualitative methods to determine the extent of silk fiber degumming, quantitative methods for sericin removal and the mechanisms by which it is removed have not been developed and proposed (ref. 20).

Impurities present in synthetic fibers are primarily oils and spin finishes used in spinning, weaving and knitting operations. These can be removed under much milder conditions than impurities in cotton and in silk. Scouring solutions for synthetic fibers contain anionic or nonionic detergents with trace amounts of sodium carbonate or ammonia; scouring temperatures for these fibers are generally 50-100°C.

Techniques and machinery for scouring different fiber types and fabrics (excluding wool) have evolved from batch processes utilizing various types of open and pressure kiers to continuous processes in which fabrics are passed through roller steamers under pressure. Continuous scouring processes are similar to those used to con-tinuously bleach fabrics (shown in Figure 1.9 in the section on **bleaching**).

Open or pressure kiers are stainless steel containers that circulate the alkali at elevated temperatures from 100°C at atmospheric pressure to over 130°C at reduced pressure. The alkaline solution is injected or introduced from the bottom of the kier, and air removed by displacement by steam. After scouring for sufficient time, the fabric is rinsed by continually introducing hot water to dilute the alkali in the bath.

In semi-continuous processes where the fabric is scoured in rope or open-width form, a device called a saturator serves as the scouring bath. The fabric is then rapidly preheated by steam and passed into a J-box storage chamber or, alternatively, the fabric is directly heated in the storage chamber. After storage in the J-box, the fabric is then washed with hot water to remove alkali and impurities. Continuous processes that rebatch the fabric or utilize roller steamers under pressure, reduce the dwell time of the fabric considerably and allow for higher processing speeds. These continuous processes are replacing batch and semi-continuous scouring processes and have been critically reviewed (ref. 19).

Solvent scouring of cotton and cotton/polyester blends has been employed, with trichloroethylene frequently used as the scouring solvent. However, only waxes are removed by this method and some form of aqueous alkaline scouring is still required. When the only impurities in fabrics are oils incorporated to assist in knitting or weaving, trichlorotrifluoroethane is claimed to be very advantageous in solvent scouring. Advantages are: (a) its low boiling point and latent heat of vaporization, which facilitates separation of solvent from waste oils, (b) high liquid density and low surface tension, which provides superior ability to remove loose particulate matter, (c) low toxicity and flammability and lack of explosive characteristics, and (d) much more effective and economical than aqueous scouring methods (ref. 21).

1.5 BLEACHING AND OPTICAL BRIGHTENING PROCESSES

1.5.1 Overview

Bleaching is the removal or lightening of colored materials. The substrate in this instance is a textile. Bleaching is accomplished chemically with oxidizing or reducing agents. It is also accomplished physically by introduction of optical brighteners or fluorescent brightening agents into the fiber. Chemical bleaching agents function by solubilizing colored substances, thus facilitating their removal from the fiber. They also function by reacting with these substances in such a manner as to alter or destroy their sites of unsaturation or conjugation. Conversely, optical or fluorescent brighteners improve distribution of light reflected from the fabric surface in the visible range and increase the total amount of light reflected from that surface. The brighteners thus make the fabric surface appear whiter and brighter by selective absorption of the ultraviolet component (300-400 nm) of visible light and reemission of energy absorbed at longer wavelengths (400-460 nm) in the visible region. Most chemical bleaching agents are oxidative in nature. Active species such as hypochlorite, chlorine dioxide, hydrogen peroxide or related peroxygen structures are responsible for the bleaching achieved. Reducing agents such as sulfur dioxide or hydrosulfite salts have also been employed for bleaching textiles. Optical brighteners encompass a variety of aromatic and heterocyclic compounds; the first commercial optical brighteners used were stilbenes containing amino- and sulfonic acid groups.

1.5.2 Chemical bleaching agents

Oxidizing agents of various types represent the largest and most widely used class of chemical bleaching agents. These can be further subdivided into those that employ or generate: (a) hypochlorite (OCl⁻), (b) chlorine dioxide (ClO_2), and (c) hydroperoxide species (OOH⁻ and/or OOH·). Hypochlorite-based bleaches are usually generated by one of three methods: reaction of water with chlorine gas, with sodium hypochlorite, or with organochloramines (ref. 22):

$$Cl_2 \quad + \quad H_2O \quad \underset{\longleftarrow}{\overset{\longrightarrow}{}} \quad HOCl \quad + \quad HCl \qquad (1.1)$$

$$NaOCl \quad + \quad H_2O \quad \underset{\longleftarrow}{\overset{\longrightarrow}{}} \quad HOCl \quad + \quad NaOH \qquad (1.2)$$

$$RR'NCl \quad + \quad H_2O \quad \underset{\longleftarrow}{\overset{\longrightarrow}{}} \quad HOCl \quad + \quad RR'NH \qquad (1.3)$$

The amount of chlorine liberated by the addition of acid to compounds such as those above is called "available chlorine." When chlorine is in the +1 oxidation state (as above), the available chlorine is twice the weight of the active chlorine in the compound, which reacts with the water to generate hypochlorite. Hypochlorite can react reversibly and shift the equilibrium back to the left. Thus, fabrics are usually treated with antichlors such as sodium bisulfite, sulfur dioxide or hydrogen peroxide to maintain equilibrium to the right and prevent color reversion in textiles.

Although hypochlorite-based bleaches were historically first used for bleaching textiles, their commercial use is now generally limited to bleaching of cellulosics and cellulosic blends in European countries. These types of bleaches are unacceptable for wool due to extensive fiber damage caused. They are also not recommended for bleaching acrylics and/or polyamides because they cause unacceptable yellowing.

pH has a dramatic effect on the concentration and availability of various species such as Cl_2, $HOCl$ and OCl^- used to bleach cotton and other cellulosics (see Figure 1.7). Cellulosics are not usually bleached with hypochlorite systems at low pH (2.0-5.0) because of the danger and nuisance of chlorine gas being generated. It has also been demonstrated in many studies that cotton or other cellulosic fibers are extensively oxidized at low pH levels by high concentrations of $HOCl$ present. Nor is it advantageous to bleach cellulosic fibers at neutral pH (5.5-7.0), since in this range, considerable oxidation and fiber degradation also occurs. The cellulose hydroxyl groups are oxidized to aldehydes, ketone and carboxylic acids, which vary in composition with temperature and pH.

However, bleaching may be readily conducted at high pH values (usually in the range of 9-11) with a minimum of fiber damage provided the system is buffered with

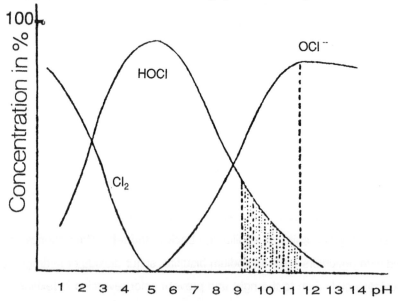

Figure 1.7. Composition of the NaOCl bleach liquor as function of the pH (shaded area = bleaching range) (ref. 23). Courtesy of Melliand Textilberichte.

compounds such as sodium carbonate or sodium borate to maintain a relatively high pH level in the bleaching bath. Cellulose acetate fabrics may also be bleached with hypochlorite provided the pH does not exceed 10. This is achieved by choice of a buffer of less alkaline character such as sodium bicarbonate or sodium borate. Hypochlorite bleaching of cotton has been conducted using batch (e.g., kier), semi-continuous (e.g., J-box) and continuous (e.g., open width) processes similar to scouring processes described earlier, and comparable to those shown in Figure 1.9 for peroxide bleaching. It has also been noted that accelerated hypochlorite bleaching may be achieved by raising the temperature from ambient to about 60°C without adverse effects to the fiber or appreciable decomposition of the bleaching agent.

A representative continuous bleaching process for cellulosic or cellulosic blend fabrics with hypochlorite under alkaline conditions consists of bleaching with a solution containing 2.5-5.0 g/l of available chlorine at 40°C for 0.25-0.50 h. The fabric is subsequently washed with water, then immersed in a bath containing an antichlor (e.g., sodium bisulfite) to prevent color reversion, fiber degradation and yellowing. The fabrics are subsequently rinsed with water, soured with dilute acetic acid to remove residual alkali, then washed again (ref. 22).

Although there have been numerous investigations and hypotheses to confirm or suggest a mechanism by which cellulosics are bleached with hypochlorite systems, no totally satisfactory mechanistic pathway has been proposed. A critical review and evaluation of possible chemical species present in hypochlorite and other types of bleaches, provide reasonable arguments to favor the dissociation of HOCl into HCl and "active oxygen" at high pH as the effective bleaching agent. Other species, such as OCl⁻, are discounted because concentrations present at different pH cannot adequately account for bleaching and fiber damage effects (ref. 23). However, there is good agreement in many studies that the hypochlorite or an active species derived from hypochlorite is effective in reacting with the colored impurities in fibers by either electrophilic or free radical mechanisms to alter or destroy chromophoric sites of unsaturation.

Chlorine dioxide (ClO_2) or related compounds ($HClO_2$ or $NaClO_2$) represent another major class of bleaching agents. In this group, the chlorine is in a higher positive oxidation state than in the hypochlorite-based bleaches. ClO_2 is usually produced by dissolving sodium chlorite in water to generate chlorous acid, which subsequently decomposes to ClO_2, or more efficiently by reacting sodium chlorite with chlorine gas (ref. 22):

$$NaClO_2 \quad + \quad H_2O \text{ --------> } \quad HClO_2 \quad + \quad NaOH \qquad (1.4)$$

$$5HClO_2 \qquad \text{--------> } \quad 4ClO_2 \quad + \quad HCl \; + \; 2H_2O \qquad (1.5)$$

$$2NaClO_2 \quad + \quad Cl_2 \text{ --------> } \quad 2ClO_2 \quad + \quad 2NaCl \qquad (1.6)$$

The compounds or species that predominate in ClO_2-based bleaches are dependent on pH. A similar dependence was discussed earlier for hypochlorite-based bleaches.

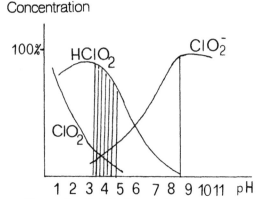

Figure 1.8. Composition of the $NaClO_2$ bleach liquor as function of the pH (ref. 23). Courtesy of Melliand Textilberichte.

The highest concentrations of ClO_2 and $HClO_2$ are present under acidic conditions, while the highest concentrations of ClO_2^- are present under alkaline conditions (Figure 1.8). In practice, chlorite bleaching is conducted in a pH range of 3.5-4.5, since this is the range in which minimum fiber damage occurs and in which a useful concentration of ClO_2 is available.

The chemistry of the chlorite bleaching solutions is complex, and the formation of various species has been shown to be dependent on the rate and order of addition of reagents. At pH values below 3, the production of excess ClO_2, a toxic and corrosive substance, presents safety and processing problems. Formation of Cl_2 and HOCl at low pH can also occur to produce fiber damage and safety hazards. Salts such as sodium nitrate or reducing agents such as formaldehyde or SO_2 are usually employed to control excess production of ClO_2 or Cl_2 gas at lower pH bleaching ranges. At higher pH values (> 5), the concentration of ClO_2 is too low to be useful, fiber damage is extensive, and bleaching times are too long to be economical.

Although no added acid is required to decrease the pH of chlorite solutions during fiber bleaching, it is often necessary to add activators to the $NaClO_2$ solutions to obtain uniform and acceptable whiteness of the fabrics. A list of frequently used activators for chlorite bleaching has been compiled and their beneficial functions described (ref. 24). These activators include various salts ($CuSO_4$), buffering agents (NaH_2PO_4), carboxylic acids and their esters (citric acid and diethyl tartrate), and reducing agents (CH_2O). The function of the activators is gradual reduction of the pH of the bleaching solution, decreased formation of chlorate and corrosive ClO_2, facilitation of the rate of bleaching at higher pH without ClO_2 present, and imparting improved fabric whiteness at lower bleaching temperatures. A one-step scouring and bleaching process has been described that utilizes an emulsifiable perchloroethylene solution with $NaClO_2$ and triethanolamine hydrochloride as the activator. Conditions and concentrations of each component have been optimized to carry out this process at a mild temperature of 60°C (ref. 25).

As with hypochlorite bleaches, chlorite bleaches are not extensively used in the United States, but are primarily employed in Europe for bleaching cotton and cotton/polyester blends. A typical chlorite bleaching process consists of immersing cotton/polyester fabrics in a solution of 2% $NaClO_2$-2%$NaNO_3$ at 100°C for 1-2 h in a J-box. In contrast to hypochlorite bleaches, chlorite bleaches are more versatile, since they may be employed to bleach various types of fibers. In addition to bleaching cellulosic and cellulosic/polyester blends, acrylics, polyamides, and cellulosic/polyamide fiber blends may also be effectively bleached with chlorite-based systems. Other advantages of ClO_2 or related bleaching systems are: (a) they are insensitive to the presence of metal impurities, (b) they are adaptable to a variety of batch and continuous processes, and (c) no prior kier boiling or scouring of cotton fabric is necessary. Their disadvantages are their high cost, the need for stainless steel or other specially fabricated processing equipment due to their corrosive behavior, and special safety precautions that need to be taken because of their toxic and explosive characteristics.

Although ClO_2 is considered by many investigators to be the active bleaching species, other species may also be active, since it has been demonstrated that $NaClO_2$ functions as an effective bleaching agent under reaction conditions where ClO_2 cannot exist (ref. 26). Agster reinforces this observation by eliminating several species that could function as the active chlorite-bleaching agent: ClO_2 (bleaching only takes place when it is in the presence of water) and ClO_2^- (high concentrations of it at pH 8.5 give no bleaching effect). He concludes that the active bleaching agent is $HClO_2$, which disproportionates into HCl and "active oxygen" (ref. 23).

Recent environmental regulations for minimizing or reducing the amount of halogenated organic compounds (AOX) have led to studies that determined the amount of such effluents for hypochlorite, chlorite and peroxide bleaching processes (ref. 27). It was concluded that even under the most favorable conditions, chlorite bleaching and especially hypochlorite bleaching would exceed desirable AOX limits. In contrast, peroxide bleaching appears to be much more environmentally acceptable even under the most unfavorable processing and disposal conditions.

Hydrogen peroxide and related peroxy compounds are the most widely used class of textile bleaching agents, particularly in the United States. These compounds always contain a peroxide bond (-O-O-) in which at least one oxygen atom is active, i. e., capable of oxidizing iodide ion to free iodine under acidic conditions. The active oxygen content of H_2O_2 or any other peroxy compound is defined as the atomic weight of active oxygen atoms times 100 divided by the molecular weight of the peroxygen compound. For example, pure H_2O_2 has an active oxygen content of 47% (16/34 x 100). With regard to peroxide reaction chemistry, the two most important reactions with regard to bleaching textiles are:

$$H_2O_2 \quad \text{-------->} \quad H^+ \quad + \quad OOH^- \tag{1.7}$$

$$H_2O_2 \quad \text{-------->} \quad H_2O \quad + \quad 1/2\ O_2 \tag{1.8}$$

The first reaction, the dissociation of H_2O_2, takes place only at high pH since hydrogen peroxide is a very weak acid ($K_{diss.}$ = 1.78 $\times 10^{-12}$ at 20°C). Thus, in practice, peroxide bleaching of textiles is most satisfactorily achieved under alkaline conditions using sodium hydroxide and sodium carbonate in conjunction with other textile auxiliaries (e.g., stabilizers and wetting agents). The second reaction is important because such decomposition may be achieved or catalyzed by the presence of metal impurities. This is true even though the decomposition of hydrogen peroxide to water and oxygen has a high activation energy (ca. 50.6 kcal/mole).

There have been several mechanisms proposed for the decomposition of peroxide in alkaline solution by metals. Most investigators favor a mechanism by which decomposition occurs through a free radical mechanism such as the one proposed in the two studies by Cates and colleagues (refs. 28, 29):

$$H_2O_2 \quad ---------> \quad H^+ \quad + \quad {}^-OOH \tag{1.9}$$

$$M^{2+} \quad + \quad H_2O_2 \quad ---------> \quad M^{3+} \quad + {}^-OH \quad + \cdot OH \tag{1.10}$$

$$\text{or } {}^-OOH + H_2O_2 \quad ---------> \quad \cdot OOH \quad + {}^-OH \quad + \cdot OH \tag{1.11}$$

$$\cdot OH + H_2O_2 \quad ---------> \quad \cdot OOH \quad + \quad H_2O \tag{1.12}$$

$$\cdot OOH \quad + \quad M^{3+} \quad ---------> \quad M^{2+} \quad + \quad O_2 \quad + \quad H^+ \tag{1.13}$$

The decomposition of the peroxide may be initiated by either (1.10) or (1.11), but there is more evidence for initiation by (1.10). The production of molecular oxygen in the last equation (1.13) has been correlated with oxidation of the cellulose and its degradation.

The use of stabilizers permits bleaching to be conducted at alkaline pH, and serves two important functions: they slow the rate of peroxide decomposition under alkaline conditions, and combine with or neutralize metal impurities (iron being the most detrimental and to a lesser extent, copper, tin, lead and brass) which may catalyze decomposition of peroxide and induce fiber damage. Sodium silicate is still the stabilizer of choice in most commercial peroxide bleaching processes because of its ability to buffer the peroxide solution against metal impurities and because it increases the pH of the bleaching bath. However, its disadvantages are that it may cause fiber tendering and uneven dyeing behavior. Therefore, a variety of organic

26

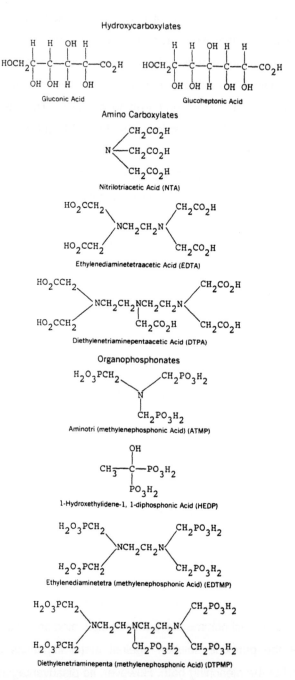

Figure 1.9. Representative organic sequestering agents in peroxide bleaching (ref. 30). Courtesy of American Association of Textile Chemists and Colorists.

stabilizers (sequestering or chelating agents) for peroxide bleaching baths are frequently utilized (Figure 1.9). Of the major classes of organic sequestering agents, organophosphonates are considered to be more suitable than amino- or hydroxycarboxylic acids (such as EDTA and its analogs). This is due to their greater resistance to oxidation in bleaching baths and because they produce whiter fabrics by preventing precipitation of silicates when used in combination with this stabilizer (refs. 30, 31).

A study was made of factors and conditions that influence damage to cotton fibers during their metal-catalyzed peroxide bleaching (ref. 32). It was determined that more damage occurred in pad-steam processes than in pad-batch bleaching processes. Use of appropriate amounts of complexing agents in the bleaching bath can minimize or totally eliminate fiber damage. Bronze powder and ferric nitrate were the most insidious metal impurities and ingredients that catalyzed fiber damage during the peroxide bleaching process.

Although sodium hydroxide or sodium hydroxide/sodium carbonate are the most frequently used and economical activators, it has been demonstrated that amides (urea, benzamide, formamide, and N,N-dimethylformamide) can be incorporated as activators into peroxide bleaching baths at temperatures from 30-90°C to obtain bleached cotton fabrics with comparable whiteness and strength to those bleached in the presence of alkali (ref. 33).

A variety of representative continuous processes employed for bleaching textiles, particularly with hydrogen peroxide, are shown in Figure 1.10. The J-box process (top figure) is still frequently used. However, the latter three systems (roller-steamer, minute-scour-rapid-bleach and pressure-scour-rapid-bleach) are considered to have several advantages over the J-box, particularly for bleaching of cotton/polyester blends (ref. 34). Short fabric contact times provide more flexibility in choosing temperature and bleach and stabilizer concentrations, and minimize or prevent fiber damage due to metal impurities and specific alkaline hydrolysis of polyester fibers. Minute and rapid open-width bleaching usually consist of a steaming time of 2-15 min at or below the boiling point of water with sodium silicate, an organic sequestering agent, caustic soda, hydrogen peroxide, tetrasodium pyrophosphate and a surfactant present in the bleaching solution in varying proportions.

A single stage process for desizing, scouring and bleaching cotton fabric with peroxide has been accomplished by using a treating bath containing sodium hydroxide, hydrogen peroxide, magnesium sulfate hexahydrate, EDTA and gluconic acid as organic sequesterants, and a mixture of anionic and nonionic wetting agents. This was done in a winch beck for 1.5 h at 95°C to obtain cotton fabrics with good whiteness, absorbency and little degradation (ref. 35).

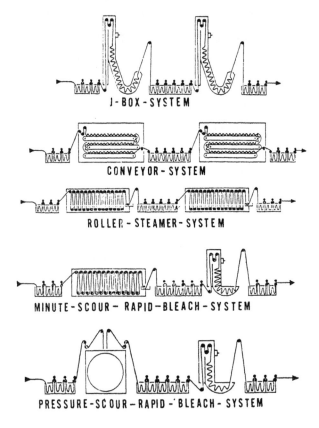

Figure 1.10. Standard and combined open-width bleaching ranges (ref. 34). Courtesy of SAF International.

Continuous bleaching techniques and open width bleaching equipment have been critically reviewed with regard to design features and advantages and disadvantages of various methods (refs. 36, 37). Topics discussed include: designs of J-boxes, preparation process steps in open width continuous bleaching, comparison of four bleaching systems (rope batch, continuous rope, continuous open width and pad-roll), use of spray drum washers for delicate fabrics and numerous other processing and cost factors. Improvements in multi-stage pad-steam peroxide bleaching processes are claimed due to newly developed nozzles in conjunction with use of multifunctional wetting agents and other textile auxiliaries (ref. 38).

It has been observed that the rate of bleaching of fabrics increases with an increase in concentration of perhydroxyl anion (-OOH$^-$). Thus, this species has been frequently proposed as the active bleaching agent. A recent study cites experimental evidence for the formation of superoxide (-OO.$^-$), the conjugate base of perhydroxyl anion, as the active species in the peroxide bleaching. The investigators reject other "active oxygen" species as the causative bleaching agent (ref. 39). However, earlier studies have proposed that bleaching action is due to the hydroperoxide radical (ref. 29, 40) and/or hydroxyl radical (ref. 29) in alkaline media by reaction of perhydroxyl anion with hydrogen peroxide:

$$^-OOH \ + \quad H_2O_2 \quad \text{--------->} \quad \ ^\cdot OOH \ + \ ^-OH \ + \ ^\cdot OH \qquad (1.14)$$

$$^\cdot OOH \ \text{or} \ ^\cdot OH \ + \ -C=C- \text{--------->} \qquad \qquad ^-OH \qquad (1.15)$$

$$_.OH \qquad + \ H_2O_2 \quad \text{--------->} \quad HO\text{-}OH^\cdot \ + \quad ^\cdot OH \qquad (1.16)$$

Additional evidence to support this mechanism is deduced from the observation that cellulosic fibers are bleached more effectively in acid media with a H_2O_2/HBr system than with other peroxide-acid systems. Under these conditions, it has been demonstrated that both $^\cdot$OH and $^\cdot$OOH radicals can also be produced to effectively decolorize impurities by free radical addition across sites of unsaturation as shown in eqs. 1.15 and 1.16 (ref. 28, 29). Additional investigations are in progress to determine the structures that actually cause peroxide bleaching in fibers.

In addition to the sizable amounts of cotton and cotton/polyester blends that are bleached with hydrogen peroxide, this agent is frequently used to bleach a variety of synthetic fibers as well as the natural protein fibers wool and silk. Lewin (ref. 24) describes the various processing conditions and formulations most commonly used for peroxide bleaching of regenerated cellulose fibers (such as rayon, cellulose acetate and triacetate), polyester, polyamide, acrylic and various cellulose/synthetic fibers and all synthetic fiber blends. Variations in time, temperature and concentrations of peroxide, stabilizers, and activators are given for each type of fabric.

The bleaching of wool with peroxide has been reviewed with regard to fiber changes occurring and representative processing conditions and formulations. When wool is bleached with peroxide, it undergoes depolymerization of its polypeptide chains, a decrease in sulfur content due to less crosslinking and cleavage of its disulfide bonds, as well as corresponding increases in alkali solubility and decreases in its weight and durability (ref. 41). Thus, the pH must be carefully controlled in bleaching wool, since it is much more susceptible to degradation under alkaline conditions than are other types of fibers. Wool and silk have been satisfactorily bleached in alkaline media with systems such as peroxide/ammonium hydroxide/sodium tripolyphosphate or with peroxide/sodium silicate/sodium carbonate, with times and temperatures varied to meet particular bleaching requirements. Alternatively, wool may also be effectively bleached in acid media (pH 4-4.5) with a solution of hydrogen peroxide and formic acid.

A new combination bleaching and dyeing process for wool involves sequential oxidation with hydrogen peroxide and reduction with thiourea in the same bath. After the bath temperature is reduced, desired dyes are added to the bath. Formation of sulfinate anions results in a strong reducing agent that promotes effective bleaching (ref. 42).

Several types of synthetic fibers are bleached with peracetic acid (CH_3COOOH) rather than with hydrogen peroxide because satisfactory reflectance values for fabrics may be obtained at a lower pH without fiber damage. Fibers subject to alkaline hydrolysis and degradation (e.g., polyester, cellulose acetate and cellulose triacetate) thus may be bleached with peracetic acid at a pH of 6.0-8.0 without appreciable fiber degradation or hydrolysis. The mechanism by which peracetic acid bleaches fibers is thought to be similar to that of hydrogen peroxide, since both are catalytically decomposed by metal ions and stabilized by sequestering agents such as the polycarboxylic acids.

The mechanism by which cotton is effectively bleached at low temperatures (ca. 30°C) with peracetic acid containing 2,2'-bipyridine (chelating agent) and a transition metal salt (preferably Co^{2+} or Fe^{2+}) has been investigated (refs. 43, 44). Initial complexation of the cellulose with the transition metal salt and 2,2'-bipyridine occurs, followed by subsequent reaction of the cellulose complex with the peracetate anion (CH_3COOO^-) to displace one unit of cellulose anion (Cell-O^-). The new complex containing one unit of cellulose and the peracetoxy group then reacts with conjugated sites by an epoxidation reaction to decolorize impurities. An alternative method of bleaching cotton at low temperatures with peracetic acid involves use of ultrasonic radiation (ref. 45). This method is claimed to be more environmentally desirable than peracetic acid/metal complex formulations described above.

In contrast to oxidative bleaching agents, bleaching agents based on the chemical reduction of colored impurities (e.g., reduction of -C=O to -CH$_2$ groups) are now only occasionally used in commercial practice. Historically, sulfur dioxide was first used for bleaching wool. Later, other bleaching agents such as sodium or zinc hydrosulfite ($Na_2S_2O_4$ or ZnS_2O_4) or bisulfite-aldehyde addition compounds (zinc formaldehyde sulfoxylate) were employed. Because of the problems that sulfur dioxide caused with regard to fading of dyes and air pollution, and the tendency for all reducing-type bleaching agents to produce color reversion in textiles, they have been almost completely replaced by oxidative bleaching agents.

1.5.3 Fluorescent brightening agents

Fluorescent or optical brighteners, in contrast to chemical bleaching agents (oxidation or reduction of colored impurities), represent a class of physical bleaching agents because these substances produce a whitening effect by exhibiting fluorescence on the surface of the fabric or fiber. These brighteners have been reviewed extensively in books, monographs, and journals with regard to principles, chemical classes, methods of application and suitability for various fibers and fiber blends (refs. 46-49).

Ideally, these compounds absorb energy in regions of short wave-length (below 400 nm in the ultraviolet region), then reemit the energy at longer wavelengths (primarily in the visible region). They should also be colorless and not absorb energy in the visible region. The theory and measurement of whiteness, reflectance and other visual characteristics commonly associated with color and appearance of fibrous surface containing optical brighteners is discussed later in this text in Chapter 5 under the subtopic of color measurement.

Figure 1.11 shows desirable absorption and emission characteristics of a representative fluorescent brightener used for cotton and other types of cellulosic fibers. Emission in the blue-violet region (400-480 nm) is preferred to give the fabric a neutral blue hue, but occasionally emission at somewhat shorter or longer wavelengths produces respectively, fabrics with a red or a green hue.

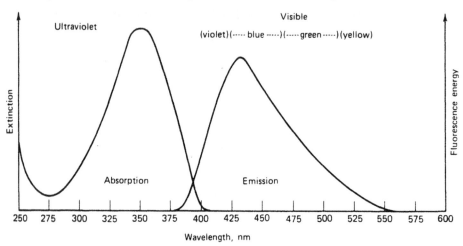

Figure 1.11. Absorption and emission spectra in solution of a bis(triazinyl)stilbene optical brightener (ref. 46). Courtesy of John Wiley & Sons Publishers.

The general structural requirements for these substances are listed and discussed by Williamson (ref. 47). The brighteners must have a planar structure and be chemically conjugated, preferentially containing electron-donating groups (e.g., -OH, -NH$_2$) and not electron-withdrawing (e.g., -NO$_2$, -N=N-), and have the capability to remit light in the region of 450 nm (blue color) to neutralize the yellow component of the visible region present in most textiles. In addition to amino, hydroxyl, alkoxy, or other electron-donating groups, brighteners may contain sulfonic acid groups (analogous to acid dyes), cationic groups (analogous to basic dyes) and water-insolubilizing groups (analogous to disperse dyes). Fiber type will dictate the particular chemical structure chosen from optical brighteners available.

The most comprehensive and current list of chemical classes and structures of these compounds and their suitability for application to specific fiber types is given by Siegrest and his co-authors (ref. 49). There are essentially three major classes:

bis(triazinyl)stilbenes

nonionic distyrl-arenes

1,3-diphenyl-2-pyrazolines

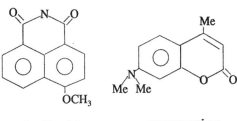

napthalimides   coumarins

Figure 1.12a. Representative chemical classes of fluorescent brightening agents.

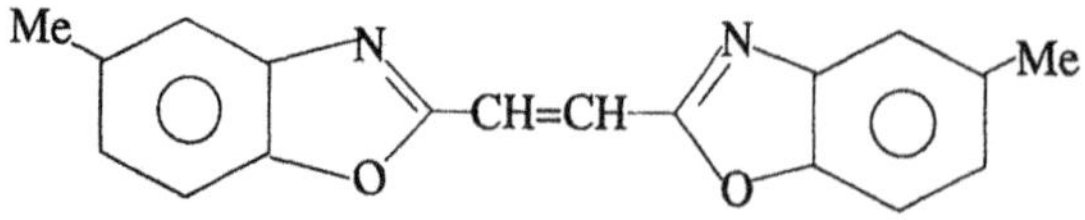

bis(benzoxazoles)

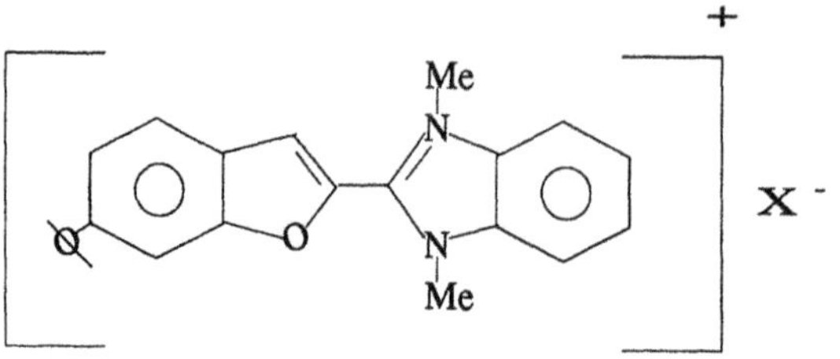

cationic azoles

Figure 1.12b. Representative chemical classes of fluorescent brightening agents.

carbocyclic, carbocyclic-heterocyclic and heterocyclic. The earliest optical brighteners were primarily of the bis(triazinyl)stilbene class (first structure in Figure 1.12a), and developed for application to cotton and other cellulosic textiles. However, these classes of brighteners were observed to be unstable to hypochlorite bleach. Thus, other types of brighteners (e.g., monoazole stilbenes containing anionic groups) were synthesized to alleviate this problem for cellulosic fabrics. The sulfonic acid groups in the first structure in Figure 1.12a are anionic groups analogous to those found in direct dyes applied to cotton. Thus, each type of fluorescent brightener has structural features similar (except for the absence of chromophoric groups) to those for classes of dyes with respect to fiber type and substantivity. Moreover, the brighteners are usually applied under conditions (e.g., pH and temperature) and processes comparable to those used for dyeing specific fiber types (as discussed in Chapter 3: **Methods of Applying Dyes to Textiles**).

The varieties of nonionic brighteners shown in Figures 1.12a and 1.12b are most frequently used with polyester, polyamide and cellulose acetate fabrics and normally applied as a disperse dye would be applied. These include stilbenes containing only nonpolar groups, and numerous nonionic heterocyclic structures such as benzofurans, azoles, 1,3-diphenyl-2-pyrazolines, coumarins, naphthalimides, bis(benzoxazoles) and bis(benzimidazoles). The anionic 1,3-diphenyl-2-pyrazoline shown in Figure 1.12a is representative of a group of brighteners that are applied to wool, silk and polyamide fabrics. The cationic azole depicted in Figure 1.12b is a brightener that is applied to acrylic fabrics.

In addition to certain structural and physicochemical aspects of optical brighteners that are required for their use and effectiveness (e.g., planarity and fluorescence), there are several other criteria and desirable properties that these agents must possess. These have been critically evaluated by Williamson (ref. 47). The fluorescent brightening agent must be substantive to the fiber, have acceptable rates of migration and build-up on the fiber, be compatible with salt solutions or electrolytes, and have stability and chemical reactivity over desired temperatures and pH ranges. The brighteners should also be compatible for coapplication with chemical bleaches (such as hypochlorite or peroxide) and/or with topical chemical finishes at a later stage of textile processing. These agents should also be amenable to a variety of processes used in dyeing (exhaust, pad-steam) and finishing (pad-dry-cure) of fabrics.

An extensive number of examples and processes by which fluorescent brightening agents are applied to fabrics are described in Williamson's book (ref. 47). For cellulosic fabrics (particularly cotton), there are three groups of brighteners that vary in their affinity and substantivity. Those with low affinity are very dependent on the presence of salt or other electrolytes and the application temperature, while those with medium affinity are less dependent on these two factors. Brighteners with high affinity are only slightly dependent on temperature and amount of added electrolytes. A typical process for application of a fluorescent brightener to cotton fabric is a single bath scour and bleach (most commonly used process for this type of fiber):

This process involves the inclusion of a sulphonated fatty alcohol (SFA) type of detergent in the bleach formula. The bath is charged with:

Hydrogen peroxide (35%)	7.5-10ml/l
Sodium silicate (1.4 g/cm^3)	3 g/l
Soda ash	1 g/l
SFA, e. g., Gardinol LP	2 g/l
Fluorescent brightener	x% (based on wt. of fabric)

The bath is set at 40-50°C and the fabric circulated for 10 min in this liquor. The temperature is then raised to the boiling point and bleaching is continued at the boil for 60 min. The material is then given a hot rinse, followed by cold rinses. In the final rinse, it is often necessary to add a suitable softening agent. (ref. 47)

For wool fabrics, fluorescent brighteners are usually coapplied with reducing agents (such as sodium hydrosulfite) to minimize yellowing that occurs as a frequent adverse side effect when these optical brighteners are affixed to wool. Proper choice of brighteners for wool fabrics is also important to minimize yellowing, e.g., use of the anionic 1,3-diphenyl-2-pyrazoline shown in Figure 1.12a. Affixing brighteners to wool fabrics is conducted under processing conditions similar to application of acid dyes to wool. Both processes use low pH baths containing acetic or formic acid. Fabrics comprised of synthetic fibers may be optically whitened by techniques and processes used to apply dyes to these types of fibers. Thus, brighteners that are nonionic (such as four of the six structures in Figures 1.12a and 1.12b) are frequently applied to polyester and polyamide fabrics by processes used to affix disperse dyes. These processes include exhaustion methods at various temperatures and continuous and semi-continuous methods (e. g., pad-roll, pad-wash, pad-steam, and pad-bake or Thermosol). A typical and frequently used method for polyester fabrics is a pad-Thermosol method that is described as follows:

A typical formula is:

Fluorescent brightener 10-40 g/l according to shade requirement
Dispersing agent   1-2 g/l
Antimigration agent
(if required)   5 g/l
Temperature of bath   25-30°C

Pad to liquor pick-up of 50% and dry at 100-120°C. Bake at recommended temperature (170-225°C) for 20-40 seconds. If an antimigration agent is used, the following scour should be applied:

Non-ionic detergent   0.5-1.0 g/l
Soda ash   0.5-1.0 g/l
Scour 20-30 min at 50-60°C, rinse well and acidify. (ref. 47)

For fiber blends, there are a variety of factors that must be considered in order to effectively apply brightening agents. In some instances, only the predominant fiber in the blend is optically bleached with the appropriate brightener, while in other situations, a mixture of two different and compatible brighteners for each type of fiber in the blend is used.

1.6 MERCERIZATION AND LIQUID AMMONIA TREATMENT OF TEXTILES

1.6.1 Mercerization of textiles

Mercerization is appropriately named after John Mercer, who in 1850, discovered that cotton yarn or fabric immersed in aqueous solutions of caustic soda (NaOH) exhibited swelling and shrinkage. Changes in fine structure, morphology and conformation of the cellulose chains occur with mercerization. More specifically, treatment of cellulosic fibers with aqueous solutions of sodium hydroxide at various

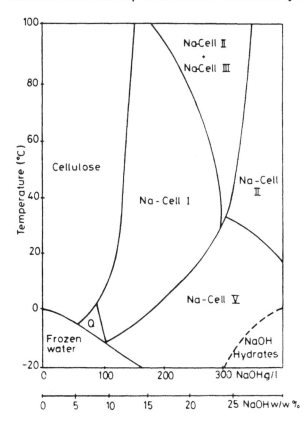

Figure 1.13. Phase diagram of soda celluloses (ref. 19). Courtesy of Oldenbourg Publishers.

temperatures (e.g., 15% or greater concentrations of aqueous NaOH at 25°C for cotton) produces a polymorphic change in the cellulose (known as Cellulose II). A concomitant increase also occurs in fiber accessibility to chemical reagents and dyes due to a decrystallization or lower degree of order by rearrangement of cellulose chains and hydrogen bonds between the chains.

The hydrodynamic diameter of various hydrates of NaOH decreases with increasing concentration of the alkali in water. Formation of cellulosate anion and the different sizes of $NaOH.xH_2O$ have been offered as explanations why certain concentrations of alkali metal hydroxides are effective in penetrating between cellulose chains to cause maximum swelling, cleavage of hydrogen bonds, and a change in the conformation of native cellulose to Cellulose II. Historical development of these observations and the composition and identification of five different soda celluloses (Na-Cell I through Na-Cell V) that are consistent with x-ray diffraction data have been reviewed (ref. 19). A phase diagram of the different soda celluloses in Figure 1.13 shows the relationship between NaOH concentration and temperature for each of the phases. Sarko and co-authors have proposed elaborate pathways by which native cellulose is converted to each of the soda celluloses that ultimately lead to the thermodynamically stable and irreversible Cellulose II structure (refs. 50-53). However, these pathways are based on the premise that Cellulose II exists in antiparallel chain conformations. These investigators argue that this antiparallel conformation occurs in initial conversion of native cellulose to Na-Cellulose I (Figure 1.14).

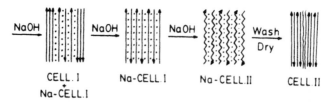

Figure 1.14. Possible mechanism of mercerization (chain directions are indicated by arrows) (ref. 51). Courtesy of John Wiley & Sons Publishers.

Mechanisms by which native cellulose (or Cellulose I) are converted into Cellulose II and the resulting structure of the alkali-treated cellulose have been and continue to be the subject of numerous theoretical and practical investigations. The most

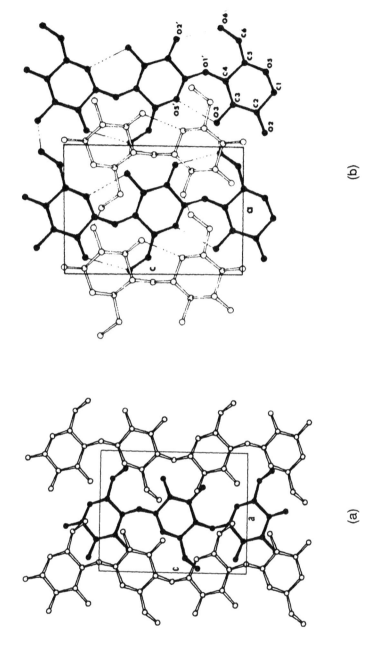

(a)

(b)

Figure 1.15. (a) ab projection for structure of Cellulose I and (b) ac projection for structure of Cellulose II (ref. 54). Courtesy of the American Chemical Society.

40

popular view that has been held for the past fifty years is that Cellulose I exists in a parallel chain conformation while Cellulose II exists in an antiparallel chain conformation. Another example of this argument is shown in the structures proposed by Blackwell and Kolpack (Figure 1.15), who based their hypotheses on extensive X-ray and electron diffraction data (ref. 54).

A more recent and simpler explanation of how Cellulose I is converted into Cellulose II and what the arrangements of the cellulose chains are in the unit cell is given by Turbak et al. (refs. 55, 56). Computer-generated models (consistent with x-ray diffraction and other experimental data) indicate that Cellulose II exists in a parallel-up chain conformation that can be achieved by initially breaking intramolecular hydrogen bonds between 6-OH and 2-OH on the adjacent glucose residue (trans/gauche or "tg" conformation). The formation of new and stronger intermolecu-

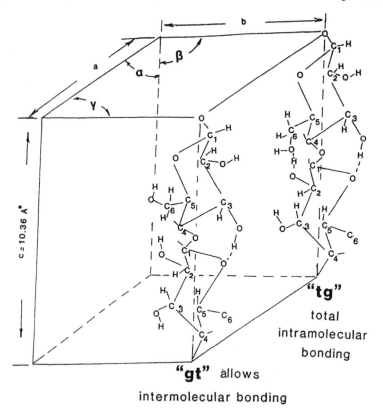

Figure 1.16. Cellulose II parallel "gt" and "tg" chains (ref. 56). Courtesy of the American Chemical Society.

lar hydrogen bonds occurs on conversion to Cellulose II. The cellulose chain is now in the less sterically hindered and more thermodynamically stable gauche/trans or "gt" conformation (see Figure 1.16).

Moreover, the computer-generated models indicate that substantial amounts of energy would be required to convert the parallel-up chains in native cellulose into the antiparallel chain conformations of Cellulose II. Although these investigators (ref. 55, 56) have not specifically discussed "soda-cellulose intermediates", their Cellulose II structural model is consistent with the presence of different degrees of solvation/hydration of sodium hydroxide between cellulose chains at various degrees of chain separation, swelling and deswelling as well as equilibria in the conversion of cellulose to cellulosate ions.

In addition to changes in crystallinity, accessibility, and unit cell structure, the orientation of cellulosic fibers, i.e., the fibrils along the fiber axis, increases with swelling in alkali due to the restraining influences of the primary fiber wall. The extent to which sodium hydroxide solutions change the crystallinity, accessibility, and orientation of the cellulose fiber depend on many factors: the concentration of the sodium hydroxide, temperature, degree of polymerization and source of the cellulose, the physical state of the cellulose, i.e., fiber, yarn or fabric, and the degree of tension employed to restrict or promote fiber shrinkage and swelling. Interrelationship of these factors and their practical significance have been critically reviewed by Zeronian (ref. 57). These variables have been effectively controlled for mercerization of cotton, other cellulosics and blends of synthetic fibers and cellulosics to improve or optimize one or more of the following properties: dimensional stability, affinity for dyes, tensile strength, higher add-on of finishing agents, luster, and fabric smoothness.

There are several commercial processes and types of machinery for mercerization of yarns and woven and knitted fabrics (cellulosic and cellulosic blends). Since maximum swelling and fiber penetration occur between 15-23% NaOH at 0-25°C for cotton cellulose, mercerization is frequently conducted at or below ambient temperatures with concentrations of NaOH at or greater than 20%. Mercerization of yarns produces more dramatic changes in properties than does mercerization of fabrics. This is due to greater penetration of the cellulosic fibers by the alkali to cause greater conversion to Cellulose II and changes in crystallite orientation. When yarns are mercerized under tension or particularly when they are mercerized in the slack state and restretched to their original length, marked increases in the tenacity of the fibers are obtained.

Woven fabrics are mercerized on machines without chains (chainless) or with stretching chains to maintain constant length of the fabric in one or both dimensions and improve the mechanical properties of the fabric by application of tension. Chainless mercerizing machines have advantages over machines with stretching chains in that fabric tears are avoided and the process is less cumbersome. However, chainless machines have the disadvantage that the amount of tension applied through bowed rollers is usually less than that achieved by machines with stretching chains (ref. 19, 58). Wetting agents compatible with the alkaline solution are used to improve mercerization of fabrics, particularly when the cotton is in the grey state, i.e., contains the natural waxes and pectins and has not been desized, scoured and bleached. Development of new high efficiency washing units in the mercerizing machines have minimized the amount of water required to adequately remove alkali from the processed fabrics (ref. 58). Newer methodologies allow for washing the fabrics in open width under tension until internal changes in the fiber and concomitant fabric dimensional stability have been achieved (ref. 59). This procedure produces a more uniform stretching effect in the fabric in the warp or long direction.

Knitted fabrics pose special problems in mercerization because of variation of the amount of shrinkage and tension from the edge to the center of the fabric when they are wet with alkali. This problem has been largely resolved by mercerization of knits in tubular machines or machines of related design that control and stabilize the shrinkage and tension of knitted fabrics (ref. 60).

Besides conventional processes mentioned above, techniques have been developed to mercerize both cotton/rayon and cotton/polyester blends. In each instance, time, temperature, and alkali concentration have to be carefully controlled. Mercerization of rayon produces more pronounced swelling and shrinkage than mercerization of cotton due to the lower degree of polymerization of the cellulose, while polyester fibers may be hydrolyzed and undergo surface etching when immersed in alkali.

Processes in which the cellulosic fabrics are immersed in hot caustic solution (called hot mercerization) afford better penetration of the alkali into the fibers than ambient temperature mercerization. However, to obtain the best improvement in dyeability, luster and tensile properties, the fabrics that are first wet with hot alkali are subsequently cooled to ambient temperature prior to final removal of caustic on washing. This two step technique allows the fabric to be stretched while in the hot alkali and results in better conversion of the fibers from native cellulose to Cellulose II. However, this does have the disadvantage of increasing autoxidation of the cellulosic hydroxyl groups when they are exposed to hot alkali (ref. 19).

1.6.2 Liquid ammonia treatment of textiles

Besides mercerization with sodium hydroxide, the only other swelling pretreatment that has gained any commercial acceptance is the treatment of cotton and other cellulosic fibers with liquid ammonia. Research studies on this subject were initially conducted in the mid 1930's when investigators such as Barry et al. (ref. 61) observed that cotton fibers became swollen and plasticized in liquid ammonia, and underwent a polymorphic change (currently denoted as Cellulose III). In contrast to cotton fibers treated with aqueous sodium hydroxide, cotton fibers treated with liquid ammonia revert to the native Cellulose I lattice when the ammonia is removed from the fibers by immersion in water.

Two comprehensive reviews (refs. 62 and 63) on the effect of liquid ammonia on cellulosic fibers indicate that substantial progress has been made in understanding the mechanism of the interaction of the ammonia with the cellulose and in developing commercial processes based on this pretreatment. The phase diagram in Figure 1.17 shows that the cellulose-ammonia complex (A-C) is the key intermediate in explaining how Cellulose III is formed, and how this metastable polymorph of cellulose is readily converted back to native cellulose. The conflicting results obtained in earlier investigations can also be rationally explained by the pronounced effect that time and temperature have on conversion of native cellulose to the cellulose-ammonia complex. Evidently the ammonia is not sterically bulky nor basic enough to cause the thermodynamically irreversible rearrangement of hydrogen bonding that occurs when native cellulose is converted to Cellulose II by alkali.

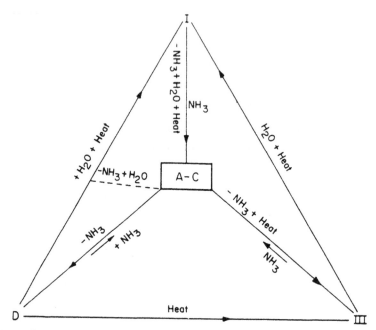

Figure 1.17. Phase diagram of ammonia cellulose (A-C), disordered cellulose (D), cellulose I (I), and cellulose III (III). A-C is the vertex of a tetrahedron and is placed above the plane of the paper (ref. 63). Courtesy of Marcel Dekker Publishers.

Although some patents were filed about the same time that fundamental studies were conducted, the swelling, shrinkage, and metastable structural changes produced by liquid ammonia on cellulosic fibers were not practically utilized until three decades later, when two commercial processes for the treatment of yarns and fabrics were described in a conference on the liquid ammonia treatment of cellulosic textiles (ref. 64). The treatment of yarns (Prograde process--schematic shown in Figure 1.18) consists of immersion in liquid ammonia at its boiling point (-33°C or -28°F) for less than a second, then subsequent immersion of the yarn under tension in hot water for about 0.1 sec to produce a 40% increase in tensile strength, improved luster and heat resistance and greater affinity for dyes (ref. 64).

Treatment of fabrics with liquid ammonia (TEDECEO process--schematic shown in Figure 1.19) is based on a patent in which cotton fabrics are exposed to liquid ammonia at -33°C for 10 sec, then the majority of the ammonia is removed from the fabric by passing it over heated cans. Final traces of ammonia are removed by subsequent steaming; the resultant fabric had improved dimensional stability, tensile

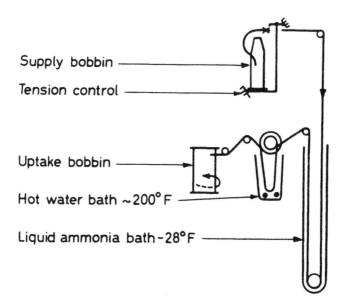

Supply bobbin

Tension control

Uptake bobbin

Hot water bath ~200° F

Liquid ammonia bath -28° F

Figure 1.18. Diagram of the Prograde process (ref. 64). Courtesy of the Shirley Institute.

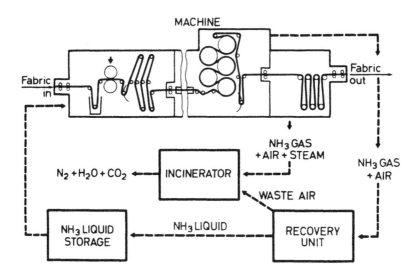

Figure 1.19. Flow diagram of the complete system for the TEDECO process (ref. 65). Courtesy of the Shirley Institute.

strength, smooth drying properties and resistance to abrasion (refs. 64, 65). Most subsequent patents and commercial practices are based on the TEDECO process for treatment of fabrics rather than the Prograde process based on treatment of yarns.

The influence of liquid ammonia treatment on the supramolecular and morphological structure of cotton has been critically reviewed and analyzed with respect to changes in pore structure, pore volume distribution and specific fiber surface area (ref. 66). It has also been reviewed from the perspective of intracrystalline swelling (ref. 57). With the aid of size-exclusion chromatography, it has been determined that the method of removal of ammonia from the fiber (by evaporation or dry process; by washing or wet process) produces distinct changes in pore structure that effect subsequent performance in dyeing and durable press finishing (ref. 66). Additional studies are in progress to further correlate morphological fiber changes on exposure to ammonia with fabric performance.

GENERAL REFERENCES FOR PREPARATORY PROCESSES

J. P. Farr, W. L. Smith and D. S. Steichen, **Bleaching agents (survey)**, in: J. I. Kroschwitz (Exec. ed.), Kirk-Othmer Encyclopedia of Chemical Technology, Vol. 4, 4th Ed., Wiley-Interscience, New York, 1992, pp. 271-300.

M. Lewin and S. B. Sello (Eds.), **Handbook of Fiber Science and Technology: Vol. I. Chemical Processing of Fibers and Fabrics. Fundamentals and Preparation. Pt. A and Pt. B.,** Marcel Dekker, New York, 1983 and 1984.

E. S. Olsen, **Textile Wet Processes. Vol I. Preparation of Fibres and Fabrics**, Noyes Publishers, Park Ridge, N. J., 1983.

E. R. Trotman, **Textile Scouring and Bleaching**, Griffin Publishers, London, 1968.

TEXT REFERENCES FOR PREPARATORY PROCESSES

1 E. Remund, **Singeing of textile webs**, Text. Praxis Intl. 39 (9) (1984) 910-912 and XXI-XXII.

2 R. S. Bhagwat, **Singeing machine**, Colourage 38 (1) (1991) 57-66.

3 R. Ebbinghaus, **Modern singeing technology - essential for all textile finishers**, Intl. Text. Bull. Dyeing, Printing Finishing Ed. 38 (2d Quarter, 1992) 29-36.

4 P. G. Drexler and G. C. Tesoro, **Materials and processes for textile warp sizing**, in: M. Lewin and S. B. Sello (Eds.), Handbook of Fiber Science and Technology: Vol. I. Chemical Processing of Fibers and Fabrics. Fundamentals and Preparation. Pt. B., Marcel Dekker, New York, 1984, pp. 1-89.

5 D. M. Hall, **Relation of fiber properties to slashing**, in: J. C. Farrow, D. M. Hall and W. S. Perkins (Eds.), Theory and Practice of Textile Slashing, ATOE Executives Monograph No. 1, Auburn Univeristy, Auburn, AL., Apr. 1972, pp. 210-262.

6 J. F. McMahon and G. H. J. van der Walt, **A review of recent developments in warp sizing**, SAWTRI Spec. Publ. No. 71, SAWTRI, Port Elizabeth, S. Africa, Aug. 1985, 35 pp.

7 K. Dickinson, **Oxidative desizing**, Rev. Prog. Color. Rel. Topics 17 (1987) 1-6.

8 N. A. Ibrahim, **Optimization of desizing water soluble sizes - Part IV: Options for washing out PVA-size**, Am. Dyest. Reptr. 80 (7) (1991) 32-48.

9 J. D. M. Gibson, W. V. Morgan and B. Robinson, **Wool scouring and effluent treatments**, Wool Sci. Rev. 57 (Feb. 1981) 2-32.

10 G. F. Wood, **Raw wool scouring, wool grease recovery and scouring wastewater disposal**, in: M. Lewin and S. B. Sello (Eds.), Handbook of Fiber Science and Technology: Vol. I. Chemical Processing of Fibers and Fabrics. Fundamentals and Preparation. Pt. A., Marcel Dekker, New York, 1983, pp. 205-257.

11 J. R. Christoe, **Developments in scouring--an Australian view**, Wool Sci. Rev. 64 (Dec. 1987) 25-43.

12 T. E. Mozes, **A review of raw wool carbonizing**, SAWTRI Spec. Pub. No. 74, SAWTRI, Port Elizabeth, S. Africa, Apr., 1986, 33 pp.

13 M. T. Pailthorpe, **Developments in wool carbonizing**, Rev. Prog. Color. Rel. Topics 21 (1991) 11-22.

14 T. Shaw and M. A. White, **The chemical technology of wool finishing**, in: M. Lewin and S. B. Sello (Eds.), Handbook of Fiber Science and Technology: Vol. II. Chemical Processing of Fibers and Fabrics. Functional Finishes. Pt. B., Marcel Dekker, New York, 1984, pp. 320-328.

15 M. S. Nossar, B. W. Edenborough and M. Chaikin, **The effectiveness of different surfactants in the 'rapid-acidification' carbonizing process**, J. Text. Inst. 70 (2) (1979) 62-71.

16 A. Kirkpatrick, **A new approach to wool carbonizing**, Text. Res. J. 56 (1) (1986) 67-71.

17 C. Wang and M. T. Pailthorpe, **An investigation of wool hydrolysis in conventional carbonizing**, Text. Res. J. 59 (1989) 232-236.

18 E. R. Trotman, **Chapter 4. Scouring**, in: Textile Scouring and Bleaching. Griffin Publishers, London, 1968, pp. 78-117.

19 R. Freytag and J.-J. Donzé, **Alkali treatment of cellulose fibers**, in: M. Lewin and S. B. Sello (Eds.), Handbook of Fiber Science and Technology: Vol. I. Chemical Processing of Fibers and Fabrics. Fundamentals and Preparation. Pt. A., Marcel Dekker, New York, 1983, pp. 94-165.

20 M. L. Gulrajani, **Degumming of silk**, Rev. Prog. Color. Rel. Topics 22 (1992) 79-89.

21 R. B. Hull, Jr., **Weighing the alternatives in scouring technology**, Am. Text. Reptr./ Bulletin 9 (6) (June 1980) 41-43.

22 B. M. Baum, J. H. Finley, J. H. Blumbergs, E. J. Elliott, F. Scholer and H. L. Wooten, **Bleaching agents**, in: M. Grayson (Exec. ed.), Kirk-Othmer Encyclopedia of Chemical Technology, Vol. 3, 3rd Ed., Wiley-Interscience, New York, 1978, pp. 938-958.

23 A. Agster, **A close look at the reaction mechanisms of the principal bleaching processes**, Mell. Textilber. 59 (11) (1978) 908-912.

24 M. Lewin, **Bleaching of cellulosic and synthetic fabrics**, in: M. Lewin and S. B. Sello (Eds.), Handbook of Fiber Science and Technology: Vol. I. Chemical Processing of Fibers and Fabrics. Fundamentals and Preparation. Pt. B., Marcel Dekker, New York, 1984, pp. 158-175, 228-241.

25 M. L. Gulrajani and R. Venkatraj, **A low temperature scouring/bleaching process for cotton using sodium chlorite**, Text. Res. J. 56 (8) (1986) 476-483.

26 I. Chesner, **Some bleaching problems and developments associated with hydrogen peroxide, peracetic acid, and sodium chlorite**, J. Soc. Dyer. Color. 79 (4) (1963) 139-146.

27 W. Sebb, **Influence of oxidative bleaching agents on the content of halogenated organic compounds (AOX) of bleaching effluents**, Text. Praxis Intl. 44 (8) (1989) 841-843.

28 A. M. Taher and D. M. Cates, **Bleaching cellulose: Part I. A free radical mechanism**, Text. Chem. Color. 7 (12) (1975) 220-224.

29 W. G. Steinmiller and D. M. Cates, **Bleaching cellulose: Part II. Effect of impurities**, Text. Chem. Color 8 (1) (1976) 14-18.

30 X. Kowalski, **Sequestering agents in bleaching and scouring**, Text. Chem. Color. 10 (8) (1978) 161-165.

31 B. J. J. Engbers and G. Dierkes, **Sequestering agents and ion exchangers used in the textile finishing industry**, Text. Praxis Intl. 47 (4) (1992) 365-368; 47 (5) (1992) 462-466; 47 (6) (1992) 557-560.

32 J. Reicher, **Decomposition kinetics of the oxidizing agent and damage to cotton cellulose during metal-catalyzed bleaching with hydrogen peroxide**, Mell. Textilber. 73 (7) (1992) 572-577.

33 T. K. Das, A. K. Mandavawalla and S. K. Datta, **Amide-activated bleaching processes**, Text. Dyer & Printer 19 (21) (Oct. 8, 1986) 21-28.

34 W. Prager, **New preparation system for cotton and blends with polyester**, Am. Dyestuff Reptr. 67 (7) (1978) 24-28, 48.

35 A. Hebeish and S. El-Bazza, **Single stage process for desizing, scouring and bleaching of cotton fabric**, Angew. Makromol. Chem. 129 (1985) 169-188.

36 B. Mahapatro, A. K. Shikarkhane and P. R. Deshpande, **Continuous bleaching of cotton, polyester/cotton blend fabrics, present practice/future trends**, Colourage 38 Annual Issue (1991) 25-39.

37 R. S. Bhagwat, **Continuous open width bleaching equipment**, Colourage 38 (7) (1991) 67-78.

38 B.-D. Bähr, J. Carbonell and P. Farber, **Development of a new preparation technology for use in the finishing industry**, Text. Praxis Intl. 46 (8) (1991) 780-785.

39 J. Dannacher and W. Schlenker, **What is "active oxygen?" The mechanisms of hydrogen peroxide bleaching in the washing process**, Textilveredlung 25 (6) (1990) 205-207.

40 P. Ney, **Chemism of the alkaline bleach of textile cellulosic fibers with hydrogen peroxide**, Mell. Textilber. 63 (6) (1982) 443-453.

41 J. Cegarra and J. Gacén, **The bleaching of wool with hydrogen peroxide**, Wool Sci. Rev. 59 (Sept. 1983) 1-44.

42 M. Arifoglu and W. N. Marmer, **Sequential oxidative/reductive bleaching and dyeing of wool in a single bath at low temperatures**, Text. Res. J. 62 (3) (1992) 123-130.

43 J. W. Rucker, **Low temperature bleaching of cotton with peracetic acid**, Text. Chem. Color. 21 (5) (1989) 19-25.

44 J. W. Rucker and S. A. Satterwhite, **2,2'-Bipyridine catalyzed bleaching of cotton fibers with peracetic acid. Pt. III. Effect of chain length of alkyl sulfate surfactants**, Text. Res. J. 61 (5) (1991) 273-279.

45 K. Poulakis, H.-J. Buschmann, U. Denter and E. Schollmeyer, **Bleaching of cotton with peracetic acid assisted by ultrasonic radiation**, Text. Praxis Intl. 46 (4) (1991) 334-335.

46 R. Zweilder and H. Hefti, **Brighteners, fluorescent**, in: M. Grayson (Exec. ed.), Encyclopedia of Chemical Technology, Vol. 4., 3rd Ed., Wiley-Interscience, New York, 1978, pp. 213-226.

50

47 R. Williamson, **Fluorescent Brightening Agents**, Elsevier, New York, 1980, 153 pp.

48 R. Levene and M. Lewin, **The fluorescent whitening of textiles**, in: M. Lewin and S. B. Sello (Eds.), Handbook of Fiber Science and Technology: Vol. I. Chemical Processing of Fibers and Fabrics. Fundamentals and Preparation. Pt. B., Marcel Dekker, New York, 1984, pp. 257-304.

49 A. E. Siegrist, H. Hefti, H. R. Meyer and E. Schmidt, **Fluorescent whitening agents 1973-1985**, Rev. Prog. Color. Rel. Topics 17 (1987) 39-55.

50 T. Okano and A. Sarko, **Mercerization of cellulose. I. X-ray diffraction evidence for intermediate structures**, J. Appl. Poly. Sci. 29 (1984) 4175-4182.

51 T. Okano and A. Sarko, **Mercerization of cellulose. II. Alkali- cellulose intermediates and a possible mercerization mechanism**, J. Appl. Poly. Sci. 30 (1985) 325-332.

52 H. Nishimura and A. Sarko, **Mercerization of cellulose. III. Changes in crystallite sizes**, J. Appl. Poly. Sci. 33 (1987) 855-866.

53 H. Nishimura and A. Sarko, **Mercerization of cellulose. IV. Mechanism of mercerization and crystallite sizes**, J. Appl. Poly. Sci. 33 (1987) 867-874.

54 J. Blackwell, K. H. Gardner, F. J. Kolpak, R. Minke and W. B. Claffey, **Refinement of cellulose and chitin structures** in: A. French and K. H. Gardner (Eds.), Fiber Diffraction Methods, ACS Symposium Series No. 141, Ch. 19, Am. Chem. Soc., Washington, D. C., 1980, pp.315-334.

55 A. Sakthivel, **Crystal structures of cellulose II derived from packing energy minimization**, Diss. Abstr. Int. B 49 (5) (1988) 1744-1745.

56 A. Turbak and A. Sakthivel, **Solving the cellulose enigma**, CHEMTECH 20 (7) (1991) 444-446.

57 S. H. Zeronian, **Intracrystalline swelling of cellulose**, in: T. P. Nevell and S. H. Zeronian (Eds.), Cellulose Chemistry and Its Applications, Ellis Horwood Ltd., Chichester, ENGLAND, 1985, pp. 159-180.

58 R. S. Bhagwat, **Section II - Desizing, scouring, mercerizing and bleaching machines**, Colourage 38 (2) (1991) 61-70.

59 S. Grief, **Addition mercerization - a new processing technology**, Mell. Textilber. 72 (9) (1991) 753-756.

60 P. F. Greenwood, **Mercerisation and liquid ammonia treatment of cotton**, J. Soc. Dyer. Color. 103 (1987) 342-349.

61 A. J. Barry, F. C. Peterson and A. J. King, **Studies of reactions of cellulose in non-aqueous systems. I. Interaction of cellulose and liquid ammonia**, J. Am. Chem. Soc. 58 (1936) 333-339.

62 F. W. Herrick, **Review of the treatment of cellulose with liquid ammonia**, in: J. Appl. Poly. Sci., A. Sarko (Ed.), Applied Polymer Symposium 37: Proceedings of the Ninth Cellulose Conference . II. Symposium on Cellulose and Wood as Future Chemical Feedstocks and Sources of Energy, and General Papers, Wiley-Interscience, New York, 1983, pp.993-1023.

63 C. V. Stevens and L. G. Roldán-González, **Liquid ammonia treatment of textiles**, in: M. Lewin and S. B. Sello (Eds.), Handbook of Fiber Science and Technology: Vol. I. Chemical Processing of Fibers and Fabrics. Fundamentals and Preparation. Pt. A., Marcel Dekker, New York, 1983, pp. 167-203.

64 R. M. Gailey, **The liquid ammonia treatment of yarns and threads. 1. Principles and practice**, Conference Proceedings on Liquid Ammonia Treatment of Cellulosic Textiles, Shirley Institute, Manchester, England, Nov. 17, 1970, pp. 9-20.

65 O. Skaathun, **The liquid ammonia treatment of fabrics. 1. Development of process and plant**, Conference Proceedings on Liquid Ammonia Treatment of Cellulosic Textiles, Shirley Institute, Manchester, England, Nov. 17, 1970, pp. 34-39.

66 K. Bredereck, **Ammonia pretreatment and its effect on finishing**, Mell. Textilber. 72 (6) (1991) 446-454.

Chapter 2

FUNDAMENTALS OF DYES AND DYEING PROCESSES FOR TEXTILES

2.1 INTRODUCTION

Since many textile materials are dyed to impart color to them for aesthetic purposes, a basic knowledge of the theory and practice of dyeing is quite important. There are several major topics that are important in this regard. The first is the historical development and current understanding of the relationship of color and color theory to the chemical constitution of dyes. Associated with the chemical constitution of dyes are their synthesis and their preparation from key intermediate compounds. A knowledge of the physical chemistry of dyeing is useful for explaining and predicting the attraction and substantivity of dyes to fibers under practical dyeing conditions. The processes of adsorption, diffusion and related phenomena markedly influence such dyeing. Classification schemes exist for dyes with regard to their chemical constitution, method of application and substantivity to fiber type. A list of terms that are frequently used in dyeing and printing and to describe various practical methods and theoretical concepts is given in Table 2.1.

2.2 RELATIONSHIP OF COLOR TO THE CHEMICAL CONSTITUTION OF DYES

2.2.1 Color and the electromagnetic spectrum

Color is associated with the region of the electromagnetic spectrum that is visible to the human eye, that is, the range between 400 and 700 nm (or 4,000 and 7,000 Å). In that region, colors (listed in the second column of Table 2.2) are absorbed at different wavelengths. Neither white nor black are considered colors, since white represents reflection of light over the entire visible region, while black represents absorption of all wavelengths in the visible range. As noted in Table 2.2, the color

TABLE 2.1

Terms used in color and dyeing theory and to characterize the chemical
structures of dyes.

Term	Definition
achromatic color *	one that is neutral, e.g., white, grey, or black, i. e., one that does not have a predominant hue.
adsorption	relationship between the quantity of an adsorbed isotherm substance and its equilibrium pressure or concentration at constant temperature.
affinity *	quantitative expression of substantivity; the difference between the chemical potential of a dye (in its standard state) in the fiber and in the dyebath.
aggregation	clustering of individual particles of a substance to give it colloidal properties.
auxochrome *	functional group (e.g., -OH, -NH$_2$) in the chromogen that enhances or alters the color of a dye.
bathochromic shift	shift of the visible absorption maximum (λ_{max}) of a substance to longer wavelengths.
chromatic color *	one that has a predominant hue, e.g., red.
chromogen *	compound that is colored or can be made colored by attachment of suitable substituents. The chromophore and auxochrome are part of the chromogen.
chromophore *	functional group present in the chromogen (e.g.,-N=N-) that is responsible for the appearance of color.
color *	that property of an object or stimulus, or quality of a visual sensation, distinguished by its appearance of chromatic rather than achromatic color.
colorfastness	resistance of the hue of an object to various agents to which it may be exposed, such as light and water, during its processing and subsequent use.
diffusion *	movement of a substance due to the existence of a concentration gradient.
dye *	a colorant that has substantivity for a substrate; this may either be inherent or induced by reactants.

TABLE 2.1 (continued)

Term	Definition
exhaustion *	proportion of dye or other substance taken up by a substrate at any stage of a process to the amount originally available.
heat of dyeing	quantity of energy (usually endothermic) thermally released due to transfer of dyes from solution to fiber.
hue *	attribute of color whereby it is recognized as being predominantly red, blue, violet, etc.
hyperchromic * effect	effect by which absorption of light by a substance is increased with no change in its concentration in solution.
hypochromic * effect	effect by which absorption of light by a substance is decreased with no change in its concentration in solution.
hypsochromic shift	shift in the visible absorption maximum (λ_{max}) of a substance to shorter wavelengths.
levelling *	migration of dye leading to uniform coloration of a substrate.
lightness *	property of a colored object by which it is judged to reflect or transmit a greater or smaller proportion of incident light relative to another object.
metamerism *	a phenomenon whereby the nature of the color difference between two similarly colored objects changes with the spectral composition of the illuminant.
migration *	movement of a dye or pigment from one part of a material to another.
photochromism *	change in hue of certain colored substrates on exposure to light that is reversible on removal of the light source.
phototropy	a reversible color change in a substrate caused by exposure to light or to other forms of radiant energy.
pigment *	an insoluble substance which is mechanically dispersed in a medium in particulate form to modify its color or light-scattering properties.
polygenetic dye *	dye which gives a different hue with different methods of application, particularly mordant dyes that use different metals to form complexes with the dye.
saturation *	nearness of a color in purity to an associated color.

TABLE 2.1 (continued)

Term	Definition
substantivity *	attraction between a substrate and a dye or other substance under precise test conditions whereby the latter is selectively extracted from the application medium by the substrate.
tristimulus *	amount of three defined primary colors (usually red, green, and blue) required to be mixed additively to match the color of an object under specific conditions.
zeta potential	difference in the electrical potential across the interface (a diffuse double layer) of a solid surface in contact with a liquid.

* These definitions based on those of Society of Dyers and Colorists (refs. 1-3).

absorbed by an object is different from the color observed by the viewer of that object. The latter color is determined by the amount of diffuse light reflected from the fabric surface. The color actually observed on a dyed substrate (such as a textile fiber, yarn or fabric) is dependent on the reflectance and absorbance characteristics of the object, the type and intensity of the light source, and visual response of the viewer. The latter two factors will be discussed later (color: measurements and standards in Chapters 5 and 6, respectively).

TABLE 2.2

The relation between colour absorbed and colour seen (ref. 4). Courtesy of Edward Arnold Publishers.

Wavelength nm	Colour absorbed	Colour seen
400-435	Violet	Yellow-green
435-480	Blue	Yellow
480-490	Green-blue	Orange
490-500	Blue-green	Red
500-560	Green	Purple
560-580	Yellow-green	Violet
580-595	Yellow	Blue
595-605	Orange	Green-blue
605-700	Red	Blue-green

56

The absorption spectra of dyes vary with their chemical structure and exhibit maxima (called λ_{max}) that generally correspond to the color or colors observed by a viewer. An example is given for the absorption spectrum of the monoazo dye (Figure 2.1) Kiton Fast Red G. When the maxima of a dye is shifted in wavelength in the visible region, this changes the intensity or hue of the dye. Such absorption shifts of the maxima may be either bathochromic (a shift to longer wavelengths) or hypsochromic (a shift to shorter wavelengths) and may be achieved by subtle alterations in the chemical structure of the dye. Absorption spectroscopy has been used in this regard to evaluate and predict desirable colors required in dyes for their application to textiles and to other substrates. For example, the absorption of dyes that follow Beer's law ($A = \varepsilon bc$, where A = absorbance, ε = molar absorptivity at λ_{max},

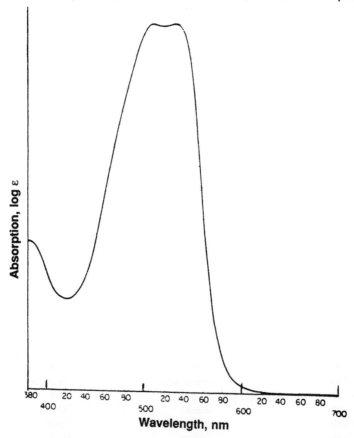

Figure 2.1. Absorption curve for monoazo dyes such as Kiton Fast Red G (CI Acid Red 18050) (ref. 4). Courtesy of Edward Arnold Publishers.

b = light path through the solution, and c = concentration of absorbing species in solution) may be calculated at different concentrations if their extinction coefficient is known. If the dyes are highly associated or aggregated in solution, they do not obey Beer's law. Thus, calculations for aggregated dyes must be made with highly dilute solutions.

2.2.2 Development of modern color theory

The historical development of modern color theory and its relation to the chemical constitution of dyes has been described and reviewed by several authors in the past twenty years (ref. 5-7). Witt recognized over 100 years ago that dyes must contain certain functional groups called chromophores and auxochromes. Chromophores impart color and auxochromes impart substantivity to substrates (fibers or fabrics). Examples of chromophores are -N=N- (azo), -N=O (nitroso), -NO$_2$ (nitro), -C=O (carbonyl) and -C=C- (alkenyl). Auxochromes are usually acidic or basic functional groups such as -SO$_3$H (sulfonic acid), -OH (phenolic), -COOH (carboxylic acid), -NR$_2$ (tertiary amino) and -NR$_4^+$ (quaternary ammonium).

The necessity of having highly unsaturated or conjugated aromatic systems for producing color by extensive delocalization of electrons was also confirmed by early investigators. In 1900, the discovery by Gomberg that the dissociation of hexaphenylethane (colorless) into two triphenylmethyl radicals (yellow color) demonstrated that delocalization of a free electron could also produce color in an aromatic system. Since that time, the advent of quantum mechanics, valence bond and molecular orbital theories, and a knowledge of the ground and excited states of molecules exposed to radiant energy, have been utilized to further refine and predict essential structural features of dyes and their absorption spectra characteristics.

From an energy standpoint, molecules may undergo rotational, vibrational, and/or electronic changes when they absorb and dissipate radiant energy. However, for colored substances or dyes, the only changes in the electronic state that are significant are those in the visible region of the electromagnetic spectrum. Differences in the electronic energy levels of the ground (E_o) and excited states (E_1) of molecules are defined by quantum mechanics as:

$$\Delta E = E_1 - E_o = h\nu \quad \text{or} \quad hc/\lambda \tag{2.1}$$

where h = Planck's constant, ν = frequency of absorbed radiation, c = velocity of the radiation, and λ = wavelength of absorbed radiation. It was noted earlier that dyes absorb in the visible wavelength region at 400-700 nm. This corresponds to ΔE values of approximately 167-297 kJ/mole required for molecules to absorb in that region.

Several theories and concepts have been advanced to explain and predict energy differences between the ground and excited states of colored substances and their absorption spectra. These include: valence bond/resonance and molecular orbital theory, the free electron model, the quantum-mechanics based Triad theory and the existence of H-chromophores. These have been summarized and critically evaluated by Griffiths (ref. 5), Coates (ref. 6) and Mason (ref. 7).

The importance of resonance or extensive delocalization of charge in dyes was demonstrated by Brooker (ref. 8) in a discussion on his comprehensive studies on the synthesis and salient structural features of cyanine dyes. He observed that an asymmetrical dye containing benzothiazole and indole ring structures [Figure 2.2(b)] had a λ_{max} 36.5 nm less than its calculated value. This value is intermediate between those of the symmetrical indole and symmetrical benzothiazole dyes shown in Figures 2.2 (a) and (c). This was attributed to the much greater basicity of the benzothiazole ring, and the inability to extensively delocalize the positive charge equally in dissimilar ring structures, thus resulting in a greater ΔE value or shorter wavelength absorption.

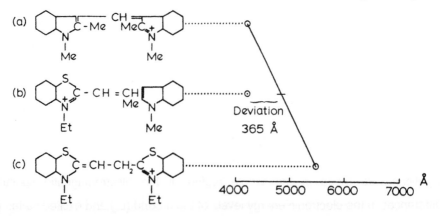

Figure 2.2. Absorption maxima of three cyanine dyes: (a) symmetrical indole, (b) unsymmetrical indole/benzothiazole, (c) symmetrical benzothiazole (ref. 8). Courtesy of American Institute of Physics.

Molecular orbital theory has been more frequently employed to explain how subtle changes in the molecular structure of dyes cause their absorption spectra to undergo shifts in the maxima and changes in their intensity. ΔE values, according to such theory, arise from differences in energy levels (σ or π) of non-bonding (n) orbitals and those of anti-bonding orbitals (σ^* or π^*) in the excited state. As shown in Figure 2.3, n -----> π^* and π -----> π^* transitions required less energy than σ -----> σ^* and n -----> σ^* transitions. Therefore, absorption occurs at longer wavelengths in transitions from nonbonding orbitals to antibonding π^* orbitals than to antibonding σ^* orbitals. Since dyes and other molecules with extensive conjugation have as many π levels as conjugated atoms, their absorption spectrum is very complex. However, bathochromic shifts in their absorption spectra (e.g., caused by replacing -C=C-C=C- conjugated systems with -N=N-C=C- or with -C=C-C=O conjugated systems) have been satisfactorily explained and predicted for lower ΔE values in the latter systems afforded by n to π^* transitions.

Refinements in the Hückel LCAO method (linear combination of atomic orbitals) that take into account electron-repulsion and antisymmertrisation have been used by Griffiths and others to satisfactorily predict the color of dyes (ref. 5). Antisymmertrisa-tion is a mathematical method that removes erroneous degeneracies in highly symmetrical molecules such as aromatic rings. The free electron model has also been used to relate the structure of dyes to their color and intensity. In this model, π electrons can move freely due to their delocalization in a field of constant potential energy. λ_{max} is computed from and equal to $8mcL^2/h$ (N + 1), where m = electron mass, c = velocity of light, L = length of the conjugated chain, h = Planck's constant, and N = number of π electrons. Such a model has been satisfactorily applied to assign absorption bands to carbocyanine and triarylmethane dyes. For example, the carbocyanine dyes in Figure 2.4 show very good agreement between their calculated and observed λ_{max} values.

The two most significant advances in the past twenty-five years of the relationship of the structure of dyes to their color have been the Triad Theory (a classification scheme discovered and devised by Dähne) and the recognition of crossed donor-acceptor systems (or H-chromophores) in dye synthesis and color prediction (ref. 5).

60

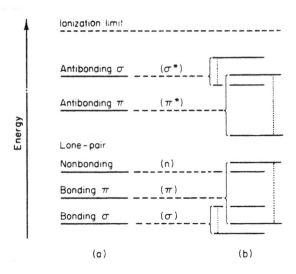

Figure 2.3. The relative energies of the σ, π, and lone-pair orbitals in (a) a simple and (b) complex unsaturated molecule (ref. 7). Courtesy of Academic Press.

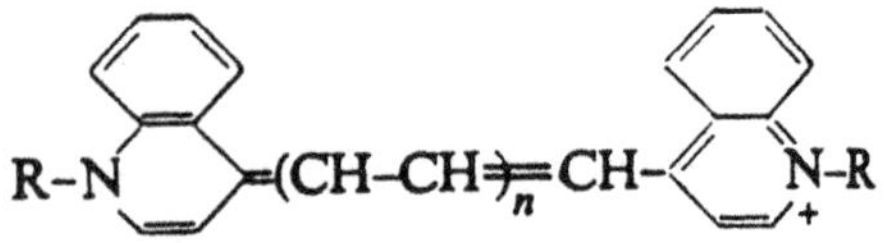

R–N⟨⟩=(CH–CH=CH–)_n⟨⟩N–R
Carbocyanines (different basic
end-groups possible)

n	N	calc.	Wavelength (mμ) obs.
0	10	579·5	590
1	12	706·3	710
2	14	832·2	820
3	16	959·6	930

Figure 2.4. Agreement between calculated and observed λ_{max} values for symmetrical carbocyanine dyes where n = number of exocyclic -C=C- and N = total number of π electrons (adapted from ref. 6). Courtesy of The Society of Dyers and Colourists.

The Triad Theory developed by Dähne (ref. 9) recognizes three electronic states of unsaturated compounds: polyenes, planar aromatics and polymethines. Polyenes have alternating π-bond orders and equal π-electron densities. The second group are aromatic structures (planar and cyclic polyunsaturated) that have equal π-bond orders and electron densities as well as high resonance energies. Polymethines, characterized by cyanine dyes with alternating π-electron densities, have equal π-bond orders and small energy differences between the highest and lowest occupied molecular orbitals.

The existence of a crossed donor-acceptor or H-chromophore system was proposed to explain the color of new indigoid and other classes of dyes, since their color could not be explained by Witt's classic donor-acceptor concept (ref. 5). Examples of new dyes having such crossed chromogens are benzodifuranones and N,N'-bisarylthioquinonedimines with their crossed donor (D)-acceptor (A) components. Other crossed chromogen structures (e.g., thiochromone-S-dioxides and indigoids) are also discussed and reviewed (ref. 9).

Structures of new chromophores for dyes used in all types of applications (non-fiber as well as fiber) have been recently compiled (ref. 10). New or modified chromophoric structures also have been synthesized to improve substantivity to most major fiber types. Azo chromophores derived from less toxic diamines may be used as direct dyes for cotton. These diazotizable amines afford viable replacements for carcinogenic benzidine structures previously used. Azo chromophores have also been synthesized that improve the fastness of disperse dyes to polyester fibers and the fastness of cationic dyes to acrylic fibers. Structures containing both styryl and azo chromophores give much deeper shades on polyester than dyes containing only one of these chromophores. Heterocyclic styryl dyes containing cationic groups have been developed for acrylic fibers. New heterocyclic systems for pigments have been synthesized and applied to various fiber types. These structures include quinacridones, fluorubine, quinophthalones, pyropyroles and isoindolenines (ref. 10).

Molecular geometry has also been observed to influence the intensity and maxima of the absorption spectra of dyes. Such spatial effects include the steric and asymmetry characteristics of molecular structures. When steric hindrance prevents extensive delocalization in a structure, e.g., many *cis* isomers of azo dyes, the substance is colorless; however, if the *cis* isomers can be converted into *trans* isomers, then there is relief from steric strain, extensive delocalization of the conju-

62

gated structure, and the substance exhibits color. When this phenomenon is associated with reversible loss or appearance of color, this is known as phototropy.

Since it has been established that radiation (in the form of light) is absorbed by molecules in the same plane as their orbitals, it is not surprising that the three-dimensional geometry of any structure profoundly influences the intensity and maxima of its absorption spectra. The example most commonly used to illustrate this point is shown in Figure 2.5. Since Crystal Violet is symmetric in both x and y directions or electric vectors of light, it exhibits only one absorption maxima; however, Malachite Green, which is not symmetric in the x and y directions, exhibits two absorption maximas. This difference illustrates the directional effect that light exerts on characteristics of a dye molecule in the visible region of the electromagnetic spectrum.

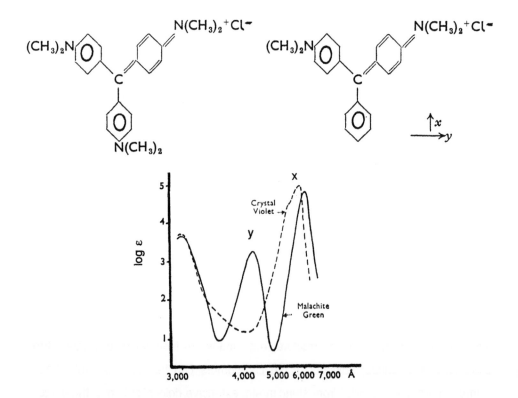

Figure 2.5. Structures and absorption curves of Crystal Violet and Malachite Green (adapted from ref. 11).

Newer and more refined theories of molecular structure will be discovered and developed. This new knowledge will be utilized in conjunction with practical discoveries to predict the color of and assist in the synthesis of dyes of desired hue and intensity.

2.3 PHYSICAL CHEMISTRY OF DYEING

2.3.1 Overview

The processes involved when fibers uptake dyes in solution have been discussed and scrutinized from several perspectives. These range from theoretical and very complex mathematical derivations and treatments based on statistical and classical thermodynamics and quantum mechanics to practical considerations of fiber, yarn, and fabric structure and operable dyebath parameters. The more precisely the dyeing processes are defined and mechanistically elucidated the more rigorous and difficult the mathematical models and assumptions become. Processes by which textile materials are dyed in solution and exhibit substantivity for a dye are normally subdivided into those (a) that are kinetically controlled by various diffusion and trans-

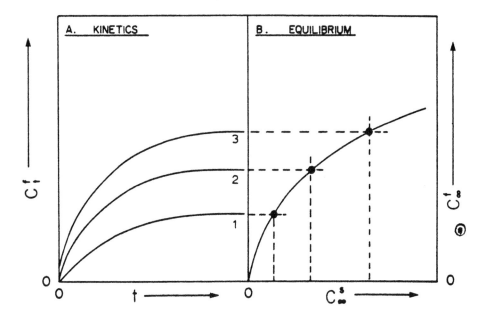

Figure 2.6. Rate of dyeing curves (A) and equilibrium sorption isotherms (B) (ref. 12). Courtesy of American Association of Textile Chemists and Colorists.

port phenomena, and (b) those that are equilibrium controlled by sorption and desorption processes that occur between a fibrous substrate and the dye in solution. The simplest relationship between the rate of dyeing a fiber (f) in a dye solution (s) at a constant dye concentration (C_s) and temperature T and its equilibrium uptake of dye from that solution is shown in Figure 2.6.

2.3.2 Convective diffusion processes

The first process that occurs is convective diffusion, that is, diffusion of the dye from solution to the fiber surface. This diffusion process is influenced by such factors as the hydrodynamic flow of the dye solution, the diffusional boundary layer at the fiber surface, and the zeta potential of the fibers and state of aggregation or association of dye molecules in solution.

When the dye solution initially makes contact with the fiber surface, its concentration is lowered near the surface. Convective transport of the solution or aqueous diffusion of the solution supplies additional dye to the surface. Models have been devised (ref. 13) to identify and differentiate two boundary layers that are present at the fiber surface. These are the hydrodynamic boundary layer and the diffusional boundary layer (Figure 2.7). At high rates of flow, the dye solutions have

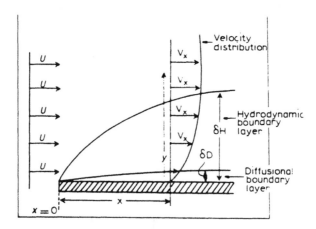

Figure 2.7. Hydrodynamic and diffusional boundary layers in steady laminar flow parallel to a plane slab. Horizontal arrows represent fluid velocity, in magnitude and direction, at the point of origin of the arrows. The diffusional boundary layer has a thickness δ_D about one-tenth that of the hydrodynamic boundary layer, δ_H. Both layers are parabolic in shape, i.e. $\delta \propto x^{1/2}$. U is the mainstream velocity (ref. 14). Courtesy of The Society of Dyers and Colourists.

decreasing rates of velocity as they approach the fiber surface. The hydrodynamic layer is approximately ten times thicker than the diffusional boundary layer. Hydrodynamic and diffusional boundary layer effects may be overcome by vigorous agitation of the fibrous substrate in the dyebath and/or increasing the velocity of the dye solution. This effectively makes it easier for the dye to reach the fiber surface and thus increases the dyeing rate.

Since the zeta potential of fibers causes electrostatic repulsion or attraction of dye molecules, it can be used to advantage to accelerate or retard dye uptake by changing the dyebath temperature and pH, or by adding electrolytes to the dyebath. An increase in dyebath temperature causes disaggregation or dissociation of dyes and usually increases their rate of diffusion to the fiber surface.

2.3.3 Sorption of dyes and sorption isotherms

The second process, chemical and/or physical sorption of the dye occurs at equilibrium. This sorption occurs in such a manner on the fiber that the dye become substantive. Mechanisms by which dyes are sorbed and become substantive to fibers are dependent on the chemical nature and morphology of the fiber and on steric and electronic structural features of dyes. Although there are various models to explain the sorption behavior of dyes onto fibers, the Nernst, Freundlich and Langmuir sorption isotherms have proven quite useful in characterizing and classifying dye-fiber sorption processes. These isotherms represent the relationship of the concentration of dye sorbed onto the fiber (C^f_∞) and its concentration in the dyebath (C^s_∞) at a constant temperature. The ratio of the former to the latter is called the partition coefficient K (Figure 2.8).

In the Nernst isotherm [Figure 2.8 (a)], the linear relationship between dye concentration on the fiber and in the bath holds except for very concentrated dye solutions. This type of sorption isotherm frequently occurs when disperse dyes are applied to hydrophobic fibers as aqueous suspensions, since their limited solubility in water produces a steady-state concentration of the dye in solution. The Freundlich isotherm [Figure 2.8 (b)] is representative of the sorption behavior of vat and some direct dyes on cellulosic fibers. It is characterized by rapid initial dye sorption on the fiber that is limited only by accessibility of fiber surface sites. This limitation occurs

66

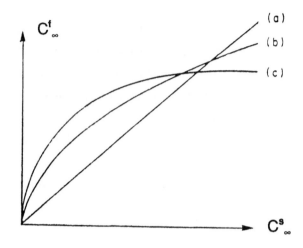

Figure 2.8. Relationship between concentration of dye in fiber (C^f_∞) vs concentration of dye in solution (C^s_∞) described by (a) Nernst, (b) Freundlich, and (c) Langmuir sorption isotherms (adapted from ref. 15). Courtesy of John Wiley and Sons Publishers.

because these types of dyes are normally bound to fibers by purely physical forces such as hydrogen bonding and van der Waals forces. The Langmuir sorption isotherm [Figure 2.8 (c)] appears to be operative in many ion-exchange processes where the fibers have or acquire a positive or negative charge that dictates attraction of dyes containing an opposite charge. Once the charged fiber sites are occupied on the fiber, sorption of additional dye levels off quite rapidly. The dyeing of wool and of nylon with acid dyes, acrylics with basic or with cationic dyes, and cellulosics with some direct dyes, are usually characterized by Langmuir sorption isotherms.

Hoffman surveys the various fundamental relationships governing the sorption of dyes onto fibers and discusses how such sorption is modified under practical conditions of levelling, temperature change, dye liquor flow and dye migration (refs. 16-18). In the first paper, he describes the kinetics of dye sorption for various fiber types and dye application classes without any consideration for dye liquor flow, fabric movement or temperature variation (ref. 16). The second paper discusses the influence of dye liquor circulation and fabric movement in the dyebath on dye bath exhaustion and uneven dyeing. The uneven dyeing is attributable to the state of the dye in the liquid state, liquor flow, and concentration of dye on the fiber surface and in the bath. Thus, the relationship between kinetics of dyebath exhaustion and uneven dyeing is not a simple, straightforward correlation (ref. 17). The final paper demonstrates the importance of differentiating initial from final uneven fiber dyeing and describes such differences in terms of fiber affinity and dyeing rate. The role of temperature differences and fiber type in uneven dyeing is also discussed. Alternative tests are also proposed to simulate migration of dyes under actual commercial production conditions (ref. 18).

Dimensionless groups and equations have been recently employed to describe a variety of dye sorption processes on fibers (refs. 19-21). Etters has developed a computational technique to approximate the dimensionless sorption time of dyes as a function of fractional equilibrium uptake of diffusant by a fibrous polymer, dimensionless dye bath exhaustion and a dimensionless boundary layer. This isothermal sorption from finite concentrations of dye in the bath is based on translational kinetics and was shown to be applicable for sorption systems based on solids of different geometric shapes (ref. 19). McGregor and Ijima have simplified equations from a generalized Donnan equilibrium model to describe equilibrium sorption of dyes with anionic groups on fibers containing both acidic (positively-charged) and basic (negatively-charged) functional groups and on fibers (such as polyamide) containing an excess of basic groups (refs. 20, 21). Thus, sorption of dyes and of inorganic ions by fibrous polymers may be described by different dimensionless groups that represents different aspects of sorption processes.

The physical and chemical mechanisms by which dyes become substantive to fibers during sorption include hydrogen bonding, van der Waals forces, covalent bond formation, ion-exchange and hydrophobic bonding. Hydrogen bonding frequently occurs between cellulosic fibers and dyes containing groups such as amino, azo, hydroxyl, and carboxyl. The substantivity of vat and some direct dyes to cellulosics has been attributed to van der Waals forces in which attraction occurs between chemically inert molecules. In this instance, the force between dye and the fiber is inversely proportional to the sixth power of the distance between them. It has been demonstrated that the planarity and molecular weight of the dye influence this type of dye-fiber bonding. Covalent bond formation occurs when reactive dyes, usually by nucleophilic substitution or nucleophilic addition reactions, are chemically bound to cellulosic and protein fibers. Ion-exchange mechanisms prevail with ionic dyes and with fibers that either have or can acquire a positive or negative charge by change in pH or by addition of an electrolyte. Such dye-fiber systems commonly include anionic or acid dyes with wool and polyamides, cationic or basic dyes with acrylics, and some direct dyes with cellulosics. Hydrophobic bonding frequently occurs when disperse dyes are applied to hydrophobic fibers as aqueous suspensions; however, this mechanism of fiber bonding has also been advanced to explain the behavior of other dye-fiber systems.

Monolayer films as models for various fiber types to explain dye-fiber bonding and characterization of such processes by the Gibbs absorption equation have also been reviewed (ref. 22). There were essentially three mechanisms observed for interaction of the dyes in solution with the film. In the first case [Figure 2.9 (a) and (b)], the monolayer A is penetrated by a nonionic solute (e.g., nonionic basic dyes with cellulosic films and monosulfonated dyes with protein films). The hydrophobic portion of the dye enters the monolayer, while the hydrophilic portion remains in solution. In the second case [Figures 2.9 (a) and (b)], a long planar solute molecule B (e.g., direct dyes with cellulosic films and disperse dyes with cellulose triacetate films) with ionic groups at either end is oriented beneath monolayer polymer chains. In the third case [Figures 2.9 (a) and (b)], C depicts crosslinking of the polymer chains by a bifunctional solute (e.g., reactive dyes with protein and cellulose-based films). Each of these three modes of interaction between dye and film can be correlated with the behavior of different classes and structures of dyes with fibers. Covalent bonds of high strength are formed between reactive dyes and cellulosic and/or wool fibers.

Acid dyes (including 1:1 metal complex) on wool and anionic and cationic dyes on cellulosics are retained by purely physical factors related to pore size and structure of the substrate. Reaction of ionic dyes with cellulosic or wool fibers appears to be determined by the configuration that the dye favors relative to the water surface [e.g., A or B, Figures 2.9 (a) and (b)] rather than by attraction to the substrate (ref. 22).

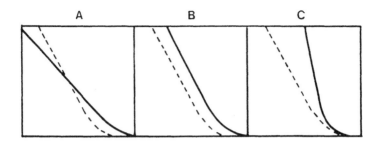

Figure 2.9 (a) - The three general types of change of slope of the surface pressure-area curve for a monolayer of a polymer when a solute is introduced into the aqueous sub-phase. The broken line represents the curve for the monolayer on pure water (ref. 22). Courtesy of the Society of Dyers and Colourists.

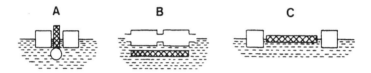

Figure 2.9 (b) - The suggested relative orientations of polymer molecules (open rectangles) and solute molecules (hatched rectangles) in the monolayer on aqueous solutions, corresponding with the three respective curve slope changes shown in Figures 2.9 (a). A shows the penetration of the monolayer by a nonionic solute, where the ionic group remains in the water and the hydrophobic polymer enters the monolayer. B shows the orientation of a long solute molecule with ionic groups at either end (e.g., a direct dye), beneath the monolayer polymer chains. C shows crosslinking of polymer chains by a bifunctional solute, whereby the viscosity of the film is greatly increased. In A and C the polymer chains are shown in cross-section, in B in side-elevation. (ref. 22). Courtesy of the Society of Dyers and Colourists.

The relevance of thermodynamics to dye-fiber behavior at equilibrium is treated at great length in many text and reference books (e.g., **general references** at the end of this chapter). It is possible, for example, to derive the various sorption isotherms (e.g., Nernst, Freundlich, and/or Langmuir) that characterize dye-fiber interactions if one knows or can measure the standard chemical potential of a dye (μ°) and related it to the activity of a dye (a) by the equation: $\mu = \mu^{\circ} + RT \ln a$. The difference in the chemical potentials of a dye in solution and in the fiber phase ($\Delta\mu^{\circ}$) is useful in calculating the heat of dyeing of fibers (ΔH°). Both these parameters can in turn be substituted in the equation $\Delta\mu^{\circ} = \Delta H^{\circ} - T\Delta S^{\circ}$ to assess the contribution that the entropy term (ΔS°) makes to the sorption of dyes onto fibers at equilibrium.

Sumner has reviewed earlier work on the determination of the thermodynamic affinity of various types of dyes to fibers and how it relates to the concentration of ions and dyes in solution and in the fiber phase (ref. 23). In earlier studies, equations were developed that related the thermodynamic affinity (i.e., the difference in the standard chemical potential of the dye (Na_zD) when it is adsorbed from the dyebath solution onto the fiber) of direct, vat and reactive dyes for cellulosic fibers. For direct and vat dyes with cellulosic fibers, this affinity can be represented by the equation:

$$- \Delta\mu° \quad = \quad RTln\ [Na]_F^z\ [D]_F/[Na]_S^z\ [D]_S \tag{2.2}$$

where z = the basicity of the dye, $[D]_F$ and $[D]_S$ are respectively, the concentration of dye ions in the fiber phase and in solution (moles/liter), and $[Na]_F^z$ and $[Na]_S^z$ are respectively, the concentration of sodium ions in the fiber phase and in solution (moles/liter). With the exception of the concentration of sodium ions in the fiber phase, the three other concentrations are directly measurable. Assuming that the dye solution contains neutral electrolytes with monovalent ions $[Na]_F^z$ can be calculated from the principles of electrical neutrality. For vat dyes, an additional term is needed in the equation (concentration of hydroxyl ions in dyebath) for this calculation (ref. 23).

Sumner has also developed generalized equations to determine the affinity of anionic dyes with cellulosics, wool and polyamides and how these could be utilized in practical dyeing situations (refs. 23, 24, 25). The affinity of anionic dyes for cellulosic fibers in a bath containing multivalent anions and over a range of alkaline pH ranges was determined by modification of the equation shown above (ref. 23). Modification of this equation from the principles of Donnan membrane equilibrium also allows one to determine the affinity of anionic dyes for wool and polyamides and of cationic dyes for acrylic fibers (ref. 24). If the affinities of the various dye-fiber combinations are determined, it is feasible to use this information to predict dyeing behavior under practical conditions and/or to calculate the amount of a particular dye to produce the desired coloration of fabrics (ref. 25).

The use of solubility parameters has been advocated for determining the compatibility of dyes and auxiliaries (such as dye carriers) for the hydrophobic-type bonding of polyester fibers (ref. 26). The solubility parameter was developed many years ago by Hildebrand, and is essentially the cohesive energy density required to completely separate molecules. This energy is equal to a mole of a substance divided by its volume. Methods for determining the solubility parameters or cohesive energy density for dyes and additives that are relevant to fibers have been compiled and reviewed by Siddiqui and Needles (ref. 27). Solubility parameters may be calculated from known or measurable physical constants, such as the heat of vaporization, coefficient of thermal expansion, van der Waals' gas constant, critical pressure and surface tension. They may also be determined from structural formulas or experimentally determined from swelling values or from various solubility measurements for polymers and dyes.

2.3.4 Diffusion of dye into fiber interior

The final process, diffusion of the dye from the fiber surface to the interior of the fiber, is kinetically controlled. Because the rate of dyeing a textile is dramatically influenced or controlled by the latter process, understanding and predicting this type of diffusion is commercially important. The three approaches most frequently utilized to explain, predict and/or obtain fundamental knowledge about this process are: (a) models that adequately explain how various dyes diffuse into different types of fibers, (b) methods of measuring the diffusion of dyes into fibers, and (c) simple and useful calculations of diffusion coefficients of dyes into fibers.

The pore model [Figure 2.10 (a)] and free volume model [Figure 2.10 (b)] are frequently used to explain, respectively, the diffusion of dyes into hydrophilic fibers such as cotton and wool and into hydrophobic fibers such as polyester. Diffusion of dyes or other molecules (pore model) occurs only through the liquids in the pores of the polymeric or fibrous substrates. These pores are formed by swelling of the amorphous regions of the substrate. The effective diffusion coefficient (D_M) is thus directly proportional to the porosity (P), partition coefficient (K) and diffusivity (D_p), and inversely proportional to the tortuosity or bending characteristics (τ) of the pores. The glass transition temperature (T_g) of the polymer chains that influence diffusion is not considered in this model. In contrast, the free volume model attributes the diffusion of dyes or other molecules into substrates to the thermal mobility of polymer chains segments. This thermal mobility is directly related to T_g (ref. 28).

72

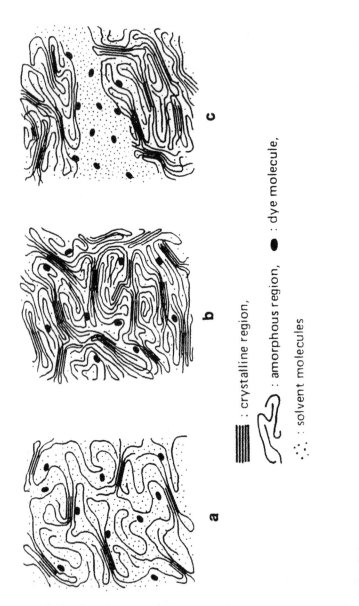

: crystalline region,

: amorphous region, ● : dye molecule,

: solvent molecules

Figure 2.10. Schematic representation of dye diffusion in a fibrous polymer: (a) pore model--dye molecules diffuse in the liquid-filled pores (porosity = P), which are bent (tortuosity = τ), and adsorb simultaneously on the pore wall (partition coefficient = K). Apparent diffusion coefficient $D_M = (P/\tau)KD_p$, where D_p is the diffusivity into the pores of the liquid; (b) free volume model--dye molecules diffuse through the amorphous region of polymer matrix, which may also contain the solvent molecules, and adsorb on the polymer chain. Diffusion coefficient is a function of the glass transition temperature of polymer. In $D = \ln D_{T_g} - A (T-T_g)/B + (T-T_g)$; (c) generalized diffusion model--dye molecules diffuse in the polymer solution (whose fractional volume and tortuosity are θ and τ, respectively), formed in the amorphous region of the polymer matrix, and adsorb on the polymer chain (partition coefficient = K). Diffusion coefficient is : $\ln D = \ln \phi_o \delta^2/6 + \ln \theta/\tau - \ln K - \Delta H_H/RT - AB/B + (T - T_g)$ (ref. 28). Courtesy of the Society of Dyers and Colourists.

A generalized diffusion model that incorporated features of both the pore and free models has been proposed [Fig. 2.10 (c)]. It purports to explain the diffusion of dyes into both hydrophilic and hydrophobic fibers. In this generalized model, the most important variables that influence diffusion of dyes into fibers are: the degree of swelling, dye affinity for the fiber, enthalpy changes necessary to produce the formation of holes (ΔH_H), and a glass transition temperature term that includes constants A and B (derived respectively from the free volume of the polymer chains at T_g and from the coefficient of thermal expansion of the polymer above T_g).

The effect that polymer morphology has on the dyeing behavior of synthetic fibers has been reviewed (ref. 29). The preference for investigators to favor structural models based on porosity or on free-volume to explain the diffusion of dyes into different types of synthetic fibers can usually be correlated to specific morphological features of the fiber chosen for the study and the conditions under which it was dyed. However, there appears to be substantial evidence that the dyeing of most acrylic and polyester fibers dyed in the presence of dye carriers occurs by the free volume mechanism of diffusion. Under these conditions good correlation between dye diffusion and segmental mobility of polymer chains and T_g has been observed and documented (ref. 29).

In a later study (ref. 30), several important observations were made about the effect that the proportion of bound and sorbed water in fibrous polymers had on the diffusion of dyes. It was surprisingly concluded (on the basis of sorption isotherms observed for various types of fibers) that even synthetic fibers such as polyester and acrylic have pore sizes and pore volume distributions comparable to those of natural hydrophilic fibers in the dry state. Although this porosity may be different for various fibers in the wet state, differences in dye diffusion may be reasonably explained by the amounts of bound and sorbed water for various fibers. Table 2.3 shows that dye diffusion in fibers represents a continuum consistent with a general composite model rather than a preference for diffusion to occur solely by a pore model or to occur solely by a free volume model.

Dyeing of hydrophilic fibers by the pore model has also been mathematically categorized into reactive (e.g., cellulosic fibers with reactive dyes) and adsorptive

TABLE 2.3

Relative contribution of the pore and free volume mechanisms to the dyeing of various fibers and films (ref. 30). Courtesy of the American Association of Textile Chemists and Colorists.

increasing weight of pore diffusion	↑	Cellulosics Highly porous acrylics Regular acrylics Polyamides Polyesters	↓	increasing weight of free volume diffusion

(e.g., polyamide fibers with acid dyes) dyeing. Equations that take into account Fick's second law of diffusion lead to derivations for both types of dyeing. It was determined that reactive and adsorptive rates of dyeing were primarily dependent on the ratio of dimensionless dyeing time required to reach fractional dye fixation to the fibers relative to the total dye fixation (ref. 31).

Modifications of the two-phase pore/free volume model have been recently proposed by Zollinger (ref. 32 and 33) and further refined by Flath in a recent critical review on this topic (ref. 34). Other investigators have essentially rejected these explanations and proposed alternate dye diffusion models. These models focus on thermal mobility of noncrystalline chain segments of a fiber, partial adsorption of the dye on the fiber surface and partial dissolution of the dye into the polymer (refs. 35, 36). The increasing recognition of the importance of polymer morphology in static and dynamic states leads the author to believe that a "template-type" model (similar to those identified in molecular biology and enzymatic reactions) will eventually be proposed as newer information is obtained. Nevertheless, this topic is important enough to warrant a more detailed discussion of the two different viewpoints on the mechanism of dye diffusion into fibers.

Zollinger demonstrated that the pore model and free volume model mechanisms could operate simultaneously by dyeing two acrylic fibers that differ only in their volume percentage of permanent pores and pore size distribution. The two acrylic fibers have identical glass transition temperatures in the dry (T_g) and wet state (T_D). However, the more porous fiber was dyed by the pore model diffusion mechanism and the less porous fiber by the free volume or segment mobility diffusion type of

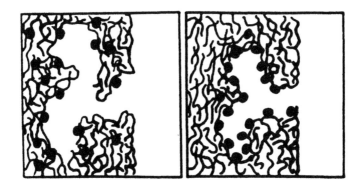

Figure 2.11. Schematic representation of dyeings of the porous acrylic fiber Dunova below (left) and above (right) T_D (ref. 32). Courtesy of Textile Research Institute.

mechanism. Figure 2.11 illustrates the results of desorption experiments to verify the hypothesis that the pore diffusion of the more porous acrylic fiber is a relatively fast reaction that occurs prior to diffusion inside the polymer matrix. Thus, the porous acrylic fiber (commercial name of Dunova) has dye absorbed on the inner pores below T_D and dye that migrated within the fiber matrix above T_D. This concept has been further refined for diverse fiber types (cellulosic, wool, polyamide and polyester) to show that different ratios of pore/free volume are appropriate to explain specific dyeing diffusion and behavior (ref. 33).

Flath has critically reviewed the various theories of dye diffusion in fibers with an emphasis on how the supermolecular structure of the fibers affects such diffusion (ref. 34). He proposes dynamic network structures (Fig. 2.12) based on viscoelasticity to show how dye diffusion differs in the relaxed and tensioned states. Practical results are offered to support this model. For example, thread tension in acrylic fabrics affects the initial dyeing rate and is inversely related to the determined value of the fiber elastic modulus.

Experimental observations by Schreiner on the distribution of dyes in polyester fibers led him to propose a totally different model for dye diffusion (ref. 35). As shown in Figure 2.13, cross-section and top views indicate the dye interacts with the surface during the initial dyeing phase. In the presence of dye carriers, the dye begins to dissolve in the polymer phase because of an increase of viscoelasticity in the amorphous regions. Maximum color depth is achieved when diffusion zones stop

Network

- without tension

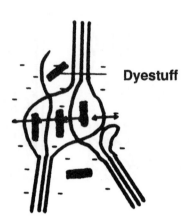

Dyestuff

- under tension

- distance decreased
- vibration amplitude limited

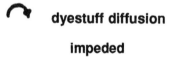
dyestuff diffusion

impeded

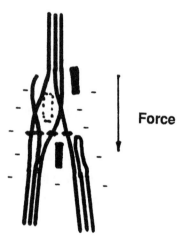

Force

Figure 2.12. Network structure sections after swelling and stressing (ref. 34). Courtesy of Melliand Textilberichte.

at crystallite boundaries. Further dissolution of the dye on the surface results in lower color strength because of transport into less optically effective interior regions of the fiber. Davis has also proposed a similar model to explain the dyeing behavior of polyester and polyamide fibers (ref. 36). He describes a method for converting

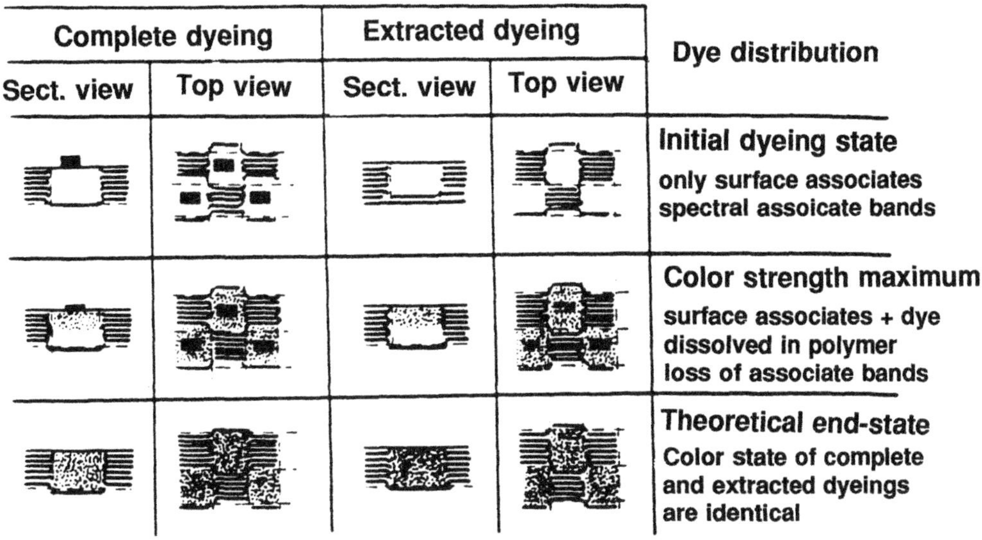

Complete dyeing		Extracted dyeing		Dye distribution
Sect. view	Top view	Sect. view	Top view	
				Initial dyeing state only surface associates spectral assoicate bands
				Color strength maximum surface associates + dye dissolved in polymer loss of associate bands
				Theoretical end-state Color state of complete and extracted dyeings are identical

Figure 2.13. Dye distribution and optical effects in the model (ref. 35). Courtesy of Melliand Textilberichte.

dynamic fiber mechanical data into two parameters that control dye diffusion, internal viscosity (α) and a mobile noncrystalline fraction (X). Thus, dye diffusion into a fiber follows the Stokes-Einstein law for Brownian diffusion of a particle and is proportional to the ratio X/α. A schematic of his model (Fig. 2.14) expresses amorphous fiber orientation as the relative ratio of end-to-end length of chain segments to their contour lengths. The model is argued to be consistent with the large effect that water has on increasing both the internal viscosity and the noncrystalline fraction X of fibers such as nylon to facilitate its dyeing.

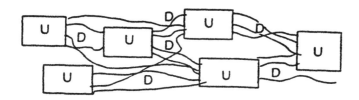

Figure 2.14. Functional fiber structure. U = undeformable solid up to 200°C and D = glass to rubber behavior over room temperature to T_g. U and D are functionally in series (ref. 36). Courtesy of American Association of Textile Chemists and Colorists.

78

Other investigators have critically reviewed barriness or streakiness that occurs during the dyeing of polyamide fibers and related this defect to the latter two theories of dye diffusion (ref. 37). The debate between Schreiner (ref. 38) and Flath (ref. 39) continues in two recent publications. Flath considers the dyeing behavior of polyurethane and polyamide fibers to be consistent with his previously proposed uniaxial, stretched dynamic polymer network (refs. 34, 39). A recent review on the diffusion of dyes into only natural fibers discusses in some length the earlier pore/free volume models of Zollinger, but does not mention the newer approaches of the dynamic polymer network or adsorption/polymer dissolution models (ref. 40). Further dialogue and experimentation are necessary to more clearly elucidate the mechanism(s) of dye diffusion into fibers.

The most frequently employed method of measuring the degree of diffusion of dyes into fibers involves rate of dyeing curves in which the time of half dyeing $(t_{1/2})$ is obtained from exhaustion of the dye at equilibrium. When the concentration of the dye on the fiber is plotted against the time or concentration of the dye in the bath is plotted against the log of the time (Figure 2.15), $t_{1/2}$ may be readily computed.

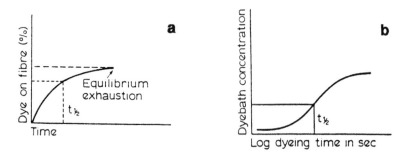

Figure 2.15. Determination of time of half dyeing $(t_{1/2})$ at equilibrium exhaustion on a (a) linear and (b) logarithmic basis (adapted from ref. 11).

Although useful and practical information may be obtained from half dyeing times, alternative methods for measuring dye diffusion are also advocated and discussed by Dawson and Todd (ref. 41). The first method is measurement of the actual penetration of dyes into fibers by microscopic examination of cross sections of dyed yarn bundles. This method has been used to explain the increased depth of color with the increasing penetration of the dye into bright or transparent fibers and the decrease in the levelling off of color depth with increasing dye penetration into fibers that have been delustered or have natural surface irregularities. The other two methods discussed are the multi-layer diffusion and microdensitometer methods. In the multi-layer method, layers of film with sealed, melted ends are tightly wound around a glass rod and a second glass rod tightly fastened outside the roll of film. Both rods are immersed in a dyebath and dyed for 1 to 2 hours, then the film is unrolled to determine the extent of dye diffusion by the number of layers that were penetrated. In the microdensitometer method, a parallel-sided block of polymer is dyed at various times, then sliced transversely with a microtome; changes in absorbance are then determined with a microdensitometer as a function of the distance from the surface of the slice (ref. 41).

McGregor notes that to achieve practical rates of dyeing, at least two sets of variables must be controlled: (a) the ratio of the partition coefficient K to the liquor ratio (L) in the dyebath that controls exhaustion of the dye in the bath at equilibrium and (b) the product of diffusion coefficient (D) times rate of dyeing (t) divided by the denier of the fibers. It is assumed that the fibers are all cylindrical in shape (ref. 12).

Theoretical equations and analytical approximations for the determination and prediction of the diffusion coefficient (D) of dyes into fibers have been devised and solved by several investigators. The most notable are Crank's complex diffusion equations (ref. 42):

$$C_t/C_\infty = 1 + \alpha/1 + 1/4\alpha \left\{ 1 - \exp 4/\alpha^2 \left(1 + \alpha/4 \right)^2 Dt/\alpha^2 . \text{erfc} \right.$$

$$\left. \left[2/\alpha \left(1 + \alpha/4 \right) \left(Dt/\alpha^2 \right)^{1/2} \right] \right\} \tag{2.3}$$

where C_t = amount of dye uptake by a fiber of cylindrical form with radius **a** in time **t**, C_∞ = amount of dye uptake at equilibrium, α = parameter related to fractional equilibrium exhaustion, D = effective diffusion coefficient, Dt/α^2 = a dimensionless quantity and erfc = the error complement function. Cegarra and co-workers claim that rank's equation and other similar types of equations are only valid at exhaustion levels of 90% or less in dyebaths (ref. 43). They have developed a series of approximations that are useful for exhaustion levels of greater than 90% to those of complete or 100% exhaustion.

Simplification of complex diffusion equations, such as those derived by Crank (ref. 42), is advocated and explained by Etters and Urbanik (ref. 44). In these equations:

$$[M_t/M_\infty = f(Dt/r^2, E] \tag{2.4}$$

where cylindrical fibers of radius r uptake dye from a finite time t (M_t) to an infinite time (M_∞) and E_∞ = fractional equilibrium exhaustion of the dyebath. Diffusion coefficients (D) may be estimated by use of semi-empirical analytical approximations with the aid of a pocket calculator (ref. 44). Because of its practical importance, methods to predict the diffusion coefficient of dyes into fibers will probably continue to be a viable area of research.

2.4 CLASSIFICATION OF DYES

2.4.1 By structure and method of application/Colour Index

Dyes may be classified or grouped in several ways. The two most common methods of classification, which form the dual basis of organization of the *Colour Index*, are: (a) chemical structure or constitution and (b) method of application. Examples and important classes for classification by chemical structure are shown in Table 2.4, while similar examples and important classes by methods of application are shown in Table 2.5. In some instances, there is duplication in the two classification methods (e.g., for azoic and for sulfur dyes), but in most instances there is little similarity between the two methods of classification. As one observes in Table 2.4, chemical classification is primarily made with regard to functional groups and/or ring systems, with the lowest numbers (10000 to 10299) assigned to nitroso or quinone oxime structures and the highest numbers (77000 to 77999) assigned to inorganic coloring matters (usually pigments). The higher numbers in the chemical classification scheme generally reflect more complex heterocyclic ring structures, while the lower numbers in this chemical classification are typically carbocyclic aromatic rings containing various types of functional groups. There are essentially twelve different dye types classified by method of application (Table 2.5) that are

TABLE 2.4

Chemical classification of dyes. Modified and condensed from *Colour Index*. Courtesy of The Society of Dyers and Colourists.

Class: CI No. range	Characteristic structural units or functional groups	Dye classes by application
1. nitroso (quinone oximes) 10000-10299	O=C-C=NOH <---> -C=C-N=O	acid, mordant, pigment
2. nitro: 10300-10999	*ortho*- and/or *para*-nitro-phenols and anilines	acid, disperse, pigment, solvent
3. azo (monoazo) 11000-19999	R-N=N-R'[a]	most major classes
azo (disazo) 20000-29999	(R-N=N--)$_2$--R'[a]	most major classes
azo (trisazo) 30000-39999	(R-N=N--)$_3$--R'[a]	acid, basic, direct, mordant
azo (polyazo) 35000-36999	(R-N=N--)$_{>4}$--R'[a]	acid, direct, mordant
4. azoic: 37000-39999	substituted aromatic -N≡N[+] (diazo component) + substituted β-naphthols (coupling components)	azoic
5. stilbene: 40000-40799	NaSO$_3$-Ar-C=C-Ar-SO$_3$Na[b]	direct
6. carotenoid: 40800-40999	--(C=C-)$_{>9}$--	-----
7. diphenylmethane (ketone imine) 41000-41999	HN=C	basic
8. triarylmethane: 42000-44999	C=Ar-X [b] (X = O or NH)	acid, basic, mordant, pigment, solvent

TABLE 2.4 (continued)

Class: CI No. range	Characteristic structural units or functional groups	Dye classes by application
9. xanthene: 45000-45999		acid, basic, mordant, pigment, solvent
10. acridine: 46000-46999		mostly basic
11. quinoline: 47000-47999		mostly acid and basic
12. methine and polymethine 48000-48999	$-\overset{+}{N}=\overset{\vert}{C}-CH=\overset{\vert}{C}-$	basic, some disperse
13. thiazole: 49000-49399		basic, direct, pigment
14. indamine and indophenol 49400-49999	X = O,N	intermediates
15. azine: 50000-50999		acid, basic, solvent
16. oxazine: 51000-51999		basic, direct, mordant, pigments
17. thiazine: 52000-52999		basic, acid, mordant, solvent

TABLE 2.4 (continued)

Class: CI No. range	Characteristic structural units or functional groups	Dye classes by application
18. sulfur: 53000-53999		sulfur and vat
19. lactone: 55000-55999		mordant
20. aminoketone and hydroxyketone: 56000-57999	chromophoric system based on -C=O with -NH$_2$, -NHR or -OH as auxochromes	mostly mordant and vat
21. anthraquinone: 58000-72999		most major classes
22. indigoid: 73000-73999	Ar-C-C=C-C-Ar [b] (-NH or -S as auxochromes)	acid, pigment, vat
23. phthalocyanine: 74000-74999	tetrabenzoporphyrazine	most major classes
24. natural: 75000-75999	various aromatic conjugated systems	natural
25. oxidation base: 76000-76999	incompletely characterized oxidation products of amines, diamines and aminophenols	_____
26. inorganic coloring matters 77000-77999	various inorganic compounds and complex salts	pigment

[a] R may or may not = R' (substituted aromatics).

[b] Ar = aromatic or heteroaromatic.

TABLE 2.5

Dyes classified by method of application

Dye type	Structure and Colour Index numbers [a]

Acid

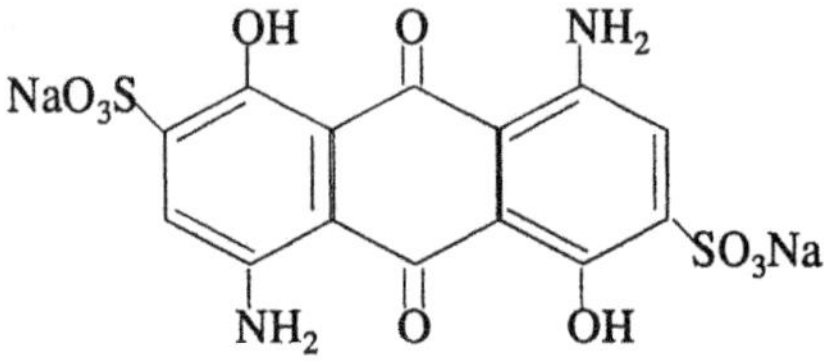

CI Acid Blue 45 (CI 63010)

Azoic

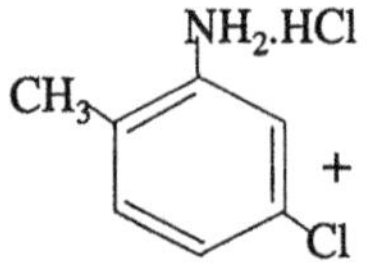

CI Diazo 32   CI Coupling 4
(CI 37090)   (CI 37560)

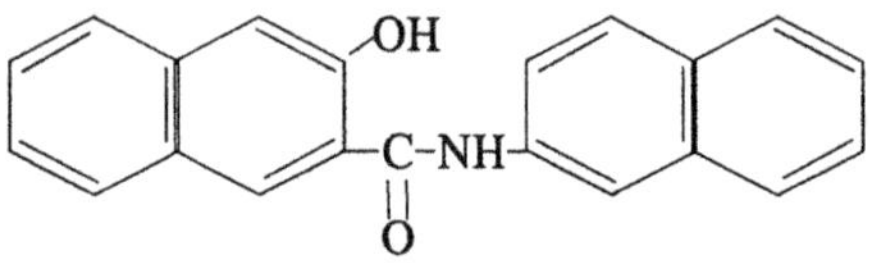

Basic

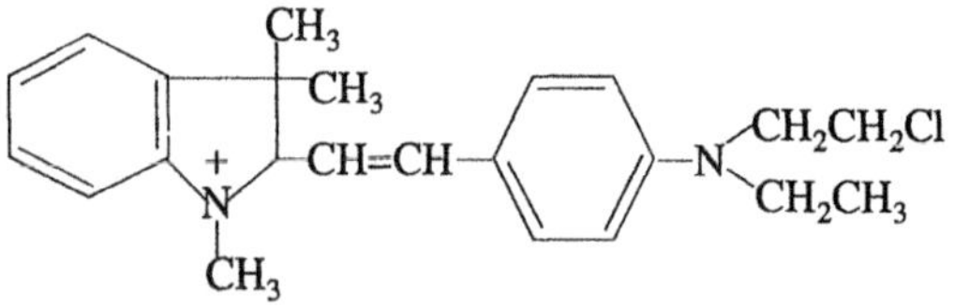

CI Basic Red 13 (CI 48015)

Direct

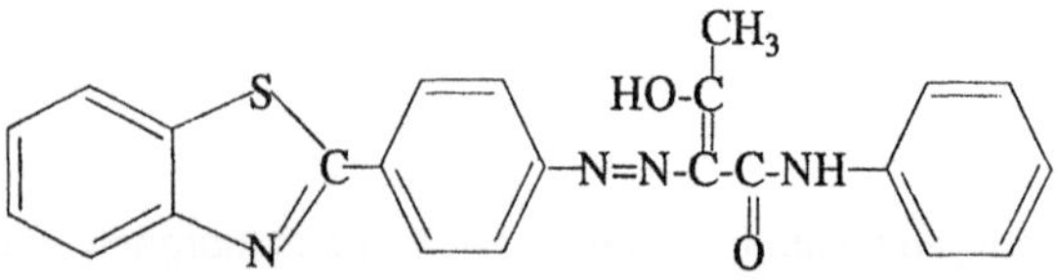

CI Direct Yellow 8 (CI 13920)

[a] Colour Index numbers according to application and chemical constitution (latter in parentheses).

TABLE 2.5 (continued)

Dye type	Structure and Colour Index numbers [a]
Disperse	CI Disperse Blue 24 (CI 61515)
Mordant	(bisulfite compound) CI Mordant Green 4 (CI 10005)
Natural	CI Natural Black 1 (CI 75290)
Pigment	(phosphotungstomolybdic acid salt) CI Pigment Green 4 (CI 42000:2)

[a] Colour Index numbers according to application and chemical constitution (latter in parentheses).

TABLE 2.5 (continued)

Dye type	Structure and Colour Index numbers [a]
Reactive	 CI Reactive Blue 4 (CI 61205)
Solvent	 CI Solvent Orange 53 (CI 10375)
Sulfur	 CI Sulfur Brown 61 (CI 53285)
Vat	 CI Vat Black 1 (CI 73670)

[a] Colour Index numbers according to application and chemical constitution (latter in parentheses).

listed in the *Colour Index* in this order: acid, azoic, basic, direct, reactive, solvent, sulfur, vat, disperse, mordant, natural and pigment.

The *Colour Index* with its dual classification of dyes, provides comprehensive and useful information that may be utilized in several different manners. In the first part of the index, dyes are classified according to their method of application. The second part of the index classifies them according to their chemical structure, while the third and last parts contain their commercial names and names of their manufacturers. Cross referencing is utilized so that one can readily retrieve or find the desired information.

2.4.2 Use of the Colour Index

To illustrate more specifically the type of information contained in the *Colour Index*, let us consider how it may be resourcefully used when certain information is known about a dye. If one knew the name and number of the acid dye listed in Table 2.5 (CI Acid Blue 45), it could be looked up in the first part of the index (under acid dyes-classified by application). When it is found under CI Acid Blue 45, there is a column of additional information about this dye (as shown in Table 2.6). It is first cross-referenced by chemical class (anthraquinone) with another *Colour Index* number related to its chemical constitution (CI 63010). Its hue in daylight and artificial daylight is provided, followed by the primary fiber type that is suitable for dyeing and preferred dyeing method. In this example, it is wool dyed at a pH of 3.5 in dilute sulfuric acid. Other types of fibers (e.g., silk in formic acid) are listed next, then information on how this dye may be used in textile printing. There is then an extensive list of colorfastness of the dye to various conditions (e.g, light, chlorine bleach, and washing) as tested by either AATCC and/or ISO methods. The higher the number rating, the better the fastness of the dyed textile is to each test (5 is excellent and 1 is poor). Other properties of the dye, such as its dischargeability and effect on metals, selected nontextile uses, and additional information are also provided.

If the dye is looked up under its chemical classification number (CI 63010), different information is provided (Table 2.7). It is cross listed under its application name or names (e.g., CI Acid Blue 45), and its structure and chemical class are shown,

TABLE 2.6

Example of information provided in *Colour Index* with regard to classification of dyes by method of application. Courtesy of The Society of Dyers and Colourists.

C.I. Acid Blue	45	
CHEMICAL CLASS	Anthraquinone	
C.I. CONSTITUTION NUMBER	**63010**	
HUE		
Daylight	Blue	
Artificial light (tungsten)	Slightly greener	
DYEING: WOOL		
Method	3	
Levelling	—	
S.D.C. migration test method/grade	I/2–3	
Staining other fibres	Acetate and cellulose—*u*, nylon—*d*	
FASTNESS PROPERTIES		
Method	AATCC	ISO
Alkali	2	2–3
Carbonising	3	4
Chlorination — alteration	—	2–3
staining wool	—	3–4
Decatising	5	4–5
Light, ½–¼ normal	4	5
normal	4	5–6
2 × normal	5	6
Milling, alkaline — alteration	1	1
staining wool	1	1
Milling, acid — alteration	—	2
staining wool	—	1–2·
Peroxide bleaching — alteration	1	1–2
staining wool	1	1–2
Perspiration	2	2–3
Potting — alteration	—	1
staining wool	—	1
Sea water — alteration	2–3	2
staining wool	2–3	1–2
Stoving	4	3–4
Washing — alteration	1–2	2
staining wool	1–2	3–4
OTHER PROPERTIES		
Dischargeability	Poor	
Effect of metals — copper	Slightly duller	
chromium	—	
iron	Slightly redder and duller	
NON-TEXTILE USAGE	Heavy metal salts as pigments for printing inks, book cloths and wallpaper Paper: beater dyeing and coating Anodised aluminium Urea, melamine and nitrocellulose plastics Soap and cosmetics See Leather Dyes section	

TABLE 2.7

Example of information provided in *Colour Index* with regard to classification of dyes by chemical structure or constitution. Courtesy of The Society of Dyers and Colourists.

63010   **C.I. Acid Blue 45** (*Blue*)

   (*a*)   Sulfonate 1,5-dihydroxyanthraquinone, nitrate the disulfonic acid formed and reduce with sodium sulfide

   (*b*)   Treat 2,6-dibromo(or dichloro)-1,5-dihydroxy-4,8-dinitro-anthraquinone with sodium bisulfite

   (*c*)   Treat 4,8-dibromo(or dichloro)-1,5-dihydroxy-2,6-anthra-quinonedisulfonic acid with ammonia in presence of copper

Discoverer — R. E. Schmidt 1897

Bayer Co., *BP* 12011/97, 19622/97, 20649/97, 7708/99, 16574/99, 7291/00; *USP* 623219, 623220; *FP* 266999; *GP* 96364, 100136, 100137, 103395, 105501, 106034, 108362, (*Fr.* 5, 246, 247, 249, 252, 250, 255, 251), 113724, 116746, (*Fr.* 6, 348, 350), 163647 (*Fr.* 8, 310), 195139 (*Fr.* 9, 713)

Agfa, *GP* 280646, 288665, 288878, (*Fr.* 12, 453, 454, 453)

I.G., *GP* 445269, 446563, (*Fr.* 15, 671, 673)

BIOS 1484, 41; *BIOS* 1661, 77; *FIAT* 1313, 2, 211

FIAT 764 — Alizarinsaphirol B

Barnett, 190, 283

Fierz-David, 520

Houben, 32, 296, 413, 447

Fierz-David, Suppl. 78

Slightly soluble in alcohol, Cellosolve

Insoluble in acetone, alcohol, benzene, carbon tetrachloride, Stoddard solvent

H_2SO_4 conc. — yellowish olive; on dilution — bluish violet

---

provided they are known or disclosed by the dye manufacturer. A representative method of synthesis, with appropriate reagents and synthetic steps, is given on the left side underneath the dye structure. In this instance, three different synthetic sequences are listed. The first sequence (discussed in more detail later in this chapter) utilizes 1,5-dihydroxyanthraquinone as the starting material. It is sulfonated, nitrated and then reduced with sodium sulfide. The other two synthetic sequences utilize dihydroxyanthraquinones containing dihalo and dinitro groups or dihalo and disulfonic acid groups that involve nucleophilic displacement reactions to obtain the desired dye. On the right side the name of the discoverer of the dye and date of discovery is provided, followed by a list of patents relating to the dye as well as literature references. Properties of the dyes, primarily their solubility in water, acid, base, and selected solvents are then listed.

At the back of this index, intermediates used to make these and similar dyes are listed. For example, the starting material for synthesizing this particular dye is 1,5-dihydroxyanthraquinone. This is listed under the dye intermediates section as anthrarufin. The melting point and molecular weight (if known) of the intermediate are given, along with the *Colour Index* chemical classification of all dyes that use this intermediate as a starting material for their synthesis (see list in Table 2.8). Acid Blue

TABLE 2.8

Example of information provided in *Colour Index* with regard to intermediates used to synthesize a dye (e.g., C. I. Acid Blue 45) or several dyes. Courtesy of The Society of Dyers and Colourists.

Anthrarufin		$C_{14}H_8O_4$	58080, 58230, 63005, 63010, 63011, 63315, 63325
		Mol. wt. 240	
		m.p. 280°	
Anthrarufin, 2,6-dibromo-4,8-dinitro-		$C_{14}H_4Br_2N_2O_8$	63005, 63010
		Mol. wt. 488	
Anthrarufin, 2,6-dichloro-4,8-dinitro-		$C_{14}H_4Cl_2N_2O_8$	63005, 63010
		Mol. wt. 399	

45 (CI 63010) can also be made from other intermediates (such as those listed in synthetic sequence (b) in Table 2.7). Starting materials such as 2,6-dibromo-4,8-dinitroanthrarufin and 2,6-dichloro-4,8-dinitroanthrarufin are also listed as intermediates in Table 2.8 for its synthesis.

TABLE 2.9

Commercial names of dyes listed under a particular dye (by application) in (a) 1973 *Colour Index* and (b) 1982 *Colour Index Supplement*. Courtesy of The Society of Dyers and Colourists.

C.I. Acid Blue 45

Acid Leather Blue CB	...	CIBA
„ Light Blue BV ...	...	Acna
Acilan Sapphirol B ...	...	FBy

C.I. Acid Blue 45 *continued*

Alizarine Azurol B ...	...	Chem
„ Blue SAP ...	...	GAF
„ „ SAPR...	...	NCC
„ Brilliant Blue B	...	LBH, YDC
„ „ „ 2 BS, G		LBH
„ Light Blue B	...	S
„ Paper Blue BP	...	NAC
„ Saphirol B ...	...	FDN, NCC
„ Saphirole BG, BLN		NCC
„ Sapphire ...		NAC
„ „ BLN	...	NAC
„ „ BN	AAP,	NAC
Bucacid Blue BL ...	...	BUC
Calcocid Quinizol Blue BP	...	ACY
Durosol Sapharol Blue BP	...	TCD
Elbenyl Brilliant Blue BL	...	LBH
Eloxone Blue RL ...	...	Gy
Erio Fast Cyanine S ...	...	Gy
Ext. D&C Blue No. 4	...	NAC
Fast Bond Blue GDS...	...	GAF
Fenazo Light Blue AS	...	GAF
Kiton Fast Blue CB ...	...	CIBA
Mitsui Alizarine Saphirol B		MDW
Pergacid Blue 2GAL ...	...	CIBA
Quinizol Blue BP ...	...	ACY
Solway Blue BN ...	...	ICI
Sulfacid Light Blue B2JL	...	Fran
Tertracid Light Blue B	...	ATS
Vondacid Light Blue BG, BN		Vond

C.I. ACID BLUE 45 (63010)

Acid Light Blue B	....	Mult
− − − BVP	......	Acna
Acilan Sapphirol B	....	BAY
Alizarine Azurol B	....	Chem
− Blue SAP	....	GAF
− − SAPR	......	NCC
− Brilliant Blue B	. LBH,	YCL
− − − BS	...	KCA
− − − FPS	..	LBH
− − − GS	...	LBH
− Saphirol B	....	FDN
− Saphirole BG, BLN	.	NCC
Atlantic Alizarine Cyanine Blue		
SAP	....	ATL
Bucacid Blue BL	....	BUC
Elbenyl Brilliant Blue BL	...	LBH
Erio Cyanine S	....	CGY
Fenazo Light Blue AS	.....	GAF
Intracid Fast Blue CB	......	CKC
Lissamine Blue B	....	ICI, ICI(O)
Mitsui Alizarine Saphirol B	..	MDW
Multacid Blue B	....	Mult
Pergacid Cyanine S	....	CGY
Quinizol Blue BP	....	ACY
Sandolan Blue E−BL	......	S
Sulfacid Light Blue B2JL	...	Fran
Sulphonol Blue E−B	......	YCL
Vondacid Light Blue BG, BN	.	Vond
Wool Fast Blue BV	....	VIO

(a) 1973 edition   (b) 1982 supplement

Finally, if one knows the commercial name of a dye (e.g., Kiton Fast Blue CB, Bucacid Blue BL, or Lissamine Blue B), these can be related to the chemical structure and class of the dye [such as Cl Acid Blue Dye--Cl 63010 (see Table 2.9 (a) and (b)]. Alternatively, one could use Table 2.9 to find out the commercial names and suppliers of this particular dye. The column on the left lists commercial dyes that were available when the 1973 edition was published, while the column on the right lists commercial dyes that were available for the 1982 updated supplement.

2.5 SYNTHESIS OF DYES

2.5.1 Relationship between structure and synthesis

Reference works and textbooks, such as those listed under **general references** at the end of this chapter, have discussed the synthesis of dyes from perspectives such as dye intermediates, organic reaction mechanisms and functional group transformations. However, since dyes represent such chemically diverse groups of substances, the approach here will be to integrate the principles of organic reaction mechanisms with synthetic strategies applicable to dye synthesis and to illustrate this approach with the aid of several examples and condensed experimental procedures.

From a structural viewpoint, there are three basic types of aromatic or highly conjugated systems that constitute the molecular architecture of dyes. The first basic type consists of substituted benzene and fused polycyclic ring systems such as anthraquinones. The second basic type is comprised of di-, tri- and polyaryl ring systems such as biphenyls and stilbenes. The third basic type consists of various heterocyclic ring systems such as benzothiazoles, xanthenes, quinolines, acridines and phthalocyanines. Examples of each system in Table 2.5 are afforded respectively by (a) the anthraquinone fused ring system of Cl Acid Blue 45, (b) the triphenylmethane structure of Cl Pigment Green 4 (Malachite Green), and (c) the benzothiazole ring system in Cl Direct Yellow 8. As one can readily ascertain by examining the structure of these and other dyes, they contain a variety of functional groups and ring structures. Incorporation of these conjugated ring systems and chemical species into dye molecules for effecting, changing, or improving fiber substantivity, chromophoric character, and colorfastness and solubility properties, thus requires a knowledge of synthetic strategies in conjunction with organic reaction mechanisms. Unless otherwise noted, synthetic sequences chosen as examples are derived from information listed in the *Colour Index*, where more detailed information may be obtained by consulting the numerous references it lists on the preparation and application of dyes.

2.5.2 Electrophilic aromatic substitution reactions in dye synthesis

A major class of organic reactions that is the most important in the synthesis of dyes is electrophilic aromatic substitution, i. e., replacement of a hydrogen atom in an aromatic or heteroaromatic ring system by other atoms or functional groups, or less frequently, replacement of functional groups by a hydrogen atom (Figure 2.16).

The presence of substituents in aromatic and heteroaromatic rings influences their reactivity towards electrophilic aromatic substitution and the orientation of the electrophilic agents into the aromatic nucleus. If the substituent already present in the ring (Y) increases the reactivity of the ring towards an electrophilic reagent or species (E^+) relative to an unsubstituted ring, it is called an activating group. Conversely, if Y causes the reactivity of the ring towards an electrophilic agent to decrease relative to an unsubstituted ring, it is called a deactivating group. This in-

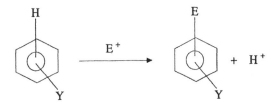

Figure 2.16. Electrophilic aromatic substitution; H = a hydrogen atom, E^+ = an electrophilic agent, and Y = other atoms or functional groups.

crease or decrease in reactivity towards electrophiles is caused by a combination of inductive (denoted by I) and resonance (denoted by R) effects, and less frequently by steric effects. If the inductive and resonance effects of the functional group Y in the ring makes the ring more reactive (relative to an unsubstituted ring), these are designated respectively as +I and +R effects; Y substituents that deactivate the ring (relative to the unsubstituted ring) towards electrophiles are designated respectively as having -I and -R effects.

The orientation effect is what position an incoming electrophilic agent (E⁺) is mostly like to react at or attack relative to functional groups already present (Y) on the aromatic nucleus. If the incoming electrophile is adjacent to Y, it occupies the *ortho-*, *o-* or 2-position; if the incoming electrophile is two atoms from Y, it occupies the *meta-*, *m-*, or 3-position; and if the incoming electrophile is three atoms from Y, it occupies the *para-*, *p-*, or 4-position.

Functional groups that exert both +R and +I activating effects cause the incoming group to substitute in the *ortho-* (or 2-) and *para-* (or 4-) positions, and the rate of reaction with an electrophile is faster than that of an unsubstituted ring (Figures 2.17 and 2.18). Groups that exert both -R and -I deactivating effects cause the incoming group to substitute in the *meta* or 3-position, and the rate of reaction with an electrophile is slower than that of an unsubstituted ring (Figures 2.19 and 2.20).

These are reinforcing effects in each case and can be explained by considering the mechanism of electrophilic aromatic substitution, in which the rate-determining or slow step is the formation of a positively-charged intermediate that precedes the second or fast step, the loss of H⁺ and restoration of the conjugated aromatic ring. In the

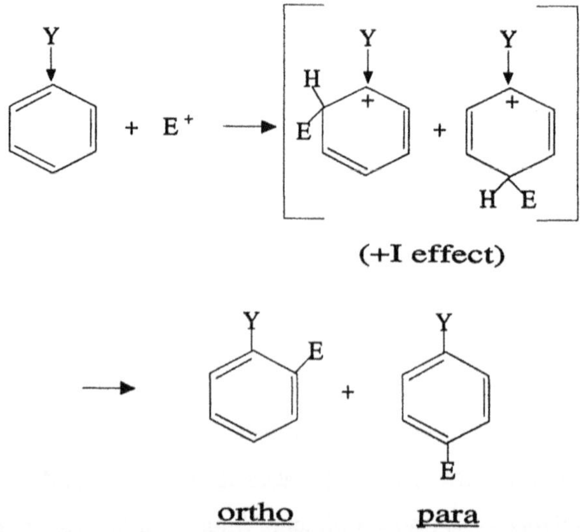

(+I effect)

ortho   para

Figure 2.17. +I effect of substituted aromatics to direct incoming substituents into *ortho* and *para* positions.

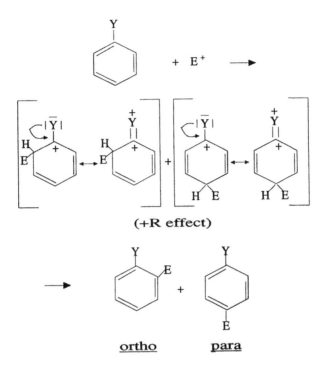

(+R effect)

ortho   para

Figure 2.18. +R effect of substituted aromatics to direct incoming substituents into *ortho* and *para* positions.

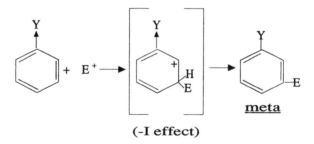

meta

(-I effect)

Figure 2.19. -I effect of substituted aromatics to direct incoming substituents into *ortho* and *para* positions.

case of directive (orientation) and ring activation, substituents that cause *ortho* and *para* substitution by resonance (+R) effects can even override a deactivating inductive effect (-I). However, electrophilic aromatic substitution is always slower and the incoming substituent always goes into the *meta* position when the resonance effect is deactivating (-R). Table 2.10 illustrates these trends and preferences and lists the

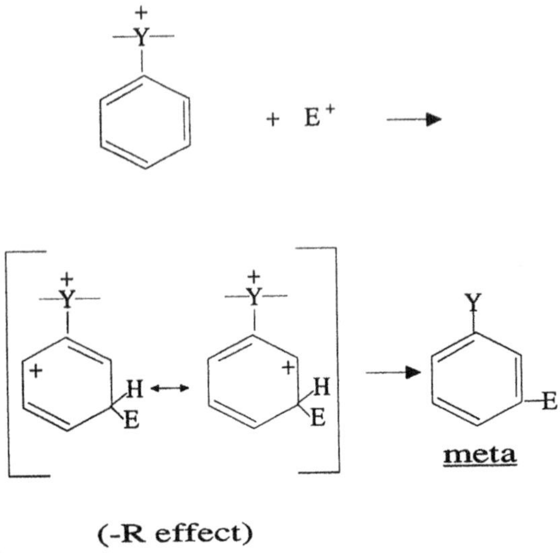

(-R effect)

Figure 2.20. -R effect of substituted aromatics to direct incoming substituents into *ortho* and *para* positions.

more common deactivating and activating groups and their directive effects. Thus, amino and hydroxyl groups make the ring extremely reactive to electrophiles because of the powerful electron-donating resonance effects that are not nullified by counterbalancing inductive effects. Alkyl substituents such as methyl have only +I effects and the ring is only moderately reactive to electrophilic aromatic substitution. In the case of most halo substituents (e.g., -Cl, Br, and -I), the +R effect is strong enough to direct substituents into the *ortho* and *para* positions, but the reaction is slow due to the -I effect. Functional groups that can withdraw electrons from the ring by resonance effects (usually groups containing heteroatom double bonds such as nitro and carbonyl) deactivate the ring towards electrophilic aromatic substitution, and as noted above, direct substituents only into the *meta* position. These well-documented effects of orientation and reactivity have been utilized in synthetic strategies for the preparation of numerous synthetic dyes.

Electrophilic aromatic substitution reactions are frequently employed for the synthesis of dyes and their intermediates. Representative reactions (Table 2.11) include: nitration, nitrosation, halogenation, sulfonation, Friedel-Crafts acylation, diazotization and coupling, carboxylation, and formylation (introduction of -HC=O).

TABLE 2.10

Effect of substituent groups on reactivity and orientation in electrophilic aromatic substitution. (adapted from ref. 45). Courtesy of Methuen and Co. Ltd.

Substituent	Polar character	Directive effect	Rate-controlling effect
O$^-$	+I, +R	o, p	Very powerfully activating
NH$_2$, NHR, NR$_2$, OH, OR	-I, +R	o, p	Powerfully activating
NHCOR, OCOR	-I, +R	o, p	Activating
phenyl	-I, +R	o, p	Moderately activating
CH$_3$, other alkyl groups	+I	o, p	Moderately activating
None	----	----	----------
F	-I, +R	o, p	Comparable to benzene
Cl, Br, I, CH=CH-CO$_2$H	-I, +R	o, p	Weakly deactivating
CO$_2$R, CO$_2$H, CHO, COR, CN, NO$_2$, SO$_3$H	-I, -R	m	Strongly deactivating
NH$_3^+$, NR$_3^+$	-I	m	Strongly deactivating

The reactions listed in Table 2.11 are representative but not at all inclusive of the numerous electrophilic substitutions that can be achieved with aromatic and a variety of heterocyclic structures. A more detailed discussion of these reactions is beyond the scope of this book, but general and advanced organic chemistry books (such as ref. 45) cover this topic in much depth.

The order of the introduction of certain functional groups in aromatic and heterocyclic rings must be carefully considered to obtain the desired product in the least number of steps and accomplished with a minimum of side reactions and low product yields. This was also noted in earlier discussions of orientation and reactivity in electrophilic aromatic substitution and in the selected examples depicted later of dye synthesis. The introduction of certain functional groups (such as -N=O, -COOH

TABLE 2.11

Electrophilic aromatic substitution reactions used to synthesize dyes and their intermediates

Type reaction	Reagents/typical conditions [a]
Nitration	HNO_3/H_2SO_4
Nitrosation	$Ar-NR_2$ + HONO ----> p-N=O-Ar-NR$_2$
Halogenation	Cl_2 or Br_2 + Lewis acids I_2 + HNO_3
Sulfonation	H_2SO_4 or $H_2SO_4.SO_3$
Friedel-Crafts acylation	Acid halides or carboxylic acids + Lewis acids or other acids/anhydrides (e.g., polyphosphoric acid). Includes intermol. ring formation and cyclizations.
Diazotization/coupling	$Ar-NH_2$ + $NaNO_2/HX$ (0-5°C) ------> $Ar-N=N + X -$ Reaction diazonium salt with Ar-OH or Ar-NH$_2$ (coupling component or reaction)
Carboxylations	Kolbe-Schmitt reaction: CO_2 + Ar-OH/NaOH -----> $ortho$-substitution (-COOH)
Formylation	CO/HCl/Lewis acid (Gatterman-Koch reaction) CO/HCN/HCl/hydrolysis(Gatterman reaction)
	Ar-OH/CHCl$_3$/OH ----->(o-hydroxybenzaldehydes) (Riemer-Tiemann reaction)

[a] Ar = aromatic or heteroaromatic rings or structures.

and -CH=O) require that other substituents first be present in the ring. Nitrosation (usually in the *para* position) requires reaction of HONO (nitrous acid) with a tertiary aromatic amine (ArNR$_2$). Phenols, particularly as the phenoxide form (Ar-O⁻) are required for several electrophilic aromatic reactions to occur. These include the Kolbe-Schmitt carboxylation with carbon dioxide, formylation under basic conditions with chloroform (Riemer-Tiemann reaction), and coupling of aromatic diazonium salts.

TABLE 2.12

Other selected reactions used to synthesize dyes and their intermediates

Type reaction	Reagents/typical conditions [a]
Nucleophilic aromatic substitution	Ar-X and subst. Ar-X + OH^- or NH_3 (X = -Cl, -Br, $-SO_3H$, etc.)/heat and/or pressure -----> Ar-OH or $ArNH_2$
Sandmeyer reaction	$Ar-N\equiv N + X - + Cu(I)Z$ -----> Ar-Z (Z = -Cl, -Br, -I, - CN, -SCN)
Reduction of nitro groups	$Ar-NO_2$ + Sn/HCl or Fe/HCl -----> $ArNH_2$
Decarboxylation	Ar-COOH + heat -----> Ar-H + CO_2 (occurs readily when -COOH group o- and or p- to two or more electron-rich groups)
Desulfonation	$Ar-SO_3H + H^+$ -----> Ar-H (reverse reaction of sulfonation)
Benzidine rearrangement	Ar-NH-NH-Ar' + $2H^+$ ------> p-NH_2-Ar-Ar'-p-NH_2 (major product)
Ullmann coupling reaction	225°C 2 Y-Ar-Cl + Cu -----> Y-Ar-Ar-Y
Oxidations	$Ar-CH_3 + KMnO_4$ -----> Ar-COOH 2 Ar-SH + H_2O_2 -----> Ar-S-S-Ar

[a] Ar = aromatic or heteroaromatic rings or structures.

In addition to electrophilic aromatic substitution reactions, there are nucleophilic aromatic substitutions reactions, ring dimerization reactions, condensations, molecular rearrangements, and the removal and/or transformation of a variety of functional groups that have and continue to be frequently utilized for dye synthesis. A few examples of these types of reactions are shown in Table 2.12.

In contrast to electrophilic aromatic substitution, nucleophilic aromatic substitution is made more facile by electron-withdrawing groups (such as $-NO_2$) *ortho* and/or *para* to ring substituents that are good leaving groups (such as -Cl or -Br). Displacement of aromatic halides with OH^- or NH_3 and moderate heating lead to the corresponding phenols (Ar-OH) or substituted anilines ($Ar-NH_2$) when this electron-deficient groups. More vigorous reaction conditions (high temperature and pressure) are required to convert aryl halides and arylsulfonic acids (Ar-SO₃H) to phenols. The latter conversion is called the alkali fusion method, and there is evidence that a different type of mechanism is require for the substitution to occur when no electron-withdrawing groups are present to facilitate nucleophilic aromatic substitution.

The Sandmeyer reaction is useful for converting primary aromatic amines into various other functional groups (-Cl, -Br, -I, -CN, -SCN) via an aromatic diazonium salt intermediate. The reaction is believed to be free radical in nature whereby an aryl radical (Ar) generated from the diazonium salt reacts with the corresponding Cu (I) salt (halide, cyanide, or thiocyanate). Variations of this reaction exist that introduce fluorine (Schiemann reaction) and other functional groups into the aromatic nucleus.

Of the numerous reductions that can be effected with various functional groups present on the aromatic ring, conversion of the nitro group to a primary amino group by use of reagents (such as Sn + HCl or Fe + HCl) has been frequently used in dye synthesis. Modern methods of catalytic hydrogenation can also be employed to reduced the nitro group to the amino group.

Decarboxylation of aromatic carboxylic acids occurs in the presence of acid and heat, and is facilitated by electron-donating groups *ortho-* and/or *para-* to the -COOH group. Similar reaction conditions with aromatic sulfonic acids using dilute acid cause desulfonation of the aromatic ring. This is the reversible reaction of sulfonation of the aromatic ring with concentrated sulfuric acid.

The benzidine rearrangement is one of several rearrangements in which functional groups in aromatic rings are transposed to different positions. It effectively converts hydrazobenzenes (Ar-NH-NH-Ar) into 4,4'-diaminodiphenyls. Some 2,4'-diamino-diphenyl product is also formed in this reaction.

The Ullmann coupling reaction is a free-radical reaction in which substituted aryl halides or halobenzenes are reacted with copper to form aryl radicals that dimerize. This effectively doubles the size and molecular weight of the starting aromatic and is used as a synthetic technique to form certain dyes of symmetrical structure.

Finally, a couple of the many oxidations are listed that are available for converting functional groups to higher oxidation states. Two relevant examples are (a) the oxidation of methyl or alkyl side chains in aromatic rings to the corresponding Ar-

COOH by use of oxidizing agents such as $KMnO_4$ or HNO_3, and (b) oxidation of thiol groups (-SH) to disulfides (-S-S-) by use of mild oxidants such as iodine, peroxide or Fe^{3+} salts. The numerous approaches and possible reactions that may be used to synthesize dyes and other aromatic and heteroaromatic structures are typified by a few selected synthetic sequences in the following pages.

The first synthetic dye, mauve, was synthesized by Sir William Henry Perkin in 1856. This accomplishment revolutionized the chemical and dye industry, and it is quite appropriate to use this as the first example of a dye synthesis. Perkin used impure aniline that contained *ortho*- and *para*-toluidine (or *o*- and *p*-aminotoluenes). The first synthetic dye mauve (azine structure in Figure 2.21) was prepared and discovered when this mixture was oxidized with potassium dichromate and cold sulfuric acid. Silk is dyed a reddish violet and cotton dyed violet with mauve.

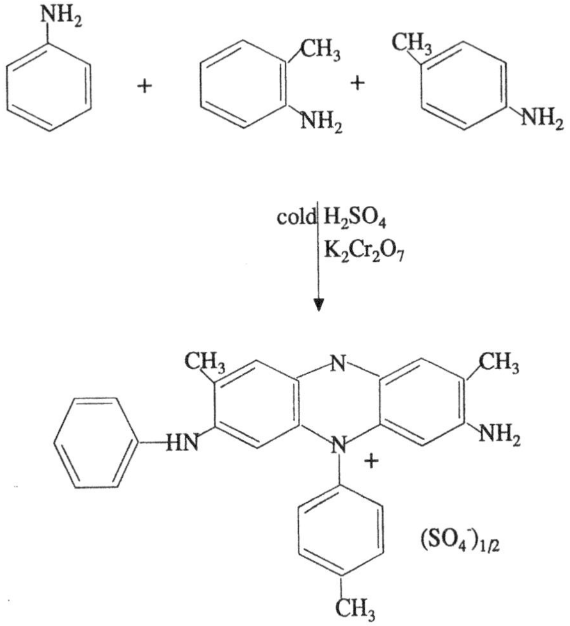

Figure 2.21. Preparation of the first synthetic dye, mauve, by Perkin in 1856.

The synthesis of CI Acid Blue 45 (depicted earlier in Table 2.5) is one example of how electrophilic aromatic substitution reactions may be utilized in sequence.

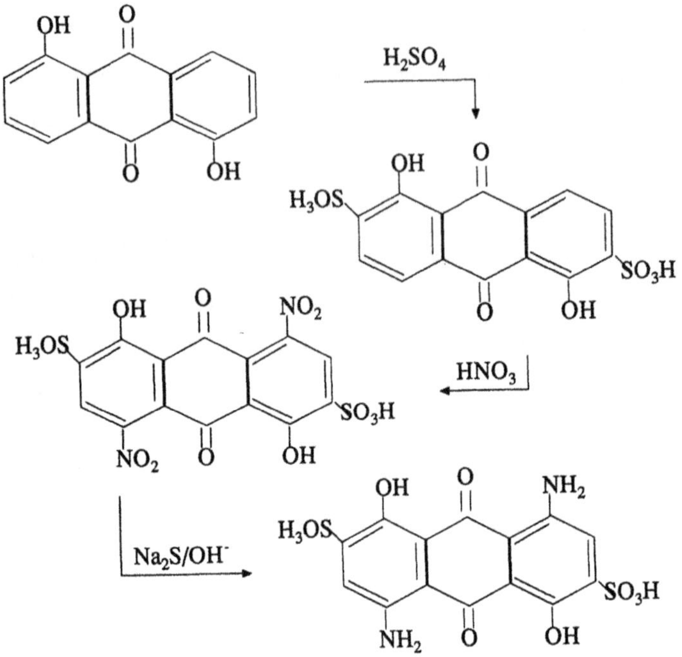

Figure 2.22. Synthesis of the anthraquinone dye CI Acid Blue 45.

The first step in the synthesis of this anthraquinone dye (Figure 2.22) consists in the sulfonation of 1,5-dihydroxyanthraquinone in the 2- and 6- positions. The activating and *ortho-* directing effect of the hydroxyl group cause this particular product to be formed. Subsequent reaction with nitric acid causes the nitro groups to be placed *meta* to an electron-withdrawing (-SO₃H) and *para* to an electron-donating group (-OH). The final product is obtained by reducing the nitro groups under basic conditions with sodium sulfide, and at the same time, making the dye water-soluble by converting the sulfonic acid groups into corresponding sodium salts.

Another example of electrophilic aromatic substitution to affect dye synthesis is the preparation of CI Pigment Green 4 (Malachite Green). This type of synthetic route (Figure 2.23) is a general method used for the preparation of a variety of triaryl-methane dyes. The protonated benzaldehyde acts as an electrophile to react with two moles of N,N-dimethylaniline in the *para* position. The central carbon atom is then oxidized with lead peroxide to form the carbinol, which is subsequently protonated with acid to produce Malachite Green.

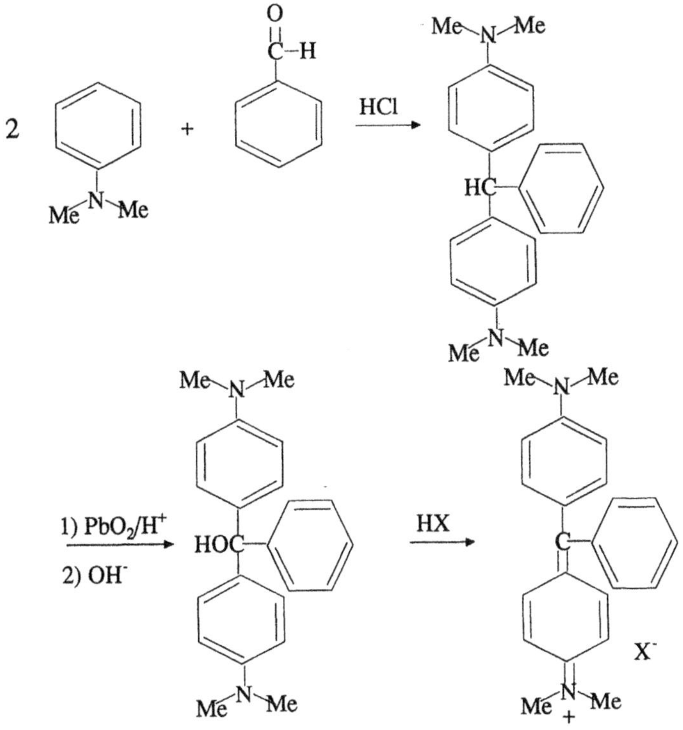

Figure 2.23. Synthesis of CI Pigment Green 4 (Malachite Green).

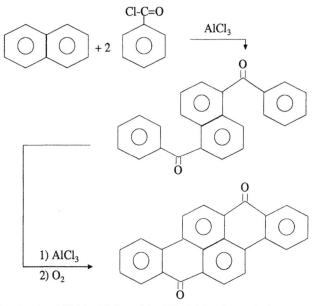

Figure 2.24. Synthesis of CI Vat Yellow 4 by Friedel-Crafts acylations.

The synthesis of a vat dye with an anthraquinone structure (Cl Vat Yellow 4/Cl 59100) is achieved by use of successive Friedel-Crafts acylation reactions (Figure 2.24). The first step in this dye synthesis is the Friedel-Crafts acylation of naphthalene with two moles of benzoyl chloride to yield 1,5-dibenzoylnaphthalene. This is followed by cyclization of the latter to the yellow vat dye by a Friedel-Crafts intramolecular cyclization and oxidation.

2.5.3 Other reactions on aromatic structures used for dye synthesis

Nucleophilic aromatic substitution reactions are usually less useful for dye and other type of organic syntheses because the presence of good leaving groups (usually halogen) *ortho* and/or *para* to electron-withdrawing groups (such as $-NO_2$ and $-SO_3H$) in the aromatic ring is required for this type of reaction to occur. Activating and directive effects are exactly the opposite of those observed in electrophilic aromatic substitution. Electron-donating groups are deactivating and *meta*-directing, while electron-withdrawing groups are activating and *ortho/para* directing towards incoming nucleophiles. However, many types of reactions are required for the synthesis of dyes of diverse structure and substantivity to many fiber types. For example, the synthesis of Cl Disperse Blue 24 (Figure 2.25) is achieved by (a) electrophilic aromatic substitution in the first step (bromination of 1-methyl-aminoanthraquinone), followed by (b) nucleophilic aromatic substitution in the second and final step (replacement of bromine with aniline). The bromo- group is *para* to an electron-donating group ($-NHCH_3$). However, the nucleophilic displacement still occurs because the carbonyl groups in the adjacent ring are available to stabilize the negatively-charged intermediate.

Although the reaction of stabilized aromatic diazonium ($Ar-N\equiv N + X -$) with phenols, aromatic tertiary amines and other suitable structures can be classified as an electrophilic aromatic substitution reaction, it is usually discussed separately because of its combination with many other types of reactions to produce dyes. Also, it deserves separate discussion because the number of dyes that are produced by these coupling reactions is so extensive and the increase in molecular weight and size of the resultant dyes is much greater than those produced by other types of electrophilic aromatic reactions.

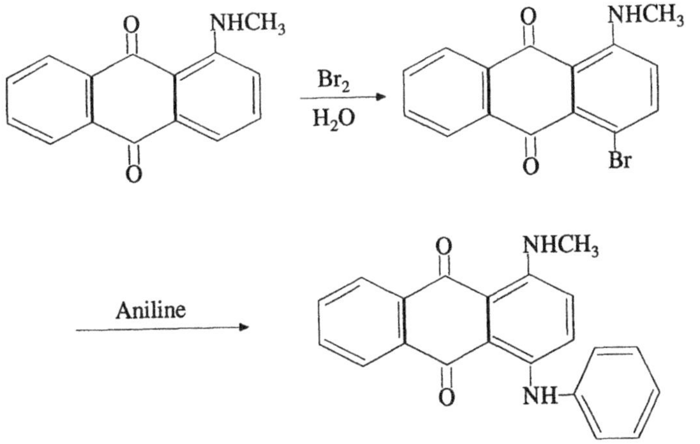

Figure 2.25. Synthesis of Cl Disperse Blue 24 by electrophilic substitution and a subsequent nucleophilic substitution reaction.

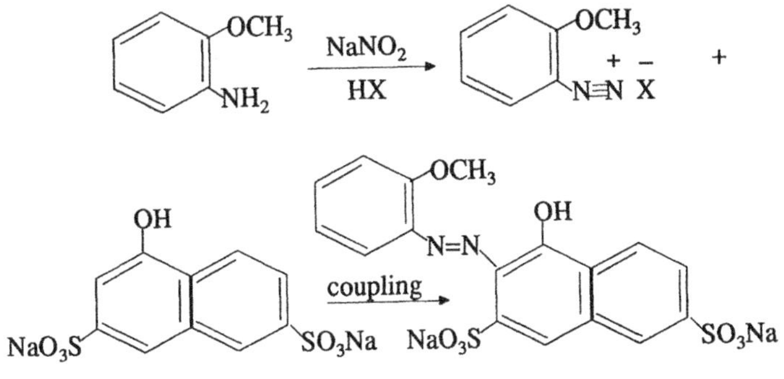

Figure 2.26. Synthesis of Cl Acid Red 5 by a diazonium/coupling sequence.

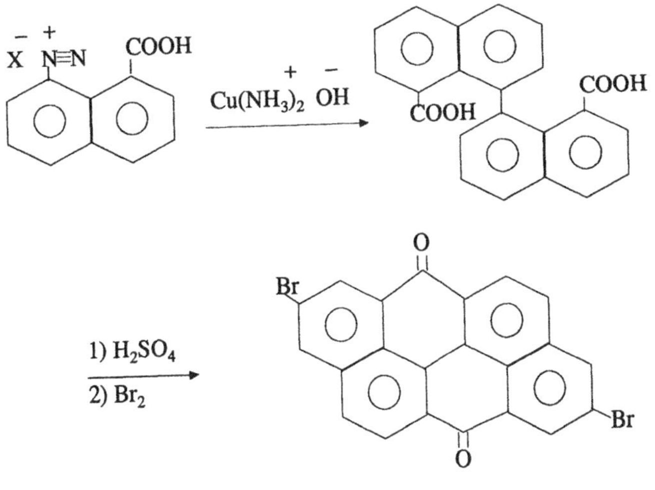

Figure 2.27. Synthesis of Cl Vat Orange 3 by a diazonium/ dimerization/ cyclization/ halogenation sequence.

The stabilization of aromatic diazonium salts and their importance as intermediate for producing dyes was discovered by Peter Griess in Germany shortly after Perkin produced the first synthetic dye. Although Griess was not the first to react his stabilized diazonium salts with coupling components, he readily recognized the commercial potential of the coupling reaction and was responsible for the discovery of many azo dyes. A relevant example of an azo dye formed by the diazonium/ coupling sequence is CI Acid Red 5 (Figure 2.26).

Other types of reactions (depicted earlier in Table 2.12) are synthetically useful for the preparation of aromatic and heteroaromatic ring structures required to synthesize dyes. These types of reactions include molecular rearrangements, condensations, dimerizations, and coupling reactions. Many of these utilize diazonium salts as one or more of the steps in the synthetic sequence. For example, the free radical coupling or dimerization of aryl radicals produced from diazonium salts is a key step in the preparation of 4,10-dibromoantaanthrone or CI Vat Orange 3 (Figure 2.27).

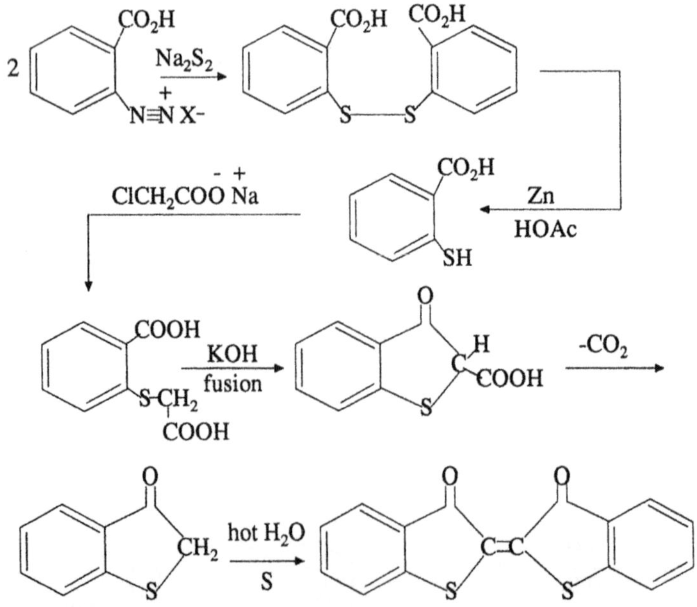

Figure 2.28. Synthesis of CI Vat Red 41 by a variety of reactions of substituents in aromatic rings.

The preparation of a thioindigo dye (CI Vat Red 41) initially involves the reaction of an aromatic diazonium salt with a good nucleophile (Na_2S_2), and in the last synthetic sequence, an oxidative coupling reaction (Figure 2.28). Intermediate synthetic sequences in the preparation of this thioindigo dye include the reduction of a disulfide to o-mercaptobenzoic acid, its subsequent reaction with sodium chloroacetate, then cyclization and decarboxylation to 3(2H)-thianaphthenone before the final oxidative coupling reaction.

Dyes with heterocyclic ring systems may be prepared by a variety of condensation reactions to form heteroaromatic nucleuses. One example of this is the reaction of N,N-dimethyl-p-nitrosoaniline hydrochloride with gallamide to form the oxazine dye commonly known as Gallamine Blue or CI Mordant Blue 45 (Figure 2.29).

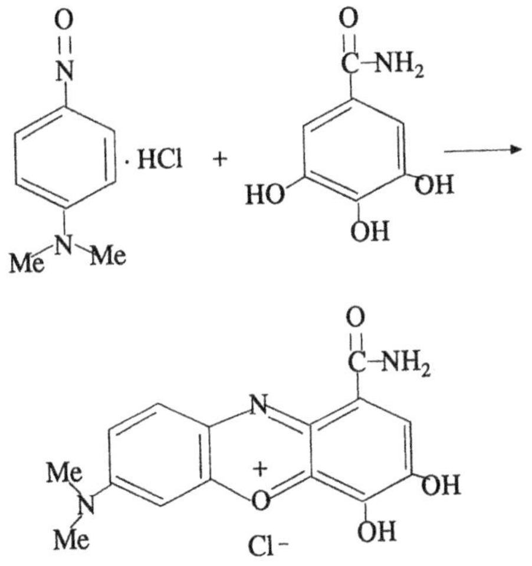

Figure 2.29. Synthesis of CI Mordant Blue 45 by heterocyclic ring formation.

As noted from several of the above synthetic sequences, oxidations, reductions, and a variety of other reactions are also employed and required for the synthesis of dyes. These advances and trends will be used to synthesize new dyes as new synthetic methods are developed and requirements for dyes change.

GENERAL REFERENCES FOR FUNDAMENTALS OF DYEING AND DYEING PROCESSES FOR TEXTILES

E. N. Abrahart, **Dyes and Their Intermediates**, Edward Arnold Ltd., London, 1977, 265 pp.

The Colour Index, Vols. I-IV and Vol. 5, Supplement, 3rd Ed., Society of Dyers and Colorists, Bradford, England, and American Association of Textile Chemists and Colorists, 1971 and 1982.

A. Johnson (Ed.), **The Theory of Coloration of Textiles**, The Textile Institute, Manchester, England, 1989, 552 pp.

H. L. Needles, **Handbook of Dyes, Finishes and Textile Auxiliaries**, 2d Ed., Noyes Publishers, Park Ridge, N. J., 1986, 210 pp.

R. H. Peters, **Textile Chemistry, Vol. III: The Physical Chemistry of Dyeing**, Elsevier, Amsterdam, 1975, 889 pp.

P. Rys and H. Zollinger, **Fundamentals of the Chemistry and Application of Dyes**, Wiley-Interscience, New York, 1972, 196 pp.

V. A. Shenai, **Fundamentals of Textile Coloration**, Text. Dyer. Print. 23 (22) (1990) 29-44.

E. R. Trotman, **Dyeing and Chemical Technology of Textile Fibres**, 6th Ed., Hodder and Stoughton Ltd., London, England, 1991, 587 pp.

K. Venkataraman (Ed.), **The Chemistry of Synthetic Dyes, Vols. I-VIII**, Academic Press, New York, 1952-1978.

D. R. Waring and G. Hallas (Eds.), **Chemistry and Application of Dyes**, Plenum Press, New York, 1990, 430 pp.

H. Zollinger, **Color Chemistry: Synthesis, Properties and Applications of Organic Dyes and Pigments**, 2nd Rev. ed., VCH Publishers, New York, 1991, 496 pp.

TEXT REFERENCES FOR FUNDAMENTALS OF DYEING AND DYEING PROCESSES FOR TEXTILES

1 Terms and Definitions Committee, J. S. Ward, Chairman, **Colour terms and definitions**, J. Soc. Dyer & Colour. 89 (11) (1973) 411-422.

2 Terms and Definitions Committee, J. S. Ward, Chairman, **New terms defined since 1973**, J. Soc. Dyer & Colour. 94 (7) (1978) 310-314.

3 Terms and Definitions Committee, **Colour terms and definitions (1985)**, J. Soc. Dyer & Colour. 101 (11) (1985) 367-371.

4 E. N. Abrahart, **Dyes and Their Intermediates**, Chapter 1: Introductory and Chapter 4: Classification of Dyes, Edward Arnold Ltd., London, 1977, pp. 2, 4 and 31.

5 J. Griffiths, **Historical development of modern colour and constitution theory**, Rev. Prog. Color. Rel. Topics 14 (1984) 21-32.

6 E. Coates, **Colour and constitution**, J. Soc. Dyer & Colour. 83 (3) (1967) 95-111.

7 S. F. Mason, **Colour and the electronic state of organic molecules**, in K. Venkataraman (Ed.), The Chemistry of Synthetic Dyes, Vol. 3, Academic Press, New York, 1970, pp. 169-221.

8 L. G. S. Brooker, **Absorption and resonance in dyes**, Rev. Modern Phys. 14 (1942) 275-293.

9 S. Dähne, **Color and constitution: one hundred years of research**, Science 199 (1978) 1163-1167.

10 D. W. Rangnekar, **Development of newer chromophores: strategies and resources**, Colourage 39 (5) (1992) 55-69.

11 E. R. Trotman, **Dyeing and Chemical Technology of Textile Fibers**, Griffin & Co. Ltd., London, 1975, pp. 336, 346, 347.

12 R. McGregor, **Kinetics and equilibria in dyeing**, Text. Chem. & Color. 12 (12) (1980) 306-310.

13 R. McGregor and R. H. Peters, **The effect of rate of flow on rate of dyeing. I. The diffusional boundary layer in dyeing**, J. Soc. Dyer & Colour. 81 (9) (1965) 393-400.

14 R. H. Peters, **Textile Chemistry, Vol. III. The Physical Chemistry of Dyeing**, Elsevier, Amsterdam, 1975, p. 777.

15 P. Rys and H. Zollinger, **Fundamentals of the Chemistry and Application of Dyes**, Wiley-Interscience, New York, 1972, p.165.

16 F. Hoffman, **Exhaust level kinetics with batch dyeing processes. I. Sorption under idealized conditions**, Textilveredlung 24 (10) (1989) 340-348.

17 F. Hoffman, **Exhaust level kinetics with batch dyeing processes. II. Effect of liquor circulation or movement of the goods**, Textilveredlung 24 (11) (1989) 381-389.

18 F. Hoffman, **Exhaust level kinetics with batch dyeing processes. III. Migration, temperature or fibre differences**, Textilveredlung 25 (2) (1990) 49-54.

19 J. N. Etters, **Isothermal sorption from finite baths: effect of the boundary layer on sorption of diffusants by solids of various geometrical shapes**, J. Appl. Poly. Sci. 42 (1991) 1519-1523.

20 R. McGregor and T. Ijima, **Dimensionless groups for the description of sorption equilibria in dyeing with anionic dyes. I. Inorganic co-ion exclusion**, J. Appl. Poly. Sci. 41 (1990) 2769-2782.

21 R. McGregor and T. Ijima, **Dimensionless groups for the sorption of dye and other ions by polymers. II. Hydrochloric acid, C. I. Acid Blue 25, and other polyamides with an excess of basic groups**, J. Appl. Poly. Sci. 45 (1992) 1011-1021.

22 C. H. Giles, **Dyeing in two dimensions. A review of the use of the monolayer method in the study of dye-fiber interactions**, J. Soc. Dyer & Colour., 94 (1) (1978) 4-12.

23 H. H. Sumner, **The development of a generalised equation to determine affinity. Part 1 - Theoretical derivation and application to dyeing cellulose with anionic dyes**, J. Soc. Dyer & Colour., 102 (10) (1986) 301-305.

24 H. H. Sumner, **The development of a generalised equation to determine affinity. Part 2 - Application to the dyeing of nylon and wool with anionic dyes and acrylic fibres with cationic dyes**, J. Soc. Dyer & Colour 102 (11) (1986) 341-349.

25 H. H. Sumner, **The development of a generalised equation to determine affinity. Part 3 - Application to practical systems**, J. Soc. Dyer & Colour 102 (12) (1986) 392-397.

26 A. Urbanik, **Relationship of the solubility parameters of dyes and auxiliaries to effects in the disperse dyeing of polyester**, Proc. AATCC 1981, Charlotte, N. C., pp. 258-265.

27 S. A. Siddiqui and H. L. Needles, **Solubility parameters**, Text. Res. J. 52 (1982) 570-579.

28 T. Hori, Y. Sato and T. Shimizu, **Contribution of swelling, dye affinity, glass transition temperature and other factors in the experimental diffusion of a dye into poly (ethylene terephthalate) from various solvents**, J. Soc. Dyer & Colour. 97 (1) (1981) 6-13.

29 K. Silkstone, **The influence of polymer morphology on the dyeing properties of synthetic fibers**, Rev. Prog. Color. Rel. Topics 12 (1982) 22-30.

30 T. Hori and H. Zollinger, **Role of water in dyeing theory**, Text. Chem. & Color. 18 (10) (1986) 19-25.

31 E. Sada and H. Kumazawa, **Kinetic aspects of dyeing processes**, J. Appl. Poly. Sci. 27 (1982) 2987-2996.

32 R. M. Rohner and H. Zollinger, **Porosity versus segment mobility in dye diffusion kinetics--a differential treatment: dyeing of acrylic fibers**, Text. Res. J. 56 (1) (1986) 1-13.

33 H. Zollinger, **Dye theories--models and reality of diffusion of dyes in textile fibers**, Textilveredlung 24 (4) (1989) 133-142.

34 H.-J. Flath, **Polymer structure and dyestuff diffusion,** Mell. Textilber. 72 (2) (1991) 132-139.

35 G. Schreiner, **Diffusion theory as a guide for the practice?-depth of shade and dye distribution in polyester dyeing**, Mell. Textilber. 71 (9) (1990) 686-688.

36 H. A. Davis, **The relationship between fiber structure and dye diffusion**, Text. Chem. Color. 24 (6) (1992) 19-22.

37 M. Duscheva, M. Itcherenska and E. Gavrilova, **The phenomenon of 'barriness' in the dyeing of polyamide fiber material. Part I. What is known about the theory of barriness?**, Mell. Textilber. 70 (5) (1989) 360-364.

38 G. Schreiner, **Dialogue on the theory of dyeing**, Mell. Textilber. 73 (4) (1992) 348-352.

39 H.-J. Flath, **Recent findings in dyeing fiber-forming polymers of polyamide and polyurethane**, Mell. Textilber. 73 (5) (1992) 423-427.

40 P. R. Brady, **Diffusion of dyes in natural fibers**, Rev. Prog. Color. Rel. Topics 22 (1992) 58-78.

41 T. L. Dawson and J. C. Todd, **Dye diffusion-the key to efficient coloration**, J. Soc. Dyer & Colour. 95 (12) (1979) 417-426.

42 J. Crank, **The Mathematics of Diffusion**, 2d Ed., Clarendon Press, Oxford, England, 1975.

43 J. Cegarra, P. Puente and F. J. Carrión, **Emperical equations of dyeing kinetics**, Textile Res. J. 52 (1982) 193-197.

44 J. N. Etters and A. Urbanik, **An automated computation of diffusion equations solutions**, Textile Res. J. 53 (1983) 598-605.

45 R. O. C. Norman, **Principles of Organic Synthesis**, 2nd Ed., Halstead Press, New York, 1978, p. 377.

Chapter 3

METHODS OF APPLYING DYES TO TEXTILES

3.1 INTRODUCTION

As indicated in Chapter 2, little similarity exists between the two methods of dye classification (chemical functionality and method of application). However, there are some useful relationships that can be discussed and developed with regard to dye application class and the six representative fiber types: wool, cotton, cellulose esters, polyamides, polyester and acrylic (Table 3.1). A further subdivision of dyes by method of application is given by Needles (ref. 1): (a) dyes containing anionic functional groups (acid, metal complex, mordant, direct and reactive); (b) dyes containing cationic functional groups (basic); (c) dyes requiring chemical reaction before application (vat, sulfur and azoic); and (c) special dyeing techniques and classes of colorants (disperse, solvent, natural and pigment). Aftertreatments that improve dye fastness, the dyeing of fiber blends, and textile printing may also be included in the latter category. Since this subdivision and its interrelationship with fiber type and dye fastness properties are generally useful, each of these classes of dyes will be discussed in the order listed, followed by the types of dyeing and printing machines that are utilized.

3.2 APPLICATION OF DYES CONTAINING ANIONIC FUNCTIONAL GROUPS

3.2.1 Acid dyes

Acid dyes are water-soluble, contain one or more anionic groups (usually $-SO_3H$), and are applied primarily to wool and polyamides. Most acid dyes contain azo groups, but there are a few that are in the anthraquinone or triphenylmethane chemical class. Adsorption of acid dyes by polyamides and by wool fibers is governed by species of polymeric functional groups that exist at different values of pH and by the hydrophobicity of the dye. In neutral solution, the functional amino- and carboxyl groups exist as zwitterions: $^+NH_3\text{--}F\text{--}COO^-$, where F = wool or polyamide. Under acidic conditions, protonation of the carboxyl group occurs, and the fiber exists as $^+NH_3\text{--}F\text{--}COOH$. Conversely, deprotonation of the positively charged amino group occurs in basic solution and the fiber exists in the form $NH_2\text{--}F\text{--}COO^-$. Sorption of acid dyes may be most appropriately characterized by the use of Langmuir-type isotherms, but Donnan equilibrium may also be utilized to explain the distribution of ions in the solid and solution phases.

TABLE 3.1

Application class and suitability for natural and synthetic fibers (ref. 2).

Class	Wool	Cotton	Cellulose derivatives	Polyamide PA	Polyester PE	Acrylics PAN
Basic	X	X	X	X	X	X
Direct		X				
Sulfur		X				
Azoic		X	X		X	
Ingrain						
Vat		X	X			
Acid levelling	X			X		
Acid milling	X			X		
Mordant	X	(X)		X		
Metal complex	X			X		X
Disperse			X	X	X	X
Reactive	(X)	X				
Pigment				X	X	X

X = suitable; (X) = of secondary importance. Pigments may be applied to any substrate by the use of adhesives.

Before acid dyeing, scouring of wool is necessary to remove natural oils and gums, while this preparatory process is not necessary for polyamides. Pale shades or tints of acid dyes with either fiber require pretreatment with peroxide or perborate. These acid dyes may be further subdivided by application into (a) levelling, (b) milling and (c) supermilling dyes. The characteristics of these three types of acid dyes are summarized in Table 3.2 (ref. 3). Metal complex dyes are derived from complexes with transition metals and dyes, and are discussed as a separate subclass containing anionic functional groups.

Levelling dyes are normally applied to wool from a strongly acidic bath (pH of 2-3 with sulfuric acid) to promote good exhaustion. Their good migration and levelling properties, due to their low molecular weight and ability to become unattached and subsequently reattached to another dye site on the fiber, are offset by their poor wetfastness. Levelling dyes applied to polyamides are usually of higher molecular weight, and are applied at a higher pH than on wool.

Milling acid dyes are applied from weakly acidic solutions (usually acetic acid), have higher molecular weights than levelling acid dyes, and contain two or more anionic groups. These dyes have good wetfastness, but possess poor levelling and migration properties. Neutral or supermilling acid dyes are applied from solutions of neutral pH,

TABLE 3.2

Characteristics of acid dyes (ref. 3).

Criteria	Levelling dyes	Milling dyes	Super-milling dyes
Fastness to wet treatment	Poor	Good	Very good
Dyeing method	Sulfuric acid	Acetic acid	Ammonium acetate
pH of dyeing	2-4	4-6	6-7
Levelling under own dyeing conditions	Good	Moderate-poor	Very poor
Dye characteristics	Low molecular wt. High solubility Molecular solns.	High molecular wt. Low solubility Colloidal solutions	High molecular wt. Low solubility Colloidal solutions
Affinity of anions	Low	High	Very high

usually contain only one anionic group. Since they are not levelling, their application must be carried out carefully; however, if they are properly applied, they exhibit good wet and lightfastness properties. Improved fastness of the supermilling and milling dyes compared to the levelling dyes may also be due to the larger molecular size of the former two types of dyes relative to the latter type of dye.

3.2.2 Metal complex dyes

Premetallized or metal complex dyes are normally classified as acid dyes, since they usually have azo groups and anionic substituents and possess good substantivity for protein and polyamide fibers. However, the anionic groups differ with the ratio of metal to dye. Numerous variations of these metal complex dyes developed over the years have been comprehensively and critically reviewed (ref. 4). Chromium or cobalt salts are most frequently used as the metal complexing agents. 1:1 complexes of the metal and the dye normally contain one or two -SO$_3$H groups) and are applied from strongly acidic solutions (usually 6-8% H$_2$SO$_4$ to attain a pH of 2.0-2.4). The dyeing procedure is quite lengthy since the dyebath is raised slowly to the boil for almost an hour, then maintained at the boil for more than an hour to achieve levelling and insolubilization of the metal complex on the fiber. 1:2 metal complex dyes usually have no ionizing groups but contain highly polar groups (such as -SO$_2$CH$_3$ or -SO$_2$NH$_2$) to confer sufficient solubility of the dye in water.

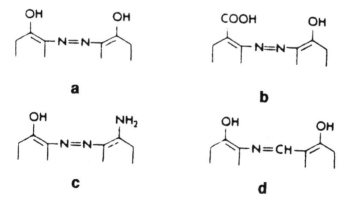

Figure 3.1. Important chromium and cobalt complexes for dyeing wool and polyamide fibers include: (a) o,o'-dihydroxyazo-, (b) o,o'-carboxyhydroxyazo-, (c) o,o'-hydroxyaminoazo- and (d) o,o'-dihydroxyazomethine structures (ref. 4). Courtesy of The Society of Dyers and Colourists.

These complexes are applied from weakly acidic or neutral solutions (pH 6-7), preferably with addition of ammonium salts. Although the dyeing period is also long, the preboil temperature is only 40-50°C. These dyes have excellent fastness properties.

Metal complexes of essentially four major types of structures comprise most of these dyes. Ligands of nucleophilic or anionic groups that are *o,o'-* to the chromophore [azo (-N=N-) and/or azomethine (-N=CH-)] can complex in the 1:1 or 1:2 state with Cr^{3+} or Co^{2+} salts to form 5, 6 and/or 7-membered fused rings. The ligands most frequently used are -hydroxyl, -amino and -carboxyl groups shown in Figures 3.1(a)-3.1(d). There are many different structures that can be obtained when these transition metals complex with these polydentate ligands. Moreover, it has been observed in the last 30 years that diastereomers of these dyes exist and can be isolated and identified by modern spectroscopic techniques. Even to the present day, 1:1 chromium complexes containing sulfonic acid groups are viewed as one of the major developments in this area. Later 1:2 chromium complexes with $-SO_3H$ groups were synthesized that generally have the advantage of a higher color yield, a color range basically of the entire visible region and minimum fiber damage relative to the corresponding 1:1 chromium complexes with these solubilizing groups.

Recent environmental regulations for substantial reduction of chromium and other heavy metal effluents have led to changes in dyeing techniques for fabrics dyed with metal complex dyes. Several strategies to minimize chromium in effluents are reviewed and discussed (ref. 5). Particular attention is given to the 1:2 metal complex dyes since this is the largest group of metal complex dyes used for textiles. Minimum amounts of added dichromate, chroming in a fresh bath and increasing dyebath exhaustion are recommended for reducing chromium-containing effluents. Addition of formic acid at optimum pH (range of 3.5-3.8) and elimination of sequestering agents in the dyebath are also beneficial. Effluent treatment by electrochemical or chemical methods is a complementary strategy that may be used with these modified dyeing techniques.

In addition to 1:1 and 1:2 metal complex structures discussed, novel structures such as iron complexes or *o*-nitrosohydroxyindazoles, formazans, and metal complex dyes of the 2'-carboxyphenyl-2-amino-naphthalene series have gained some commercial acceptance (ref. 4).

3.2.3 Mordant dyes

Mordant dyes, as classified in the *Colour Index*, include only acid dyes whose fastness properties are markedly improved by insolubilization when they are complexed with various inorganic salts. Mordant dyes were used to dye wool fibers before the discovery and application of premetallized acid dyes. Thus, these mordant dyes differ from premetallized acid dyes in that they are not isolated as stoichiometric compounds, and are affixed by applying the metal salts before the dye is in the bath (premordanting), after the dye is in the bath (postmordanting) or coapplied (comordanting) with the acid dye to the fiber. As with acid dyes, mordant dyes are applied primarily to wool and polyamide fibers. Metal salts normally used to mordant fibers are chromium, cobalt and aluminum, and to a lesser extent, copper, tin and iron. It is probably appropriate to also include under mordant dyes the following two groups of dyes: (a) natural dyes (such as alizarin or 1,2-dihydroxyanthraquinone) that are complexed with metal salts to dye proteinaceous fibers and (b) basic dyes that normally have no substantivity to cellulosics, but possess such substantivity when they are mordanted with antimony salts and tannic acid. Before the advent of many of the synthetic dyes, most mordanting was achieved by using natural dyes such as alizarin and cochineal (both have anthraquinone structures) with various metal salts to obtain different colors (Table 3.3). Presently, only chromium and to a lesser extent cobalt are used to produce mordant dyes.

Of the three mordanting techniques, postmordanting is usually preferred for acid dyes applied to polyamide and proteinaceous fibers, while premordanting appears to be the most suitable process for affixing basic dyes to cellulosics. Premordanting is rarely used in current dyeing operations because it is difficult to achieve levelling when the chromium salt is applied first. Comordanting, more commonly called the metachrome process, is useful with some acid dyes. In this process, dichromate is reduced to chromate and the dyes that are suitable for metachroming must be substantive in the range of pH 5-7. This technique has been largely superseded by the synthesis of stoichiometric 1:1 and 1:2 metal complex dyes that can be applied

TABLE 3.3

Mordanting of natural dyes (ref. 3).

Mordant	Colours of chelate compounds of	
	Alizarin	Cochineal
Aluminum	Red	Crimson
Tin	Pink	Scarlet
Chromium	Puce Brown	Purple
Iron	Brown	Purple
Copper	Yellow Brown	

under a variety of bath conditions and pH. The afterchrome process, whereby the chromium or other metal salt is applied after the dye is applied to the fiber in the bath, is the only one of the three processes still commercially used as an alternative to the application of metal complex dyes. This method has the advantage of obtaining a level dyeing before addition of the chromium salt with subsequent fixation by the metal to prevent migration. The technique of afterchroming has been extensively reviewed from both fundamental and practical perspectives by Maasdorp (ref. 6).

3.2.4 Direct dyes

Direct dyes are water-soluble, contain anionic substituents, but differ from acid dyes because of their good substantivity to cellulosics (especially cotton) and sorption characteristics (usually governed by accessibility of the fibers and/or Langmuir-type isotherms). Recent developments in direct dyes have been critically reviewed with regard to newer structures, theoretical investigations, pretreatments and after-treatments and simultaneous dyeing and finishing of cellulosic fabrics (ref. 7). Most direct dyes have azo structures, particularly in the disazo and trisazo chemical classes. However, concern over exposure to benzidine-derived azo dyes and related structures have led to the synthesis of amino intermediates with low genotoxicity and mutagenicity as alternatives. These intermediates include diaminoquinolines and diaminobipyridines such as 7-amino-3-(4'-aminophenyl)quinoline and 5,5'-diamino-

2,2'-bipyridine.   Direct dyes derived from non-azo chromogens such as metal phthalocyanines containing sulfonamide groups have also been synthesized (ref. 7).

Cotton fibers, rayon staple fibers and yarns of both of these types of fibers require little purification before application of direct dyes.   Mercerization of cotton only substantially increases uptake of direct dyes when the alkaline pretreatment is conducted in the relaxed state.   However, woven cotton fabrics are normally desized and bleached, and woven rayon fabrics are desized before application of these dyes. Addition of salts or electrolytes, such as sodium sulfate or sodium chloride, promotes exhaustion of these dyes on cotton.   Various cationic pretreatments with monomeric or polymeric agents have been used to improve the dyeability of cellulosic fabrics with direct dyes.   Aftertreatment of fabrics treated with direct dyes are used to improve their washfastness (cationic surfactants, formaldehyde-containing resins, and diazotization and coupling) and their lightfastness (copper or chromium salts).

There are basically three subdivisions of direct dyes with respect to their method of application: (a) those that migrate and level well even in the presence of excess electrolyte, (b) those that level poorly, but whose exhaustion may be controlled by

TABLE 3.4

Salt required to produce 50% exhaustion of direct dyes (adapted from ref. 3).

Dye	NaCl addition for 50% exhaustion (% on wt. of fibre)
Chlorazol Brown MS	0
Diazo Black OT	1.0
Primuline AS	2.0
Diazo Brilliant Orange GR	4.0
Chlorazol Fast Eosin B	7.0
Benzo Fast Yellow 4GL	8.5
Chyrosophenine G	13.0
Rosanthrene Pink	16.0
Rosanthrene Violet 5R	30.0

proper addition of electrolyte, and (c) those that have poor levelling properties, but are satisfactorily exhausted by increasing the dyebath temperature without addition of an electrolyte. The method of application of direct dyes is also somewhat related to their structural aspects, i.e., their molecular weight, linearity and coplanarity. As noted in Table 3.4, the amount of salt or electrolyte used with direct dyes varies with the fabric to liquor ratio and the exhaustion characteristics of the dye at equilibrium.

The substantivity of direct dyes for cellulosic fibers is facilitated by linear, coplanar aromatic structures of dyes that can align themselves parallel to the cellulose chains and by the volume term (basically accessibility of the fiber to the dye). However, there is still some dispute whether or not the sorption isotherm of these dyes is due to hydrogen bonding or is due to van der Waals' forces. Added electrolytes neutralize the negative charge of the cellulosic fiber that would repel a direct dye anion, and thus promote exhaustion, binding of the dye onto the fiber surface and facilitate diffusion of it into the fiber pores. Saturation of direct dyes on the fiber may be related to formation of a monolayer that is typical of a Langmuir-type sorption isotherm. This saturation may be estimated directly using large concentrations of salt and dye or by extrapolation of reciprocal plots of adsorption isotherms (ref. 3). Sumner's work on a generalized equation to determine dye affinity (see **Chapter 2**) and its extension to direct dyes for cellulosics is considered an important new development in this area (ref. 7). Other recent theoretical studies in this area are noted such as the influence of electrolytes on dye sorption, effect of dye disaggregation of sorption and ultrasonic fields to enhance dyeability.

3.2.5 Reactive dyes

Reactive dyes also contain anionic substituents, but differ markedly from acid, direct, basic and mordant dyes because (a) their negatively charged groups are primarily for solubilizing the dye and (b) they become substantive to fibers by covalent bond formation. Methods for producing dyed cotton fabrics by covalent bond formation through esterification and etherification were developed in the 1920's. However, it was not until the mid-1950's that dichlorotriazine structures were utilized to produce the first group of commercially acceptable reactive dyes.

The structures, development and application of reactive dyes to textiles have been critically reviewed and discussed by several authors (refs. 8-10). As shown below (Figure 3.2), reactive dyes contain sulfonic acid groups to increase their solubility in water, chromophoric groups or systems, and a linking group to an electrophilic struct-

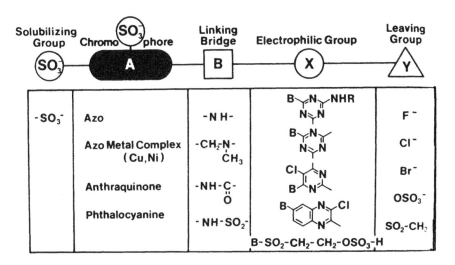

Figure 3.2. General structure of reactive dyes (ref. 8). Courtesy of Dr. Dietrich Hildebrand.

ure that contains a good leaving group (halogen, sulfate, or methanesulfonate). Most of the electrophilic structures are heterocyclic rings (triazines, pyrimidines, quinoxalines and benzothiazoles) containing one or more halogens as leaving groups; these react with sodium cellulosate or any other negatively charged fiber substituent (-NH groups in wool or polyamides) by nucleophilic aromatic substitution to form covalent bonds. There are also reactive dyes with a hydroxyethylsulfone structure ($B-SO_2-CH_2-CH_2-OSO_3H$) that, in the presence of alkali, are converted into a vinyl sulfone structure ($B-SO_2-CH=CH_2$). This latter structure then reacts with cellulosics by nucleophilic addition across the double bond to form cellulose ethers (Cell-O-CH_2-CH_2-SO_2-B). A major class of reactive dyes that has been gaining commercial acceptance because of their excellent alkaline stability and lightfastness are those based on triphenodioxazines (ref. 11). A typical structure is shown in Figure 3.3, where A and B = -H or SO_3H, R = amino- or aminoethylamino- and X = halogen.

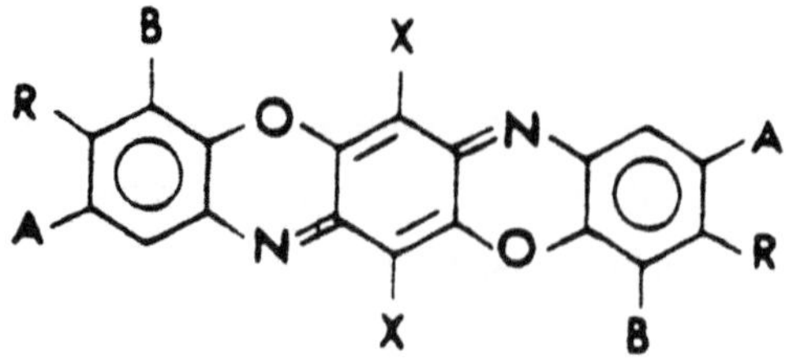

Figure 3.3. Representative structure of triphenodioxazine reactive dyes (ref. 11). Courtesy of The Society of Dyers and Colourists.

Reactive dyes based on the reaction of cellulosic fibers with phosphonic acid groups have also received some attention. These dyes form covalent bonds with cellulose to give phosphonate monoesters that are fixed by addition of cyanamide or dicyanamide. A possible mechanism for such fixation is shown in Figure 3.4.

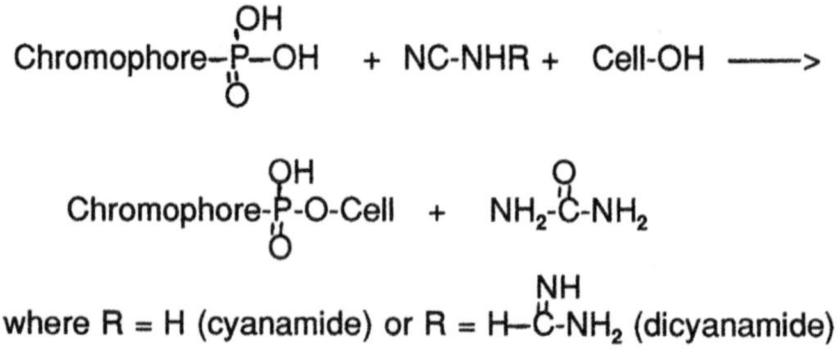

$$\text{Chromophore–}\underset{\underset{O}{|}}{\overset{\overset{OH}{|}}{P}}\text{–OH} \quad + \quad \text{NC-NHR} \quad + \quad \text{Cell-OH} \quad \longrightarrow$$

$$\text{Chromophore–}\underset{\underset{O}{|}}{\overset{\overset{OH}{|}}{P}}\text{-O-Cell} \quad + \quad \text{NH}_2\text{-}\overset{\overset{O}{\|}}{C}\text{-NH}_2$$

where R = H (cyanamide) or R = H–$\overset{\overset{NH}{\|}}{C}$-NH$_2$ (dicyanamide)

Figure 3.4. Proposed mechanism for fixation of phosphonic acid reactive dyes on cellulosic fibers (adapted from ref. 12). Courtesy of The Society of Dyers and Colourists.

These dyes are attractive because they can be applied to cotton and cotton/polyester blends at a slightly acidic pH (5-6) and the dye-fiber is exceptionally stable to strong acid or base. However, high fixation temperatures and expense of synthesizing these types of reactive dyes have impeded their wide acceptance in commercial dyeing processes (ref. 12).

For cotton and for other fibers, it is necessary to remove sizes such as starch and poly(vinyl alcohol) before application of reactive dyes, since they can react with the hydroxyl groups in these types of sizing materials. It is also important to thoroughly wash the fabrics after desizing and bleaching to prevent oxidation or reduction of the reactive dyes. Reactive dyes applied to cellulosics are characterized by their excellent bright shades and washfastness; however, their washfastness is dependent on the effective removal of chemically unreacted dye from the fiber and on the stability of the covalent bond between the dye and the fiber to alkaline and to acid hydrolysis. Although there is some hydrolysis of the reactive dye that competes with its reaction with the fiber, at least 70% of the dye affixed to the fiber is usually acceptable.

Reactive dyes are most frequently applied to cotton and other cellulosic fibers at an alkaline pH of 9-12, but have also been applied to wool, silk, and polyamide fibers in weakly acidic dyebaths. In earlier processes, it was believed that it was necessary to use strongly alkaline solutions such as sodium carbonate or even sodium hydroxide to covalent bond these dyes to the fibers. However, later studies and processes frequently use milder alkaline conditions (such as sodium bicarbonate) in reactive dye fixation processes.

Since reactive dyes cannot migrate once they are affixed to the fiber, the amount of unacceptable dyed fabric tends to be higher with these classes of dyes than for other dye classes. Level dyeings can be obtained by matching the rate of exhaustion of the reactive dye with the two stages of exhaustion (primary and secondary). In the primary exhaustion stage, physical absorption occurs because of dye addition and presence of inorganic salts. In this stage, migration or levelling is possible. Controlling rate of fixation to form covalent bonds with the fiber is important in the secondary exhaustion stage. Methods have been developed to obtain high reproducibility of dyeing in both shade and depth and better levelness by predicting and controlling the pH of the dyebath to obtain optimum exhaustion and fixation curves for specific reactive dyes (ref. 13).

There are numerous batch, semi-continuous and continuous variations (see **Dyeing Machines** in this chapter for details of these techniques) in which reactive dyes have been and are being affixed to textiles. In exhaust dyeing, the most commonly used process, the dye and salt are normally added to the dyebath first to exhaust it, then alkali added to react the dye with the fiber. In cold pad-batch dyeing (a semi-continuous method), the alkali and the dye are padded onto the textile, then batched up to a day slightly above ambient temperature to render the dye substantive to the fiber. Continuous dyeing methods for affixing reactive dyes to fibers and to fabrics include both single and double padding methods. In single padding methods, possible sequences are: (a) pad-dry-wash, (b) pad dry-bake or cure (thermofixation)-wash, and (c) pad-dry-steam-wash. For double padding methods, possible sequences are: (a) pad-dry-pad-pad alkali/salt-batch-wash, (b) pad-dry-pad alkali/salt-steam-wash and (c) pad-dry-impregnate alkali/salt-wash (ref. 10).

3.3. DYES CONTAINING CATIONIC FUNCTIONAL GROUPS

3.3.1 Basic dyes

Basic dyes are water-soluble and contain cationic groups. The positive charge on the dyes may be localized on a quaternary ammonium group or delocalized over the entire aromatic structure (such as in triarylmethanes, methines, xanthenes and other heterocyclic systems). These dyes are applied primarily to acrylics and occasionally to polyester and polyamide fibers. Although basic dyes gave brilliant colors on wool, silk and cellulosic fibers, they had poor fastness properties on these types of fibers. Thus, they fell into disuse until it was discovered that they were quite suitable for application to acrylic fibers.

As noted earlier, these dyes are of historical interest because the first synthetic dye mauve (prepared by Sir William Henry Perkin) is a basic dye with an azine ring structure. Other cationic dyes of historical interest include the triphenylmethane dyes Crystal Violet and Malachite Green. Important chemical classes of basic dyes include: methines, eneamines and hydrazones, diazacyanines and cyanines, di- and triarylmethanes, xanthenes, azines (phenazines, oxazines and thiazenes) and azo dyes with cationic groups (ref. 14).

It is usually necessary to scour acrylic fabrics to remove additives present from knitting or weaving processing to obtain uniform dyeing. Acrylic fabrics dye slowly below their glass transition temperature (generally 70-80°C), then extremely rapidly above it and are difficult to level by migration unless cationic retarders are used as auxiliaries in the bath with the basic dye. These retarders are usually quaternary ammonium salts and are more accessible to dye sites than the basic dye itself; thus, they are gradually exchanged with the cationic dye at higher temperatures to promote the desired rate of dye adsorption and levelling. Dimensional stability of fabrics above the glass transition temperature of the acrylic fabrics also has to be carefully controlled, and thus dyeing on a winch or paddle machine is preferred to jig dyeing. Basic dyes exhibit Langmuir-type isotherms (analogous to that of polyamide fibers with acid dyes) because of the binding of anionic acrylic groups on the fiber with cationic dye structures.

A typical formulation for dyeing acrylics with basic dyes contains a chelating agent to prevent off-shade dyeing due to metal ion effects, a salt such as sodium sulfate to facilitate levelling, a nonionic wetting agent, a cationic retarder (usually a colorless basic dye structure) that competes with dye sites in the fiber to provide more uniform dyeing, and acetic acid to reduce the pH to that required to conduct dyeing with these systems (ref. 3).

3.4. DYES REQUIRING CHEMICAL MODIFICATION BEFORE APPLICATION

3.4.1 Introductory remarks

In contrast to dyes that contain anionic and cationic groups, there are three important classes of dyes that must undergo chemical reaction before their application

(a) $\quad$ D $\quad$ + $\quad$ 2 (H) $\quad\xrightarrow{\text{OH}^-}\quad$ DH$_2$ $\quad\xrightarrow{\text{(O)}}\quad$ D

$\qquad$ Vat $\qquad$ Reducing $\qquad\qquad$ Leuco form $\quad$ (oxidation) $\quad$ Vat dye on
$\qquad$ dye $\qquad$ agent $\qquad\qquad\qquad\qquad\qquad\qquad\qquad\qquad$ fiber

(b) D-S$_x$ $\quad$ + $\quad$ (H) $\quad\xrightarrow{\text{OH}^-}\quad$ D-S$_x$H $\quad\xrightarrow{\text{(O)}}\quad$ D-S$_x$

$\qquad$ Sulfur $\quad$ Reducing $\qquad\qquad$ Reduced dye $\quad$ (oxidation) $\quad$ Sulfur dye
$\qquad$ dye $\qquad$ agent $\qquad\qquad\qquad\qquad\qquad\qquad\qquad\qquad$ on fiber

(c) D-NH$_2$ $\quad$ + $\quad$ HONO $\quad\xrightarrow{\text{H}^+}\quad$ D-N≡N$^+$ $\quad\xrightarrow{\text{D'H}}\quad$ D-N=N-D'

$\qquad$ Diazo $\qquad\qquad\qquad\qquad\qquad$ Diazonium $\quad$ Coupling $\quad$ Azoic dye
$\qquad$ component $\qquad\qquad\qquad\qquad$ salt $\qquad\quad$ component $\quad$ on fiber

Figure 3.5. Dyes requiring chemical reaction before application: (a) vat, (b) sulfur and (c) azoic (adapted from ref. 1). Courtesy of Noyes Publishers.

on textiles: vat, sulfur, and azoic (Figure 3.5). Both vat and sulfur dyes must usually be available as their reduced or solubilized forms (usually called leuco for the vat dyes) for initial application to fabrics. Subsequent oxidation renders these dyes substantive to the fibers. In the case of azoic dyes, this consists of a coupling component padded onto the fabric that is subsequently reacted with a diazo component to render the dye substantive to the fiber. These three dye classes were in major use for dyeing cellulosic fibers before the turn of the century. They are still very important application classes for the dyeing of cotton and other cellulosics.

3.4.2 Vat dyes

Vat dyes are water-insoluble compounds that contain anthraquinone or indigoid (such as indigo or thioindigo) ring structures. Important subclasses of vat dyes include: indanthrones, flavanthrones, pyranthrones, dibenzanthrones, benzathrone acridones, anthraquinone carbazoles and anthraquinone oxazoles (refs. 15, 16).

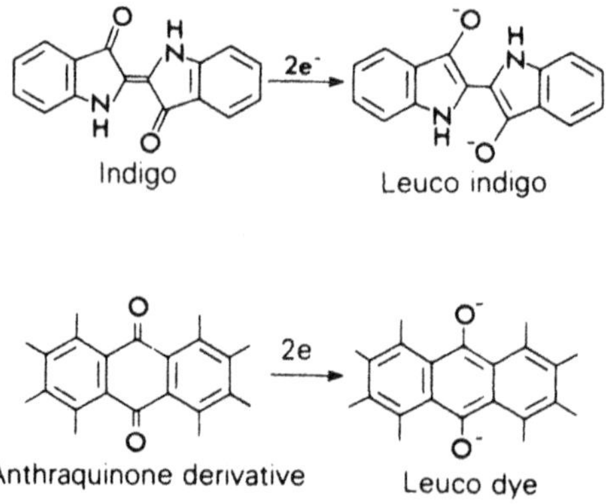

Figure 3.6. Oxidized and leuco forms of representative vat dyes (ref. 15). Courtesy of The Society of Dyers and Colourists.

Carbonyl groups in these structures are reduced to their soluble or leuco form in the dyebath with sodium hydroxide or with sodium hydrosulfite, then insolubilized in the fiber by oxidation with air or with various oxidizing agents. This redox system is depicted (Figure 3.6) for indigo and anthraquinone-type vat dyes and occurs by a two electron transfer mechanism. Solubilized leuco esters of vat dyes that do not require reduction before application to fibers have also been developed.

The important processing steps and measurements for application of vat dyes are shown in Figure 3.7 and are concerned with effective reduction of the vat dye, exhaustion and its reoxidation (refs. 17, 18). For suitable dyeing results, it is necessary to know the reduction potential of the particular vat dye that is being applied. Most vat dyes have reduction potentials in the range of -650 to -1,000 mv and require use of reducing agents with a more negative reduction potential to convert the dye into its leuco form. Dyeing kinetics indicates that vat dyes are ex-

Reduction to the Leuco Dye

1. Measurements of the potential
2. Kinetics of vatting, over-reduction
3. Secondary reactions (oxygen, substrate)
4. Practical aspects, analysis

Exhaustion of the Leuco Dye

Reooxidation to the Initial Dye

1. Measurements of the potential
2. Kinetics of oxidation, over-oxidation
3. Washing or soaping

Figure 3.7. Reduction and oxidation processes in dyeing with vat dyes (ref. 17). Courtesy of BASF.

hausted in two phases. Exhaustion in the first phase occurs rapidly and results in the majority of dye adsorption, while the second phase of exhaustion is much slower. Care must be taken to avoid over-reduction of these dyes because byproducts formed cannot readily be reooxidized to the desired quinoid structure, and thus are likely to result in poor and/or off-shade dyeings. Appropriate amounts of reducing agents such as hydrosulfite must also be present in the dye bath until exhaustion is completed. This can be complicated by side reactions such as the oxidation of hydrosulfite to sulfite and sulfate. Over-oxidation, while not as serious or as common as over-reduction, can lead to undesirable shade changes due to the formation of side reactions. The final operation, soaping or treatment with a surfactant at the boil (ref. 15) is very important since it removes unaffixed dye and the properties of the dyed fabric are generally made more washfast and occasionally undergo a color or shade change in this final step.

Vat dyes are particularly substantive to cellulosics, but may also be applied to other fibers such as polyamides. Before application of vat dyes, cotton fabrics are usually desized, scoured, bleached and mercerized. The latter process affords deeper shades with the vats. Although these dyes are noted for their excellent fastness to washing and to chlorine, certain shades on fabric accelerate degradation on exposure to light (called phototendering). This phenomenon will be addressed in the section on mechanisms of photodegradation of dyed fibers in **Chapter 6.**

Most vat dyes are currently applied to textiles by continuous processes, although various exhaust dyeing processes are still employed. In exhaust dyeing, dyebath temperature, concentration of alkali or of hydrosulfite to solubilize the dye, and the amounts of salt or electrolyte vary with the particular vat dye employed. An example of a frequently used continuous process consists of padding the fabric with the vat dye, drying, padding it into a solution containing alkali and salt, steaming, rinsing, oxidizing, then soaping, rinsing and drying.

Substantivity of vat dyes to fibers has been attributed to hydrogen bonding and van der Waals' forces, but enough anomalies exist to preclude a simple relationship between vat dye substantivity and molecular structure. However, there is some indication that the more planar and to a lesser degree the more linear that the vat dye structures are the more substantive they become to fibers.

3.4.3 Sulfur dyes

Sulfur dyes are prepared by fusing sulfur and alkali or sodium polysulfide with aromatic compounds containing nitro-, amino-, and/or hydroxyl groups. Their chemical structure is very complex. The comparatively few sulfur dyes whose structures have been satisfactorily elucidated are characterized by disulfide linkages with either poly(thiazine)--six-membered rings or poly(thiazole)--five-membered rings as repeating units in the main polymer chain.

Production of sulfur dyes occurs by reaction of sulfur with organic aromatic structures by one of four general methods: (a) sulfur bake in which aromatic intermediates are heated with sulfur at 160-320°C, (b) polysulfide bake, same as process (a), except sodium sulfide is used instead of sulfur, (c) polysulfide melt using aqueous sodium sulfide + aromatic compound under reflux conditions, and (d) solvent melt, a variation of (c) in which another solvent is used partially or totally to replace water, e.g., butanol or cellosolve. Dyes in the sulfur bake and polysulfide bake classes probably have benzothiazoles as the chromophore. Polysulfide melt and solvent melt sulfur dyes are somewhat better characterized and are prepared from 2,4-dinitrophenols and indophenols (refs. 19, 20).

Since sulfur dyes that are classified chemically include two different dye classes by application (vat and sulfur), some confusion and disagreement exist as to what constitutes a sulfur dye with regard to application. However, if one uses the definition of the *Colour Index*, sulfur dyes are defined as those that have to be reduced by some form of sulfur, then subsequently oxidized on the fiber to render them substantive. The four major subdivisions, according to the *Colour Index* (ref. 21) are:

TABLE 3.5

Characteristics of subclasses of sulfur dyes (adapted from ref. 21). Courtesy of The Society of Dyers and Colourists.

Generic name	H_2O solubility	H_2O substantivity	Descriptive class	Dyeing method
C.I. Sulfur	Insoluble or partially soluble	Variable;slight substantivity	Conventional	1
			Dispersed/spec -ial convent.	2
C.I. Leuco Sulfur	Soluble	Substantive	Liquid mixtures & Na_2S/NaSH	3
			Dry mixts. with Na_2S-reducing agent	4
			Dry mixts. with CH_2O-type red. agents	5
C.I. Solubilized Sulfur		Non-substantive	Thiosulfonic acid derivatives	6
C.I.Condensed Sulfur	Soluble	Non-substantive	S-aryl & S-alkyl thiosulfates	7

(a) sulfur, (b) leuco sulfur, (c) solubilized sulfur and (d) condensed sulfur. Each of these subclasses differ in water solubility, substantivity in water and method of application and may be differentiated by these characteristics (Table 3.5).

The conventional C.I. Sulfur dyes are water-insoluble and are made substantive to the fiber by one of two methods (Methods 1 and 2 in Table 3.5). In the first method, sodium sulfide is used as the reducing and solubilizing agent, while in the second method, a 2/1 mixture of sodium carbonate/sodium formaldehyde sulfoxylate is used. Oxidation and fixing occur when the fabric is dyed at the boil in both methods. The C. I. Leuco Sulfur dyes are already solubilized, and are available as liquid (Method 3) or dry mixtures (Methods 4 and 5) with a little reducing agent present. Procedure for dyeing and fixation is essentially the same as that for the C.I. Sulfur dyes. C. I. Solubilized Sulfur dyes are dissolved in boiling water and reducing agents may be added before or after introduction of these dyes into the dyebath (Method 6). C. I. Condensed Sulfur dyes were previously classified as Ingrain dyes and are totally different from the other subclasses of sulfur dyes with regard to their structure and method of application. These dyes are applied to fabrics by padding and drying, then repadding the fabric in an aqueous alkaline solution of a reducing agent (Method 7). This causes a condensation reaction between divalent sulfur atoms and produces an insoluble pigment in the fiber. It is also possible to apply C. I. Solubilized Sulfur dyes by this method or coapply them with C. I. Condensed Sulfur dyes (ref. 21).

These dyes are applied primarily to cotton and rayon fabrics (cellulosics). If the sizing is poly(vinyl alcohol), it is necessary to remove the size to obtain good dyeing results with the sulfur dyes. Desizing, bleaching and particularly mercerizing afford brighter and deeper shades on subsequent application of sulfur dyes even if this is not required. Although sulfur dyes on cellulosics have good washfastness, they have poor fastness to chlorine (presumably due to the instability of the disulfide groups), and in some instances cause the fabric to tender or degrade at high temperatures and humidities.

Although reduction of water-insoluble sulfur dyes may be achieved by pasting them with reducing agents and a little water, deposition of insoluble material on the fabric surface after oxidation may be problematic. Water-soluble sulfur dyes, either prereduced (e.g, Leuco Sulfur) or thiosulfonic acid derivatives (C. I. Solubilized Sulfur) are thus preferred and are more frequently applied (preferentially in liquid rather than in dry form) by continuous rather than by batch processes. Continuous processes commonly used are pad-steam, pad-dry-develop and pad-dry-chemical-pad-steam. Although little information exists with regard to the structures of the sulfur dyes and their fiber substantivity, it would appear that the same characteristics that govern fixation of vat dyes (i.e,. coplanarity and linearity) would also influence the fixation of sulfur dyes onto fibers.

3.4.4 Azoic dyes

For azoic dyes, the fabric is first padded with the coupling component, then reacted with the diazo component to form an insoluble dye in the fiber. The earliest C. I. Azoic Coupling Components were based on derivatives of 2-hydroxy-3-naphthoic acid (Structure I, Figure 3.8) because of their substantivity to cellulosics when they were applied as the sodium salt. This led to the development of three other general structures for azoic coupling components (Structures II-IV, Figure 3.8). Structures II and III are variations of Structure I (arylamides of o-hydroxycarboyxlic acids); all three

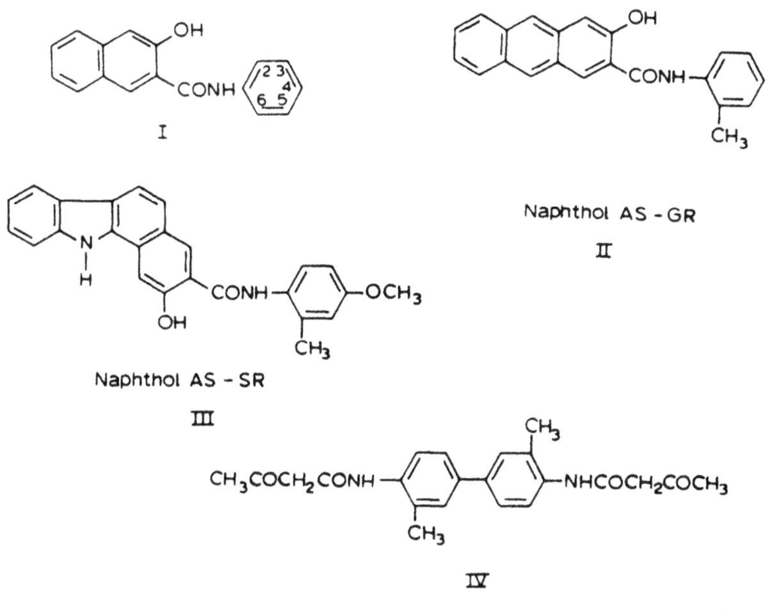

Figure 3.8. Structures of commonly used C. I. Azoic Coupling Components (ref. 3).

of these structures are suitable for all colors but yellow. The linear diarylamide IV is used as the coupling component to produce yellow colors on fabrics (ref. 3).

There are also several subclasses of C. I. Diazo Components. Although conventional diazonium salts are stable at cold temperatures, problems of prolonged storage and reproducible dyeing results led to the use of more stable forms of the diazo component of the azoic dyes. One stable form of diazonium salts is the *anti*-diazotate (Ar-N=N-OH), formed by the reaction of an aromatic diazonium halide with silver hydroxide . The potassium salt of this *anti*-diazotate is stable, and can be acidified as needed to regenerate the aromatic diazonium salt during the dyeing process. Another method of stabilization is conversion of the aromatic diazonium salt to an aromatic diazosulfonate (Ar-N=N-SO_3Na) by reaction with sodium sulfite; the diazosulfonate is stable under alkaline conditions and can be coupled with the azoic component at this pH. A third method involves conversion of the aromatic diazonium salt to a diazoamino compound (Ar-N=N-NR_1R_2) at pH 5-7 by reaction with primary or secondary aliphatic or aromatic amines; the active diazonium salt can be regenerated as needed by reducing the pH below 5. Aromatic diazonium double salts formed by reaction with $ZnCl_2$ or HBF_4 are more stable than aromatic diazonium salts with simple anions and can be used more effectively as diazo components (ref. 3).

Azoic dyes are primarily applied to cellulosics, but have also been applied to polyester fibers. Before dyeing with azoics, the fabric should be scoured and/or bleached, then rinsed until its aqueous extracts have a neutral pH. Textiles dyed with azoics have good fastness to washing, light and chlorine, but have poor fastness to crocking. As with most dyes, continuous processes with azoics are more widely practiced than are exhaust or batch processes.

In a typical continuous process for application of azoic dyes, the fabric is padded through a hot solution (80°C) containing the coupling component and alkali. The fabric is then subsequently dried, cooled, then immersed for less than a minute in two to three baths containing the cold buffered diazonium salt (diazo component) at a pH of 4.0-8.0 (varies with effectiveness of particular diazonium salt) and a temperature below 5°C. This is followed by a rinse in cold water, washing with hot alkaline soap, washing in water and drying.

Exhaust processes differ from continuous processes in that the bath temperature for the azoic component must be much lower (usually around 30°C) for the former than for the latter. Excess salt or electrolyte solution is also used to remove unfixed coupling agent from the fabric in an exhaust process, and the fabric is kept in the coupling bath at least 20 minutes to form an insoluble dye in the fiber. Generally, coupling components with moderate to high fiber substantivity are used in exhaust processes, while coupling components with low fiber substantivity (less than 25%) are used in continuous processes.

It appears that the substantivity of coupling components or substituted hydroxynaphthoic acid derivatives on fibers are enhanced by the same structural factors that increase the substantivity of direct dyes. Thus, linear, coplanar aromatic structures that can hydrogen bond with functional groups of the fibers are the most substantive of the azoic dyes.

3.5 SPECIAL DYEING TECHNIQUES AND CLASSES OF COLORANTS

3.5.1 Disperse dyes

Disperse dyes are very slightly water-soluble compounds whose fixation properties to fibers depend on their particle size and uniformity and on the nature of the dye dispersing agent. The vast majority of chemical types of disperse dyes that are used are monoazo-, anthraquinone and disazo- structures shown in Figure 3.9. Three types of chemical structures make up 85% of the total of all disperse dyes used. The remaining 15% are methines, styryl derivatives, aroylenebenzimidazoles, quino-naphthonones, aminonaphthylimides and naphtholquinoneimines (ref. 22).

In addition to these structures, there are a variety of disperse azo dyes containing heterocyclic diazo and coupling components or miscellaneous chromophores that are actively being evaluated to replace a variety of currently used disperse anthraquinone dyes. These disperse azo dyes give brilliant blue and red colors and are considered advantageous when compared to the tinctorially weak disperse anthraquinones. Research on these disperse azo dyes have been reviewed and major structural types discussed include: amino-disubstituted benzothiazoles; heteroannelated-aminothia-

Type	Common Structure	Percent of total
Monoazo	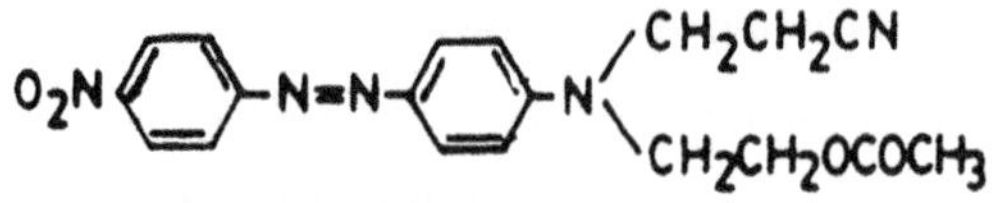 O_2N—⟨⟩—N=N—⟨⟩—N⟨$\begin{array}{l}CH_2CH_2CN\\CH_2CH_2OCOCH_3\end{array}$	50
Anthraquinone	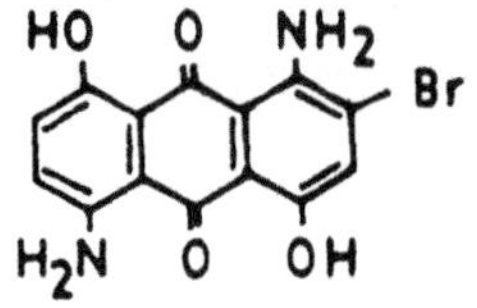 (structure with HO, O, NH₂, Br, H₂N, O, OH substituents)	25
Disazo	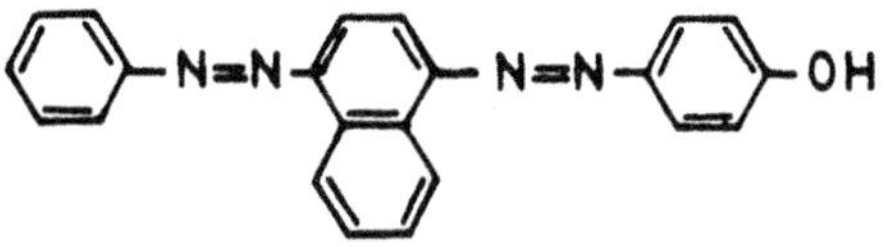 ⟨⟩—N=N—⟨naphthalene⟩—N=N—⟨⟩—OH	10

Figure 3.9. Major chemical types of disperse dyes used in commercial dyeing (adapted from ref. 22). Courtesy of Prof. M. L. Gulrajani, Indian Institute of Technology.

zoles; aminoisothiazoles; amino-dicyanoimidazoles, diamino-cyanopyridines, and related heterocyclic structures (ref. 23).

A new class of disperse dyes based on the benzodifuranone structure (Fig. 3.10) have been commercialized for obtaining bright red shades for all polyester microfibers and for cotton/polyester blends (ref. 24). These dyes are suitable for exhaust dyeing, do not stain other fibers in blends (such as cotton) and are characterized by outstanding wetfastness relative to other types of disperse dyes.

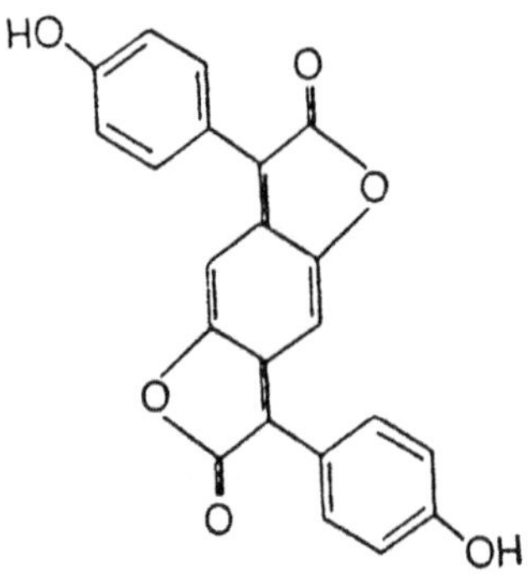

Figure 3.10. The first benzodifuranone dye (ref. 24). Courtesy of American Association of Textile Chemists and Colorists.

Disperse dyes, irrespective of their chemical structure, may also be classified in four groups (A through D) according to their fastness to light and heat/sublimation and their overall dyeing properties. This classification was proposed by ICI in 1973 and is extensively used at the present time; these groups of dyes are also referred to as low to high energy types. Group A has poor sublimation characteristics, but reasonably good lightfastness and excellent dyeing properties. Group B has good heat and lightfastness and are very suitable for carrier dyeing. Group C is similar to Group B but has superior heat or sublimation fastness, and Group D has poor dyeing properties, but extremely good sublimation characteristics (ref. 25).

Although disperse dyes were initially developed in the 1920's for dyeing cellulose acetate and cellulose triacetate fibers, these dyes are now primarily used for dyeing of polyester and polyester fiber blends (ref. 26). They represent the most important class of dyes for synthetic fibers, and are also applied to some extent to polyamides, acrylics, and modacrylics, polyolefins and poly(vinyl chloride). Dye carriers are necessary for affixing disperse dyes to polyester and cellulose triacetate fibers, but other types of synthetic fibers do not usually require this type of auxiliary or additive.

The mechanisms by which synthetic fibers are dyed by disperse dyes, the kinetics involved in their diffusion and sorption, and models that explain their diffusion into and fixation on fibers have been critically and comprehensively reviewed (refs. 22, 26). Dyeing of fibers with disperse dyes are generally believed to occur by the model shown in Figure 3.11. In this model, the insoluble and suspended dye particles slowly dissolve and form smaller crystals that eventually form a dilute dye solution that is transported to the fiber by diffusion, is absorbed on the exterior surface of the fiber, diffuses into the fiber and is made substantive. The kinetics of diffusion is noted to be consistent with the free volume theory of dyeing because rapid dyeing of the polyester occurs above its T_g (ref. 22). However, alternative dye diffusion theories based on dye distribution and polymer morphology may also explain this behavior (see **Chapter 2** for additional discussion of dye diffusion models). Initial studies were postulated that insoluble disperse dyes were attracted to the fiber by absorption similar to that of direct dyes on cellulosics. However, later studies provided evidence that the Nernst isotherms observed for the dyeing of fibers with disperse dyes were

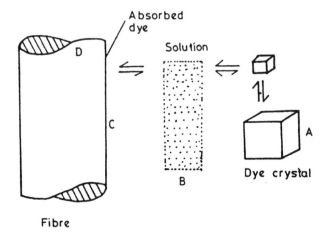

Figure 3.11. Simple model of a disperse dyeing system (ref. 22). Courtesy of Prof. M. L. Gulrajani, Indian Institute of Technology.

consistent with dyeing to form a solid solution with the fiber. Abrupt changes to a horizontal plateau at saturation are consistent with a mechanism characterized essentially as partition of solid solute (dye) between two immiscible solvents (water and fiber). Current views are that neither the term "adsorption" nor "solid solution" is totally appropriate to describe the disperse dyeing process because of the difficulty in determining the extent of solubilization of dye in crystalline and amorphous regions of synthetic fibers (refs. 22, 26). Alternative dye diffusion models that focus on polymer morphology should be considered to explain disperse dyeing behavior of synthetic fibers. Although hydrophobic bonding between disperse dyes and fibers account for some of the substantivity to the fibers, there is sufficient evidence that hydrogen bonding and van der Waals' forces are of major importance (as these bonding mechanisms are for vat and direct dyes) in the fixation of disperse dyes to fibers.

It is essential that fabrics, particularly knits, be thoroughly scoured before application of disperse dyes to remove any dirt, sizing, or machine oils present in the fibers. The fastness properties of these dyes vary widely with their molecular structure, the type of fiber to which they are applied and are particularly dependent on the particle size, stability and other dispersion properties of these dyes. With regard to structural factors of the disperse dyes, good substantivity is achieved if the dye has a specific balance of hydrophobic and hydrophilic groups, exists in a coplanar structure and is of sufficient shape and size to promote a strong dye-fiber interaction. In some instances, fastness properties are improved by diazotizing disperse dyes containing primary aromatic amino groups, then immersing the diazotized fabric into a developer (a coupling component) bath. Fastness to sublimation is important for dyes that are used in high temperature dyeing processes. Gas fading, especially by nitrogen oxides, may be exacerbated with cellulose acetate and polyamide fibers that are dyed with disperse anthraquinone structures.

The fiber type dictates the temperature, pressure and use of a dye carrier for the various batch and continuous processes used to apply disperse dyes. Disperse dyes are nonionic and can only be made substantive to hydrophobic fibers by diffusion into their pores. Thus, dye carriers are essential for dyeing of polyester at temperatures below 120°C and are also usually required for dyeing cellulose triacetate fibers. The important classes of dye carriers and their suitability for various disperse dye/synthetic fiber combinations have been summarized and discussed (refs. 27, 28). Chemical types and typical examples of dye carriers are: phenolics (o-phenyl and p-phenylphenols), chlorinated aromatics (o-dichlorobenzene and 1,3,5-trichlorobenzene), aromatic hydrocarbons and ethers (diphenyl ether and 2-hydroxydiphenyl), aromatic esters (n-butyl benzoate, dimethyl phthalate and methyl salicylate).

Acceptable dye carriers are those that have a suitable boiling point, produce acceptable color yield and dye migration properties, and have both shelf stability and emulsion stability in the dyebath. The major disadvantages of using dye carriers are the difficulty in completely removing them from the fabrics after dyeing and processing, their volatility and disposal as effluents, and their tendency to cause fabric shrinkage when they are applied at temperatures above 115°C. The various theories that have been advanced to explain how dye carriers function have been critically reviewed (refs. 3, 29). None of these theories are totally satisfactory for explaining the effectiveness of the carriers in facilitating dyeing of polyester and other synthetic fibers. A critique of earlier theories indicates that adsorption of the carrier by the fiber is more important than solubilization of the dye by the carrier. This is consistent with hypotheses that the fibers are coated with the carrier to form a layer through which the disperse dye is transported and that the carriers plasticize the polymer chains to allow penetration of the dye into the fibers (ref. 3). In a more recent study on how carriers facilitate dyeing of synthetic fibers with disperse dyes, the associative and disperse forces of solubility parameters of the fiber and various carriers were determined; reduction in the glass transition temperature of the fiber by the carrier should result in plasticization of the fiber with optimum affinity of the dye occurring when its solubility parameters are modified and match those of the fibrous polymer (ref. 29). An alternative view is that the polyester fibers have improved wettability even after the carrier is removed from the fiber, and thus are more accessible to dyes or any other reagent (ref. 30).

In exhaust or batch dyeing of synthetic fibers with disperse dyes, the dyebath contains a dispersing agent (usually a nonionic surfactant) and a dye carrier if needed. In a typical procedure, the dyebath is adjusted to a pH of 4.0-6.0 with acetic acid before introduction of the fabric into the bath. The dye is then added, the bath heated for 1-3 hr, and the fabric washed and dried. Bath temperature varies with fiber type, being less than 100°C for acetate and polyamides, and in the range of 100-120°C for cellulose triacetate and polyester fibers. A sizable quantity of polyester is batch-dyed as loose fiber and as yarn in bale, hank and package dyeing machines. For batch processes, beam and jet dyeing machines are replacing winch and jig dyeing machines because the polyester fabrics and their blends have improved dimensional stability and can be dyed more rapidly with this machinery than they can by older machinery used in batch processes. Rapid dyeing methods are also being used with jet and package-dyeing machines that have programmed heating rates and improved liquor flow through the fabric. A variety of dyeing machines are described that will be discussed in detail later in the section on **Dyeing Machines and Processes**.

High temperature dyeing, e.g. of polyester, at 120-130°C or greater, usually requires no dye carrier and the dyebath can be exhausted much more rapidly than it is when using lower dyeing temperatures. Batch dyeing processes are carried out in pressurized high-temperature equipment around 130°C after dispersion of the dyes and pH adjustments are made at lower bath temperatures (50-70°C). Continuous dyeing processes for triacetate and for polyester fibers are typified by the Thermosol method of Dupont or by other comparable techniques. These are now used extensively in the textile industry whenever possible instead of batch processes. In a typical thermofixation process, the dyes are padded onto the fabric with thickeners, and the fabrics passed through hot cans or an infrared oven to dry them at 135°C. The dye is then fixed to the fiber by heating it for 0.3-1 min at 200°C or higher before washing and drying. Thermofixation is a prototype continuous dyeing process that will also be discussed in more detail in the section on **Dyeing Machines and Processes.** Another frequently used continuous dyeing process for applying disperse dyes is high temperature steaming. In this process, superheated steam is used to fix the dye at atmospheric pressure. This steaming process has proved to be useful for improving the fastness and fixation of mixtures of dyes applied to cellulosic/polyester blends.

As is frequently observed for the carrier dyeing of textiles, the sorption of disperse dyes onto fibers applied by vapor-phase processes (such as high-temperature thermofixation) has also been observed to occur by a Nernst-type isotherm. As noted earlier, in this type of sorption the dye is in partition between two immiscible solvents, the dyebath liquor or vapor and the fiber. A simple mathematical model has been developed for the transient heating of polyester fabrics dyed by three popular continuous processes: thermofixation, high temperature steaming and heat transfer printing (refs. 31, 32). A general expression for heat transmission to a body (fabric) under commercial dyeing conditions is given by :

$$\frac{A_s \cdot h_T \cdot t}{C_T} = \ln \frac{T - T_o}{T - T_m} \tag{3.1}$$

where A_s is the specific heat transfer for different body geometries (e.g., semiinfinite plate--fabric thickness; semiinfinite cylinder--fiber radius; and spherical dye particle radius); h_T is an overall heat transfer coefficient for convection, conduction and radiation for all heating modes from heat sources (processing equipment); C_T is the heat capacity of the body, and T_o, T_m, T are the initial and mean temperatures of the fabric and of the heating medium, respectively during dyeing processes (refs. 31, 32).

The most recent method for fixing disperse dyes to polyester and other fabrics composed of synthetic fibers is the vapor phase transfer of the dye to the fiber under low pressure or vacuum conditions. This may be considered a variation of vacuum transfer printing. Satisfactory results were obtained by this sublimation or vapor phase transfer of disperse dyes in specially constructed machinery. Since the dye uptake in the vapor phase is much more rapid than in aqueous phase disperse dyeing, this method offers promise as a commercial alternative to other methods for dyeing polyester at high temperatures (ref. 33).

3.5.2 Pigment dyeing

Pigment dyeing is a mass coloration technique that strictly speaking, is not considered a form of dyeing, but a technique by which a finely-dispersed, water-insoluble substance (usually 0.5-5.0 μm in diameter) is applied onto a material such as a textile from an oil-in-water emulsion containing a resinous thermosetting and/or thermoplastic binder. For synthetic fibers, the pigments are frequently incorporated during the polymerization or melt spinning of the fibers. Most modern day pigments applied to textiles are organic compounds (azo compounds, insoluble vat dyes, phthalocyanines, quinacridones) that do not contain salt-forming groups. However, the Colour Index also considers and lists the following subclasses as pigments: toners (concentrated coloring matters formed by the reaction of water-soluble dyes with a precipitant such as phosphomolybdic acid), lakes (similar to toners except that precipitation is effected by alumina), extended pigments (pigments or toners diluted with alumina or other substances), and inorganic pigments (such as zinc oxide or titanium dioxide). Pigments may be applied to any type of fiber or substrate by use of a suitable binder. Although they initially were applied to cotton, cotton/polyester and polyester fabrics, they are increasingly used for polyamide and particularly polypropylene fibers and fabrics. Resistance to crocking and production of stiffness in fabrics are two disadvantages of pigments applied to fabrics despite their excellent washfastness and lightfastness. When pigments are applied to fabrics, this is usually from a cold water bath containing a binder and a catalyst, followed by curing at high temperatures, and washing and drying to remove the unbound pigment from the surface of the fabric.

For polyamides, such as Nylon 6, mass coloration may be conducted during the polymerization, tumbling of the pigment onto the polyamide chips during drying or dyeing the polyamide chips prior to washing, or by injecting the pigment during the melt spinning process. There are advantages and disadvantages to each of these methods. However, the light and washfastness of Nylon 6 yarns produced from fibers with pigments incorporated was comparable to those yarns dyed by conventional techniques for polyamides (such as acid, disperse or metal complex dyes) (ref. 34).

The vast majority of polypropylene textiles are pigment-dyed by incorporation of the pigments during the melt spinning of these fibers (refs. 35, 36). Both organic (azo compounds and anthraquinones) and inorganic (metal oxides, sulfides and chromates) pigments are suitable for dyeing polypropylene fibers. However, there are other factors that must be considered: the physical properties of the pigments, their compatibility with other additives and the polypropylene fiber, and the cumulative fastness properties of the colored polypropylene textiles.

Important physical properties of the pigments include their heat stability and dispersibility. The pigments should be thermally stable under melt spinning conditions, i.e., at processing temperatures that generally range from 230-300°C. Dispersibility of the pigments into finely divided particles is desirable to avoid clogging of filtration devices in the spinnerets and to effectively impart the desired shade and tinctorial strength to the dyed polypropylene fibers (ref. 35).

The pigments must be compatible with the fiber and additives such as ultraviolet stabilizers. Compatibility of the pigment with the fiber includes the ability of the added pigment to retain the melt flow characteristics of the polypropylene fibers for proper extrusion and for the pigment not to act as a prodegradant of the fiber during its melt spinning. Since many polypropylene fibers contain ultraviolet stabilizers for outdoor applications, it is also important that the pigment not produce antagonistic effects to reduce the ultraviolet stability of the polypropylene material (ref. 36).

More recent studies have focused on the pigment dyeing polypropylene fibers by pretreatment with binders (such as comonomers of butadiene-acrylates that can crosslink under certain conditions) and by use of certain solvent and disperse dyes with lipophilic additives (refs. 37, 38).

3.5.3 Solvent dyeing

Solvent dyes originally included only water-insoluble aromatics containing azo or amino groups soluble in and applied from non-aqueous media. However, this group now also includes application of disperse and other types of dyes that are partially or completely soluble in non-aqueous solvents or in water/non-aqueous solvent emulsions. Solvents commonly used are: alcohols, esters, ethers, ketones, aliphatic and aromatic hydrocarbons, chlorinated hydrocarbons and various oils, fats and waxes. For synthetic fibers, disperse dyes in solvents such as perchloroethylene are employed, while for natural fibers and polyamides, ionic dyes in water-perchloroethylene emulsions are used. Continuous processes are more frequently used in solvent dyeing of fabrics than are batch processes. Although solvent dyeing has been considered commercially promising for textiles because high dyeing speeds may be attained, it has not received widespread acceptance for several reasons. The greatest hindrance to commercialization is that special processing equipment is needed and precautions have to be taken for the disposal, safety and storage of various solvents used in the dyeing processes.

The behaviors of dyes in non-aqueous solvents with regard to their sorption characteristics at equilibrium and their mode of diffusion have been critically analyzed and reviewed (ref. 39, 40). The following observations were made: (a) sorption isotherms of most dyes are linear for solvent dyeing for many fiber types, (b) diffusion of the dye varies with the solvent, but is constant for a given solvent, is much more rapid in non-aqueous solvents than in water, and decreases linearly with the molecular weight of the dye, and (c) sorption of solvent by fiber may adversely change or effect the fiber/fabric mechanical properties (e.g., stiffness, modulus, etc.).

The solvent dyeing of polyester and of cellulose acetate fibers has also been reviewed with emphasis on the use of appropriate solvents, rate of dyeing, the effect of the solvent on T_g and physical properties of these synthetic fibers, and the importance of solubility parameters and trace amounts of water in the mechanism of such dyeing (ref. 41). It was concluded that chlorinated solvents, particularly perchloroethylene (solvent commonly used in dry cleaning) are the most suitable because (a) they can be applied in other aspects of textile processing (such as scouring and finishing), (b) have a high boiling point (122°C) that allow dyeing of the fibers under atmospheric conditions, (c) are not flammable, corrosive or toxic, (d) have low latent and specific heats and thus use less energy than water, and (e) have good wettability on the fibers (ref. 41).

The rate of dyeing of polyester fibers has been reported to be at least 100 times faster in perchloroethylene than in water due to the absence of a diffusional boundary layer effect or similar factors. Moreover, diffusion of dyes from perchloroethylene follows Fick's law and thus gives constant diffusion coefficients that are independent of time and of the concentration of the dye in the polyester. Various studies conducted differ in the magnitude of reduction of the glass transition temperature of the polyester fiber when it is solvent dyed in perchloroethylene; this may be due to differences in fiber history and experimental techniques (ref. 41). Measurements indicate that perchloroethylene and polyester fibers have similar solubility parameters that correctly predict high solubility of dyes in that solvent, high diffusion coefficients and swelling of the fiber, and low partition coefficients and exhaustion; the relationship between solubility parameter and diffusion and partition coefficients for polyester fibers and disperse dyes is shown in Figure 3.12. (ref. 42). Addition of small amounts of water to solvents containing disperse dyes markedly increases rate of dye diffusion, and suggests that water acts similar to dye carriers to promote disperse dyeing of polyester (ref. 41).

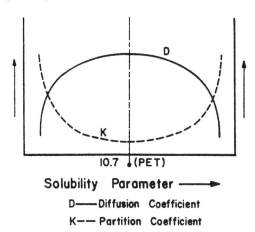

Figure 3.12. Correlation between solubility parameter and diffusion and partition coefficients (ref. 42). Courtesy of Textile Research Institute.

3.5.4 Natural dyes

Before the discovery of synthetic dyes by Perkin in 1856, only natural dyes were available for the coloration of textiles. These dyes are rarely used today on a commercial basis and are usually complex mixtures with similar structures as the synthetic mordant dyes, and to a lesser extent, those of vat, direct, acid, and/or pigment dyes. An informative review of these dyes (ref. 43) subdivides them into several classes. These include: plant anthraquinone dyes (such as madder or alizarin), animal anthraquinone dyes (kermes or kermesic acid, cochineal and lac--all derived from insects), α-naphthoquinones (juglone from walnut shells), flavones (morin from the wood of *Osage Orange* and quercetin from Persian berries), dihydropyrans (logwood), anthocyanidins or flavylium salts (carjurin from the leaves of *Bignonia chica*) and the historically famous indigoid dyes (Tyrian purple from shellfish and indigo from the leaf of a leguminous plant).

3.5.5 Miscellaneous dye application classes

Miscellaneous dyes that are presently applied to textiles in only very small quantities include: (a) ingrain dyes (dyes formed *in situ* in the fiber by development and coupling of an intermediate that now only include phthalocyanines, but also used to include azoics and oxidation bases); and (b) oxidation bases (aromatic amines oxidized on the fibers that have complex structures not generally elucidated). Neither reducing agents nor fluorescent brighteners are technically dyes, but they are also listed in separate application sections in the *Colour Index*. Fluorescent brighteners have previously been discussed in the first chapter of the book in the context of preparatory and purification processes for textiles.

In their broadest sense, ingrain dyes are defined as any type of dye that is formed *in situ* in the fiber by development and coupling of one or more chemical intermediates. However, since azoic dyes and oxidation bases are no longer listed in the Ingrain section of the *Colour Index*, there are only a few ingrain dyes listed that are characterized by phthalocyanine structures.

Oxidation bases are used only sparingly in the current dyeing of textiles, but are of historical significance. The initial experiments by Sir William Henry Perkin in 1856 to produce the first synthetic dye mauve were quickly followed by the preparation of the first oxidation base (Aniline Black). This oxidation base was prepared by reaction of aniline sulfate with potassium dichromate, and used extensively in early commercial dyeing practice for cotton fabrics. The primary current use for oxidation bases, such as the oxidation products of p-phenylenediamine, is for coloring hair, feathers and fur.

Although reducing agents are not dyes, they are essential in many textile dyeing processes and are listed in a separate application section in the *Colour Index*. Their major uses are for application of vat dyes (to reduce them to leuco or soluble forms), as stripping agents and for obtaining discharge printing effects. Some of the more commonly used reducing agents include: sodium dithionite or sodium hydrosulfite ($Na_2S_2O_4$ or $Na_2S_2O_4 \cdot 2H_2O$), thiourea dioxide and salts of formaldehyde sulfoxylate [$M(HSO_2 \cdot CH_2O \cdot nH_2O)_x$], where M = Na and x = 1 or M = Zn and x = 2 and n = 0-2.

3.5.6 Dyeing of fiber blends

Dyeing of fiber blends requires proper choice of processing conditions (pH, temperature, continuous or batch techniques) and use of one or more dyes in a particular application class. This selection will depend on the color desired and the similarity or dissimilarity of the fiber types with regard to their substantivity for the dyes. The technique is known as union or tone-on-white dyeing when one of the fibers remains colorless or is only tinted (e.g., wool/cellulosic blends). Union dyeing may also be defined as dyeing of two different types of fibers to the same shade. When different fiber types are dyed the same color or shade, this is referred to as solid shade or tone-on-tone dyeing. When different types of fibers are dyed different shades or colors, this is known as cross dyeing. Important fiber blends whose representative dyeing processes will be described include: cellulosic/polyester, cellulosic/wool, cellulosic/polyamide, cellulosic/acrylic, wool/polyester, wool/polyamide, wool/acrylic and polyester with other synthetic fibers. Some representative fiber blends and the selection of dyes, machinery and/or processes for dyeing these blends are listed in Table 3.6.

Dyeing of cellulosic/polyester blends is now more commercially important than the dyeing of either fiber alone. This is due to the extensive use of cellulosic/polyester

TABLE 3.6

Dye selection and processes for fiber blends

Fiber blend	Dye classes used	Processes and/or machinery
Polyester/cellulosic	Disperse/reactive disperse/direct;disperse/vat disperse/sulfur	Beam,jet, high temperature winch or jig for bath/exhaust
Cellulosic/polyester	Reactive/disperse	Inverse dyeing (cellulosic first) on beam
Polyester/cellulosic	Disperse/vat disperse/reactive	Semi-continuous (pad/ batch, pad/thermofix or pad/ develop)
Polyester/cellulosic	Disperse/reactive disperse/vat;disperse/azoic pigment	Continuous (pad/dry/ thermofix or cure)
Cellulosic/wool	Direct/acid reactive/reactive reactive/acid	Exhaust (one or two bath) pad/batch
Cellulosic/polyamide	Direct/acid reactive/reactive	Exhaust (one or two bath)
Cellulosic/acrylic	Direct/basic Reactive/basic	Exhaust (two bath) continuous pad-steam
Polyester/wool	Disperse/acid disperse/metal complex special dye class	Exhaust/batch jet or beam
Acrylic/wool	Basic/reactive basic/metal complex	Exhaust (one or two bath)
Polyamide/wool	Acid or metal complex	Exhaust (levelling acid dyes or neutral pH metal complex
Polyester/polyamide	Disperse/disperse disperse/acid	Exhaust (usually two bath)

blends including regenerated cellulose and linen as well as cotton for the cellulosic component) in apparel and other textile applications throughout the world.

A variety of dye combinations may be applied in the batch or exhaust dyeing of polyester/cellulosic blends. Suitable machinery includes high temperature winches, jigs, and beam and jet dyeing apparatuses. Irrespective of whether a single or double bath is used to affix both dyes to the cellulosic/polyester blends, the dyeing is usually conducted in such a manner that the disperse dye is the first affixed to the polyester component of the blend. Disperse/reactive and particularly disperse/vat dye combinations are most suitable for dyeing these blends in a single bath (ref. 44, 45). In a typical process using a disperse and vat dye, the fabric is dyed with a uniformly dispersed mixture at 130°C for 1-1.5 hr. to fix the disperse dye on the polyester component. The dyebath is then cooled to about 80°C and the cellulosic component is dyed by converting the vat dye to its leuco form (addition of caustic and a reducing agent). After dyeing about 15 minutes at this temperature, the dyebath is further cooled to 50°C and further exhausted by addition of an electrolyte or salt. The fabric is then rinsed and oxidized and soaped at the same time to fix the vat dye to the cellulosic fiber of the fabric blend. With disperse/direct dye combinations, aftertreatment may be required to produce good fastness properties for the direct dye.

One of the most frequent problems with dyeing any blend is the cross staining that occurs on the fiber that is not substantive to that particular class of dye. For polyester/cellulosic blends, the cotton component of the blend is frequently stained by the disperse dye that is fixed to the polyester. Classes of disperse dyes suitable for the exhaust dyeing cellulosic/polyester blends without staining the cellulosic fiber have been developed (ref. 46). There are basically two types of disperse dyes that do not cross stain the cellulose (shown in Figure 3.13): (a) those that cleave under alkaline conditions by hydrolysis of diester groups to water-soluble carboxylate salts and (b) by loss of a thiophene unit attached to an azo linkage. Thus, the cellulosic component is not stained and suitable for subsequent fixation of a reactive dye at high pH.

Semi-continuous dyeing of polyester/cellulosic blends is generally defined as a process in which one or both of the component fibers are dyed by a padding method delivering onto a batch. This type of dyeing has the advantage of splitting large batches of fabrics with a fixed disperse dye into smaller batches to produce different colors or cross-dyeing effects by subsequent fixation with different dyes substantive to cellulose. There are three major variations for semi-continuous processes: (a) pad-batch (reactive/disperse or disperse/reactive), (b) pad-thermofix (first fix disperse, then

148

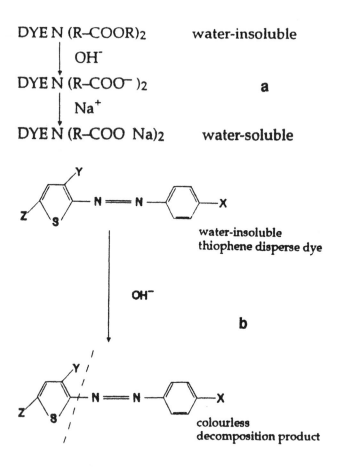

Figure 3.13. Novel disperse dyes that do not cross stain cellulosic fibers: (a) with hydrolyzable diester groups and (b) hydrolyzable thiophene components (adapted from ref. 46). Courtesy of The Society of Dyers and Colourists.

reactive), and (c) pad-batchwise develop (all disperse/other dye combinations) (ref. 44). The latter method has the advantage of better quality control in reproducing the amount of dye padded and thus leads to less shade variations.

A novel method of dyeing cotton/polyester fabrics uses a pad-dry-steam technique that was previously successful for dyeing all polyester fabrics. Mixtures of reactive/disperse and vat/disperse dyes were use in this method. Polyester components were dyed first with disperse dyes and the cellulosics components next with either reactive or vat dyes. Various colorfastness and tensile properties of the dyed blend fabrics compared favorably to those dyed by conventional high temperature methods (ref. 45).

Continuous thermofixation processes with a bath containing a disperse dye for the polyester and either a vat or a reactive dye for the cellulosic fiber are quite efficient combinations that produce good fastness properties. After the fabric is padded and dried, it is passed through a high-temperature thermofixation unit that usually renders only the disperse dye substantive. The vat or reactive dye is then usually made substantive to the cellulosic fiber by subsequent padding of the fabric in alkali, steaming, then washing the fabric. Alkali causes reduction of a vat dye to its leuco form. Thus, an oxidation step is also required for the fixation of vat dyes. In addition to these dye combinations, pigments with appropriate binders may also be employed by a continuous pad/dry/cure process to obtain pale shades on cellulosic/ polyester blends. Compatibility and stability of both types of dyes (e.g., disperse/vat and disperse/reactive) to alkali and high temperatures are essential to obtain uniform and reproducible dyeing of cellulosic/polyester blends. In addition to these considerations, staining of one type of fiber by the dye for which it has no substantivity (e.g., cellulose by the disperse dye) must also be carefully controlled for the successful dyeing of these blends.

Cellulosic/wool fiber blends may be dyed in a single bath containing direct (for cellulosics) and acid (for wool) dyes at a weakly acidic pH by allowing the cellulosic component to be dyed first. Normally acid milling or 2:1 metal complex dyes are used to dye the wool component of the blend. A dye retarder or reserving agent is used to prevent dyeing of the wool by the direct dye. Because direct dyes are used to dye the cellulose, it is customary to improve their fastness by using an appropriate aftertreatment. It is also possible to dye cotton/wool blends by a two bath method utilizing two different reactive dyes or a reactive/acid dye combination. Caution must be taken not to damage the wool under the alkaline conditions required to fix the reactive dye to the cellulose (refs. 48, 49).

Cellulosic/polyamide blends can be dyed by many of the techniques and combinations of dyes used for dyeing cellulosics/wool blend fabrics. There is somewhat more flexibility in methods for dyeing cellulosic/polyamides than for dyeing cellulosic/wool fabrics since disperse dyes can be used to dye the polyamide component and the pH of the dyebath can be slightly alkaline as well as neutral to selectively fix dyes to the cellulosic component. For two bath systems, it is common to first dye the polyamide component with a disperse, acid or metal complex dye under conditions that will not stain the cellulosic fiber, then raise the pH and dyeing temperature for application of a direct dye to the cellulosic component. Single bath methods are available whereby two different reactive dyes (one substantive to the polyamide and the other substantive to the cellulose) can be used to dye cotton/polyamide fabrics.

Cellulosic/acrylic (primarily cotton/acrylic) fabric blends are usually batch-dyed by single or two bath processes with the proper combination of dye classes (refs. 50, 51). The acrylic component of the fiber blend is always dyed with cationic or basic dyes. For single bath processes, only direct dyes are suitable for the cotton component, but anti-precipitating agents must be included in the dyebath to avoid precipitation due to the interaction of anionic groups in the direct dyes and cationic groups in the basic dyes (Fig. 3.14). To avoid staining of the cellulose by the basic

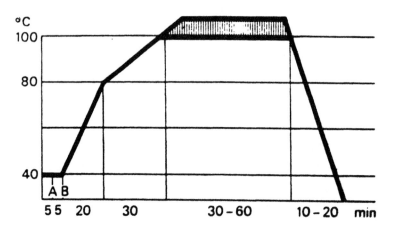

A. 2% precipitation inhibitor
    5% Na_2SO_4 or NaCl added at 80°C
    x% direct dye
    pH 4-4.5 with NaOAc

B. y% basic dye
    0-0.3% cationic retarder

Figure 3.14. Dyeing of cotton/acrylic blends with direct/basic dyes (ref. 50). Courtesy of Textilveredlung.

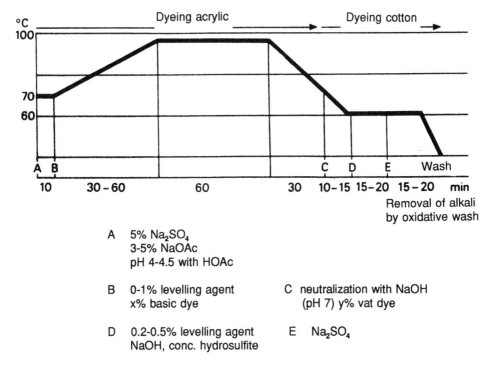

Figure 3.15. Dyeing of cotton/acrylic blends with vat/basic dyes (ref. 50). Courtesy of Textilveredlung.

dye in one or in two bath systems, a retarder is usually added. In two bath processes, the acrylic fiber is always dyed first at temperatures near the boiling point of water. The pH of the dyebath is subsequently made alkaline, then a reactive or vat dye is introduced to dye the cotton or cellulosic component at ca. 60°C. Figure 3.15 shows a representative two bath process for dyeing acrylic/cotton blends with basic/vat dyes.

As in the case of cellulosic/polyester blends for cellulosics, wool/polyester blends have become the most dominant of the wool blends and are thus the most frequently dyed. These blends are usually dyed with acid or neutral premetallized dyes (wool)/disperse dyes (polyester) in a single bath for light or medium shades; however, for heavier shades, a two bath process is recommended. In the latter process, the polyester is dyed first with a disperse dye to prevent staining of the wool by the disperse dye.

A variety of auxiliaries and additives are important to obtain the desired dyeing effect for wool/polyester blends. The most important additives are dispersing agents, dye carriers and agents that protect the wool fiber from damage (refs. 52, 53). Dispersing agents serve to maintain a fine dispersion of the disperse dye affixed to the polyester. The use of appropriate carriers is critical to allow commercial dyeing of the polyester at low temperatures, with carrier selection highly dependent on the type of dyeing machinery used. For closed and covered machinery, volatile carriers such as mixtures of diphenyl/trichlorobenzene and o-phenylphenol/biphenyl may be used, while for open machinery, it is preferable to use less volatile carriers such as o-phenylphenol or benzylphenol. To protect the wool from damage at dyeing temperatures above 110°C, the pH of the dyebath should be acidic and generally in the range of 3.5-5.5. Formaldehyde or N,N'-dimethylolethyleneurea (also known as DMEU), a controlled release source of formaldehyde, has been demonstrated to be effective in protecting the wool fiber from damage when it is dyed under these conditions (ref. 53).

A typical one bath dyeing process to produce pale or medium shades of polyester/wool blends consists of setting the dyebath at about 50°C with acetic acid (pH 5-6), 3-5% formaldehyde (based on fabric weight) and a dispersing agent. The disperse and metal complex dyes are added after a short pretreatment. The bath is then brought to a temperature of 110-120°C over an hour and dyed an additional hour at this temperature. It is usually necessary to aftertreat the dyed fabric to obtain good fastness properties and to remove any disperse dyes that stains the wool fiber or that is not bound to the polyester fiber (ref. 52).

In addition to the conventional mixtures of dye classes that are substantive to wool and polyester (e.g., metal complex and disperse, respectively), there have been some experimental dyes synthesized that function both as disperse and acid dyes and are thus suitable for dyeing wool/polyester blends. These dyes have the basic general structure: Chromophore--X--CH_2-CH_2-$SC(NH_2)_2$ $^+$ I $^-$ or Br $^-$, where X is either -NHCO-- or a different covalent bond. These isothiuronium salt dyes can act as cationic dyes on wool by being applied at acidic pH, then fixed by raising the pH to 9-10. The same dye can also be successfully applied as a disperse dye to the polyester component by thermolysis of the dye before application in the presence of dispersing agents (ref. 54).

A variety of dye machinery and processes are suitable for dyeing wool/polyester blends. Beam dyeing machinery is much more effective than conventional winches for attaining temperatures necessary to exhaust disperse dyes onto the polyester component of the blend (usually above 110°C). However, most current jet dyeing machinery is far superior to any type of atmospheric winch or beam dyeing machinery for coloration of wool/polyester blends. Jet dyeing is not only more economical, but produces fabrics with better color yield and shades, and has several other technical advantages over other dyeing methods.

Wool/polyamide blends may be dyed with the same application classes of dyes (acid and/or metal complex) or even the same dye. This type of dyeing is possible because of the similar affinity and substantivity of both fiber types. Thus, only minor variations of conditions normally used for dyeing either type of fiber (typically an exhaust, single bath process) are usually required to dye wool/polyamide blends. Polyamide fibers have a tendency to dye darker shades than the wool for pale shades of dyeing. However, this difference in dyeability is barely noticeable when the fabrics are dyed darker or heavier shades. Levelling acid dyes that contain disulfonic acid groups or premetallized acid dyes are usually used to dye wool/polyamide blends. The latter are preferred, because it is easier to dye both fiber types the same desired shade.

Wool/acrylic blends can be batch-dyed in a single or two bath process with combinations of reactive/basic or metal complex/basic dyes. However, as with cotton/acrylic blends, the use of a precipitation inhibitor is usually required for a single bath process for wool/acrylic blends because all dye classes substantive to wool have anionic groups. These negatively-charged substituents tend to form agglomerates with the cationic groups of basic dyes used for the acrylic component (refs. 51, 52). Another precaution that must be taken is minimizing or eliminating degradation products formed when the wool is boiled at or near neutral pH, since these byproducts are capable of reducing the basic dyes. Thus, careful selection of dyestuff combinations and bath conditions is required. Single bath dyeing procedures that use reactive dyes for wool and basic dyes for acrylics have been devised to produce fabrics of medium to dark shades when the dyeing is performed at pH of 5 or less; under these conditions, there is good stability of the basic dyes in the bath (refs. 51, 52).

Although all synthetic fiber blends are not nearly as common as cellulose/synthetic fiber or wool/synthetic fiber blends, there are at least three all synthetic fiber blend combinations that are commonly dyed: polyester/polyamide, polyester/acrylic and polyamide/acrylic. As in dyeing of other blends, these usually require two dissimilar dye application classes and consideration of factors such as compatibility, prevention of cross staining and related problems to produce uniform and acceptable pale to dark shades.

Although it is possible in principle to use only a single disperse dye or mixtures of disperse dyes for both polyester and polyamide components, most disperse dyes will produce unacceptable staining rather than acceptable color shades on the polyamide fiber. Thus, in a single bath process, the polyester/polyamide blends are first dyed with a disperse dye, the stained polyamide part is then stripped with a reducing agent such as sodium hydrosulfite, and the polyamide component dyed with an acid or metal complex dye.

There are similar considerations with regard to cross staining when one dyes polyester/acrylic and polyamide/acrylic blends. For polyester/acrylic blends, it is customary to first dye the acrylic component with a basic dye, cool the dyebath, then apply the disperse dyes with a carrier to dye the polyester component. For polyamide/acrylic blends, the acrylic component is also dyed first with a basic dye. If the polyamide is stained in the process, this staining may be removed or cleared by using appropriate reagents such as dilute permanganate in acetic acid followed by an aftertreatment with sodium hydrosulfite. The polyamide fiber is then subsequently dyed at a slightly acidic pH with appropriate acid or metal complex dyes.

3.5.7 Aftertreatments for dyeing of textiles

Although many of the aftertreatments for improving the fastness of dyes were developed for direct dyes, most dye application classes have improved fastness properties when fabrics treated with them are given an appropriate aftertreatment.

This topic has been extensively and critically reviewed for a 100 year period (1880-1980); there have been diverse and complex approaches used to improve the wash-fastness and lightfastness of dyed textiles (ref. 55).

Early approaches to improve the fastness of dyes were of course only directed to natural fibers and usually included some form of mordanting with natural tannins and/or transition metal salts. As factors that enhanced the stability of the dyed fabrics to washing, light and rubbing were better understood, more sophisticated aftertreatments were devised that were related to both fiber types and dye application classes. For cellulosic fibers, a variety of approaches have been developed and are still used to improve fastness of these dyed fabrics. Aftertreatments include crosslinking with formaldehyde and other aldehydes to form a new dye with decreased solubility, use of various transition metal salts (nickel, zinc and particularly copper), cationic agents such as quaternary ammonium salts, reaction products and/or resinous products derived from cyanamides and biguanides, polyfunctional crosslinking agents such as vinylsulfonyltriazine derivatives and similar agents. Currently, quaternary ammonium salts and cationic resins are used as aftertreatments to improve the overall fastness properties of direct dyes to cellulosic fibers.

A specific type of aftertreatment to improve the fastness properties of reactive dyes is not usually suitable for all fiber types (cellulosics, wool and polyamides). However, cationic agents used to improve the fastness properties of direct dyes on cellulosics have also been satisfactorily used as aftertreatments for both cellulosic and wool fibers. The primary problem with the reactive dyes is either removal or fixation (insolubilization) of unreacted and/or hydrolyzed reactive dye from the fabric. Many of the approaches suitable for aftreatment of direct dyes on cellulosics are also suitable for aftreatment of reactive dyes on cellulosics. For wool dyed with reactives, a variety of reducing agents (such as sodium sulfite) are also considered appropriate aftertreatments for improving fastness properties.

Aftertreatment of wool and polyamide fibers with a variety of metal salts such as copper and chromium (latter described earlier as afterchroming or postmordanting in the section on **Mordant Dyes**) is an effective technique that is still used to improve the fastness properties of acid dyes applied to these fibers.

A considerable amount of effort has been expended in using natural tanning agents (originally defined as water-soluble cellulosics precipitated by gelatin from solution and those materials that were irreversible absorbed by collagen in animal hides to preserve leather from microbial attack and decay), followed later by full backtanning treatments and synthetic tanning agents or syntans as aftertreatments to improve the fastness of wool, polyamide and even other synthetic fibers. A more useful and modern classification of polyphenolic compounds that constitute tanning agents is: (a) gallolylated saccharides that hydrolyze to gallic acid, (b) hydrolyzable ellagitannins that give ellagic acid and (c) condensed or catechol tannins that contain little carbo-hydrate structures but are converted into insoluble, amorphous polymers (ref. 55).

Full backtanning treatments (successive treatments with tannic acid, potassium antimony tartrate and stannous chloride) were developed to improve washfastness of polyamide fibers dyed with acid dyes. Although not as effective as the complete backtan process, at least two major classes of compounds have been used as syntan aftertreatments to improved the fastness properties of polyamides dyed with acid dyes: (a) phenol-formaldehyde condensation products containing solubilizing sulfonic acid groups and (b) thiophenols. For all fiber types, a variety of compounds and polymers have been used as aftertreatments to improve overall fastness properties. Irrespective of the structure of the agents used in the aftertreatment, the vast majority function by decreasing the solubility of the class of dye applied to the fiber and thus enhance its substantivity.

3.5.8 Garment dyeing

Garment dyeing is not new but has emerged from a specialty type of dyeing technique to one of moderate commercial importance because of the need of "quick response" to fashion demands and streamlined inventories of textile goods. A critical review of this topic subdivides garment dyeing into four categories: (a) fully fashioned garment dyeing by major commission dyers and finishers; (b) cut and sew garments derived from woven and knitted fabrics to meet high colorfastness requirements; (c) dyeing of all cotton with low colorfastness suitable only for hand washing and (d) specialized effects such as stone-washing, overdyeing or highlighting (ref. 56). There have been specialized machines constructed for garment dyeing that are essentially perforated rotary drums with external circulation, high speed centrifuge, flexible control systems, variable cylinder speed during the dye cycle and dye-liquor recovery. Most garment dyeing has been conducted on all cotton or all wool fabrics with a single color or with overdyed pigment prints. It is likely that cross-dyeing and other more specialized dyeing techniques will be continually adapted to garment dyeing processes.

3.5.9 Dyeing of carpets

Carpets composed of one type of fiber and those composed of fiber blends are dyed both by batch (usually in rope form) and continuous processes. Because of the three dimensional nature and more complex physical structure of carpets, it is more difficult to obtain level dyeing on carpets than it is for conventional fabrics. Since up to 75% of all carpet fibers are polyamides or variants of polyamides, most dyeing processes involve the use of acid or disperse dyes. In addition to the normal considerations for obtaining uniform and reproducible dyeing of fabrics, there are several other fundamental aspects that should be addressed in the dyeing of carpets (refs. 57, 58). Most are concerned with dyeing by exhaust or batch processes, but may be equally applicable to continuous dyeing processes.

Heat setting of synthetic carpet fibers (such as polyamides) can either decrease their dyeability (in the dry state that raises their T_g) or increase their dyeability (in the presence of steam because water acts as a plasticizer and carrier to lower T_g). The levelness of dyeing carpets in batch dyeing may also be improved by using consistent methods for cooling down the dyebath, monitoring and adjusting the pH during dyeing and by controlling the liquor-fabric ratio (ref. 57). Moreover, piece dyeing of carpets in open width super becks (Figure 3.16) can minimize side-center-side shading problems by uniform spray distribution of the dye liquor onto the carpet face and by maintaining a linear speed sufficient to turn the carpet over every 2-3 minutes. Nevertheless, rope form dyeing of carpets is still much more prevalent than beck dyeing of carpets in batch processes.

The description of a carpet as a structure composed of two capillary systems connected to each other is used to determine the ideal conditions for dyeing by both exhaust and continuous processes (ref. 58). One capillary system consists of the pile

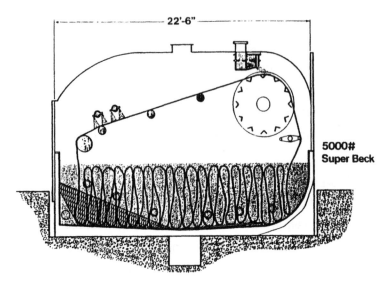

Figure 3.16. Gaston County Carpet Machinery open width super beck (ref. 57). Courtesy of Gaston County Dyeing Machine Co.

because of its interstitial volume between single fibers while the other capillary system is the carpet backing that extends over the total area of the carpet with liquor being retained at crossings of the backing structure. It is thus not surprising that carpets retain approximately twice the level of liquor when they are held in a horizontal rather than a vertical position. In view of the laws of capillarity, suggestions to improve levelness include mechanical redistribution of liquor, use of foaming agents, higher wet pickup, increasing dyebath temperature, use of wetting and prewetting agents and proper choice of dye (ref. 58).

The dyeing of carpets derived from fiber blends has been reviewed with respect to dye selection and criteria, important fiber blends used and at what stage the fiber blends are dyed, i.e., as fibers (loose stock), yarns or in batch or dyeing processes for the fabricated carpet (ref. 59). A substantial amount of dyeing for tufted carpets is conducted at the fiber stage (loose stock dyeing) by using appropriate combinations of dye classes (such as acid/basic dyes for polyamide/acrylic fiber blends); most loose stock dyeing is done by batch rather than continuous processes. For smaller scale runs, dyeing of carpet yarns in hank form is popular for Axminster-type carpets but rarely used for dyeing tufted carpets; package dyeing of carpets yarns is also gaining more commercial acceptance. As in the case of carpets derived from one fiber type, both exhaust (such as winch and beck) and continuous (primarily pad-steam) methods are used for dyeing carpets comprised of blended fibers. The major fiber blends of interest are wool/polyamide, wool/rayon, polyamide/acrylic and polyamide/polyester. As discussed earlier in the dyeing of fiber blends for flat fabrics, dyeing of carpets made of blended fibers requires similar considerations with regard to choice of combinations of dyes, order of application and use of auxiliaries.

3.5.10 Dyeing of nonwovens

Since nonwovens represent the fastest growth area of all types of textile constructions, it is not surprising that methods of dyeing or coloring them have recently been reviewed (ref. 60). Basically, nonwovens can be colored before and after web formation. The more commonly used methods of dyeing nonwovens before or during web formation are dyeing of cellulosic fibers in a wet-laid web forming process (such as that shown in Figure 3.17) in which the dye can be added at different stages of the process and incorporation of pigments or selected dyes before spinning in the polymer melt (Figure 3.18).

Dyeing or coloration methods after the nonwoven web is formed include application of pigments to nonwoven webs with latex binders, dyeing of polyester webs by continuous Thermosol processes similar to those used to dye woven and knit polyester fabrics, and printing methods that will be discussed later for all types of textiles.

3.5.11 Low wet pickup methods for dyeing textiles

The use of low-liquor or low wet pickup methods for dyeing and finishing textiles has been reviewed (refs. 61, 62). Most of the effort has been directed to finishing techniques as opposed to dyeing processes, but the same principles and techniques applicable to finishing textiles have generally been utilized. Since low wet pickup methods will be discussed in much more detail for finishing processes later (**Chapter 4**), this topic will now only be discussed briefly with regard to dyeing processes.

Techniques for reducing the amount of liquor and/or wet pickup of fabrics are generally based on expression methods (removal of liquor by squeezing or other mechanical means), topical methods (limiting amount of liquor initially contacting the fabric) and foam application (specialized method where majority of liquor is replaced

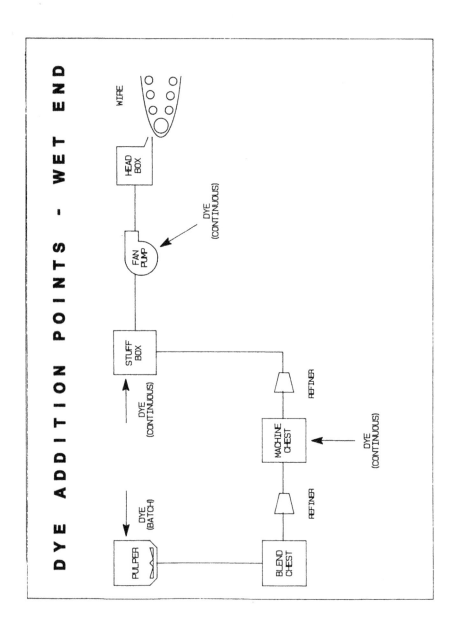

Figure 3.17. Dye addition points: wet end (ref. 51). Courtesy of American Association of Textile Chemists and Colorists.

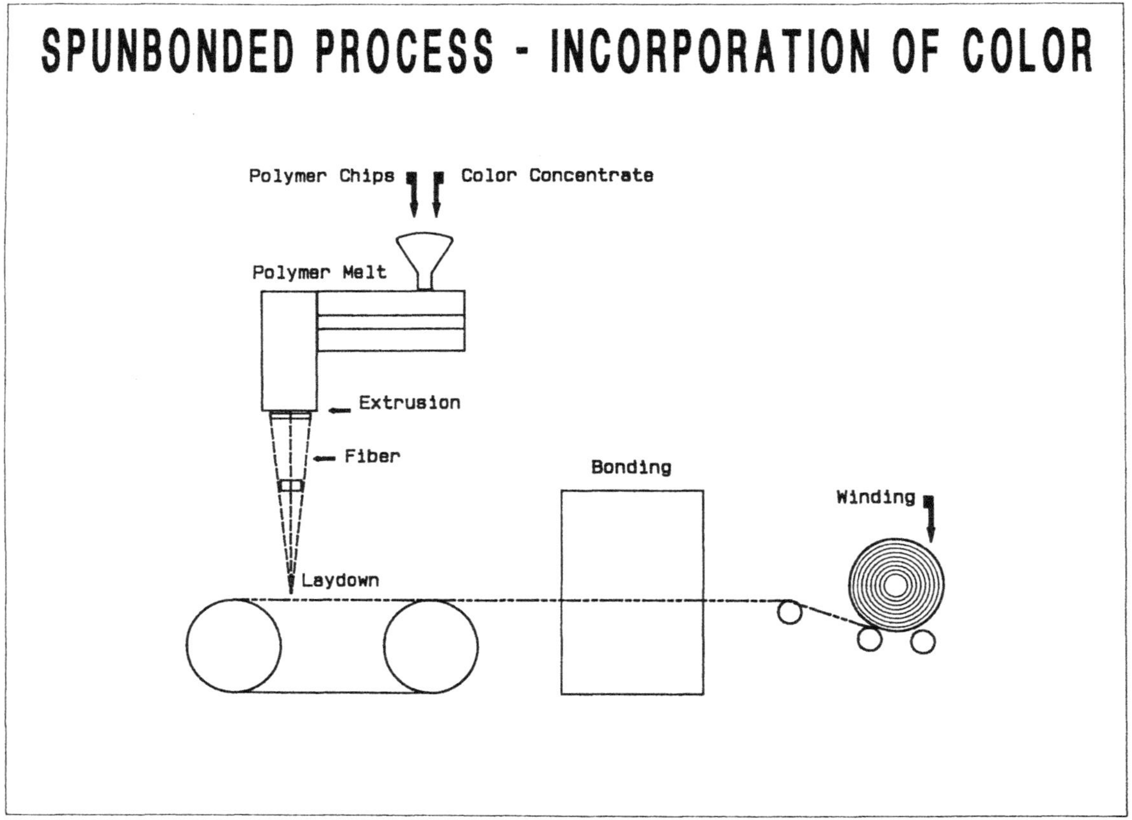

SPUNBONDED PROCESS - INCORPORATION OF COLOR

Polymer Chips   Color Concentrate

Polymer Melt

Extrusion

Fiber

Laydown

Bonding

Winding

Figure 3.18. Spunbonded process: incorporation of color (ref. 60). Courtesy of American Association of Textile Chemists and Colorists.

by air). With regard to expression techniques, there are a variety of rollers composed of special materials that will remove excess liquor from fabrics. Also, vacuum extractors and air-jet ejectors (such as the Machnozzle) have been used to reduce the wet pickup of fabrics to as low as 5% for some synthetic fabrics to about 50-60% for natural fabrics such as cotton and wool. Topical methods include the use of engraved rolls, kiss or lick rolls, nip padders with doctoring devices, wicking systems based on rotary screen printing methodology, pad-transfer and loop-transfer systems and special spraying devices (ref. 61).

Foam application techniques are the most extensively used for low liquor dyeing and many finishing processes and have been reviewed with regard to stability, structure and properties of the foam and various applications in dyeing and printing (ref. 62). Although foam application methods are suitable for both batch and continuous dyeing processes, they are particular cost and energy efficient for various continuous processes. Pad-batch continuous processes are clearly two to four times less expensive than other continuous processes such as pad-dry, pad-dry-steam and pad-steam.

Advantages of dyeing by foam application methods include less use in energy (less liquor to evaporate), high running speeds, choice of one-sided or two-sided application methods, less deformation of piles in towels and carpets, savings in chemicals and auxiliaries, lower dye migration and less streaking. Conversely, disadvantages of dyeing by foam application are inadequate or careful control of foam stability and reproducibility, incomplete wetting of fibers in some instances, variability in wetting due to difference in moisture content in the fabric and/or uneven foam drainage and lower liquor stability due to higher concentrations of dye in the liquor (ref. 62). A variety of techniques can be used to apply dye to one or both sides of the fabrics as shown in Figure 3.19. One-sided methods include the use of doctor, roller and flow coaters, and vacuum and pressure methods. Two-sided methods include the use of twin foam transfer rolls or two-bowl pad apparatuses. There are also numerous methods of removing the foam by drainage such as self-drainage, pressure and vacuum drainage and the use of squeeze rolls.

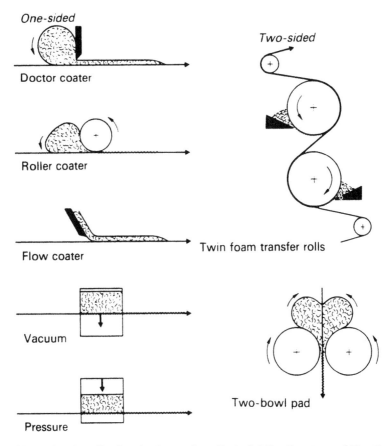

One-sided

Doctor coater

Roller coater

Flow coater

Two-sided

Twin foam transfer rolls

Vacuum

Pressure

Two-bowl pad

Figure 3.19. Methods of application for foamed media (ref. 62). Courtesy of The Society of Dyers and Colourists.

3.6 DYEING MACHINES AND PROCESSES

3.6.1 Introductory remarks

There are three major processing or mechanical principles normally employed for dyeing textiles: (a) the dye liquor is moved and the textile is stationary; (b) the textile is moved and there is no mechanical movement of dye liquor or (c) both the dye liquor and the textile undergo movement or mechanical agitation (ref. 63). A second and ancillary classification that will be discussed within the context of liquor and fabric mobility is the dyeing of textiles by batch or continuous processes.

3.6.2 Machines in which only dye liquor is mobile

Raw stock, package and beam dyeing are representative batch processes in which only the dye liquor is mobile. In raw stock dyeing, the textile may be in fiber, yarn or fabric form and is tightly packed around one or more perforated spindles in a large stainless steel kier. Dye liquor is pumped in one direction through the textile, then flows to the bottom of the machine and into the return side of the pump (Figure 3.20).

When level dyeing of textiles is required, package or beam dyeing machines are used because they have a unique automatic reversing mechanism that changes the direction of the flow of the dye liquor from inside to outside and reverses the flow of liquor at preset times. These types of dye machines are versatile because they afford a constant rate of temperature rise, time of dye application, volume of dye liquor flow, and can accommodate many different fiber types and blends and textile constructions.

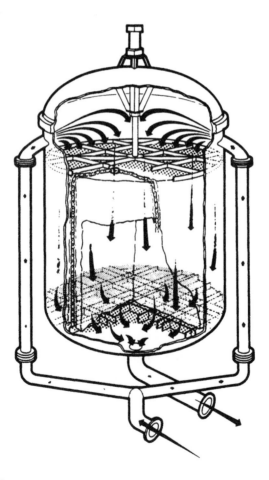

Figure 3.20. Representative raw stock dyeing machine (ref. 63). Courtesy of Gaston County Dyeing Machine Co.

Package dyeing machines are particularly versatile (Figure 3.21) since they can be readily fitted with interchangeable carriers to dye loose fibers and yarns in the form of cheeses, cones or bobbins either in conventional or rapid low liquor ratio dyeing systems (ref. 64).

Beam dyeing machines can be considered variations of package dyeing machines in which the fabric is rolled onto a perforated, hollow cylinder and held in a horizontal position in an enclosed chamber through which the liquor is pumped in a radial direction throughout the beam. Most of the advances in beam dyeing machines have

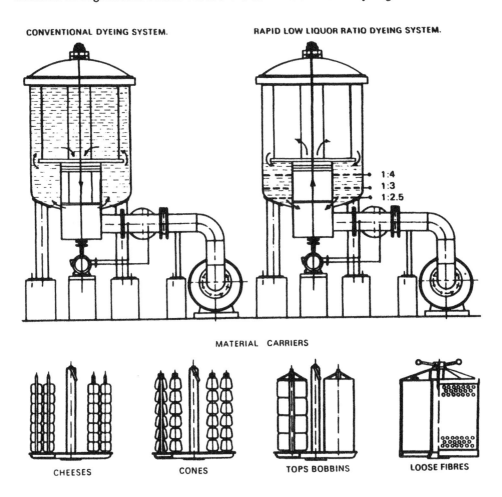

CONVENTIONAL DYEING SYSTEM.   RAPID LOW LIQUOR RATIO DYEING SYSTEM.

1:4
1:3
1:2.5

MATERIAL CARRIERS

CHEESES   CONES   TOPS BOBBINS   LOOSE FIBRES

Figure 3.21. Versatility of package dyeing machines (adapted from ref. 64). Courtesy of Prof. M. L. Gulrajani, Indian Institute of Technology.

been in the development of high temperature, high pressure apparatuses in which the fabrics could be end-loaded and have four way flow valves in contrast to earlier beam machines that could only be top-loaded with only one way flow. Prerequisites, essential features, operation and recent developments of beam dyeing machines have been summarized and reviewed (ref. 65). These machines should have pumps (centrifugal, axial, turbo or other design) suitable for liquor penetration into various types of fabrics and can add dyestuffs and auxiliaries to them at temperature below and above 100°C. Newer beam dyeing developments include the use of horizontal beam machines to dye cross wound packages of various types of yarns, the use of low liquor ratios and the use of an air injection device that avoids the undesirable "moire" effect (wavy pattern) on knits and that gives more flexibility in control of tension during winding of batches on the perforated beams.

3.6.3 Machines in which only textile is mobile

In batch processes using jigs, winches or becks and in continuous processes, the textile is moved and the liquor is stationary. Jig dyeing is a very old batch process in which long lengths of fabric are wound around a beam roller on one side of the jig, then run off the beam through a small volume V-shaped dyebath and wound on the opposite beam. The winding direction is reversed until the desired color shade is achieved. The major disadvantage of jig dyeing is that irregularity of color uniformity or shade occurs, particularly at the edges of the fabric roll where the temperature is different from the temperature in the center of the fabric. However, jig dyeing machines afford an inexpensive and simple method of batch dyeing when crease-free fabrics are required.

In winch dyeing, fabrics that have poor dimensional stability (such as knits and/or wool) have their ends stitched together to make an endless rope that is move continuously in the dyebath over a winch (Figure 3.22). One disadvantage of the winch is that as the fabric leaves it, the fabric tends to pile up for a brief time into the dyebath at the back of the machine. This causes undesirable creases in fabrics made of rayon or cellulose acetate and has been resolved to some extent by using elliptical drums for the winch and/or shallow dye vessels. Another major disadvantage is that winches require long liquor ratios (as high as 40:1) and thus can

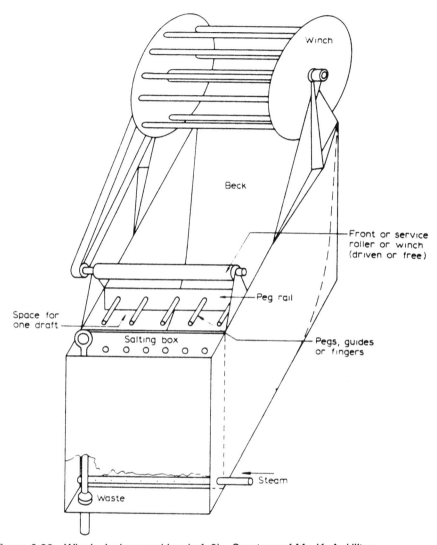

Figure 3.22. Winch dyeing machine (ref. 3). Courtesy of Mr. K. A. Hilton.

consume large quantities of dyes for systems that exhaust poorly (ref. 3). Beck dyeing is an older form of winch dyeing in which the fabric in open-width or rope form open width or rope form is continually moved through the dye/liquor in a shallow U-shaped box by means of an elliptical reel mechanism that transports the fabric from the front and allows it to slide down back and return to the front.

In continuous dyeing, the fabric is first immersed in the dyebath and then fed through a padder or comparable device to remove excess liquor, followed by partial or complete dye fixation (usually by dry heat and/or steam). If necessary, the fabric is then immersed in solutions containing agent to further fix or chemically modified the dye (such as alkali for reactive dyes or alkali/hydrosulfite for vat dyes), fed through a padder, then given an appropriate aftertreatment to remove excess dye. When conducted properly, continuous dyeing has numerous advantages over batch dyeing for large commercial operations with regard to processing speed, uniformity and minimum use of auxiliaries, dyes and water.

A multipurpose continuous dye line can be constructed to process many different types of fiber types/fabric constructions and dye application classes if the machinery is sequenced as described in the preceding paragraph. Schematics of such multipurpose lines are described as well as the historical development and improvement of continuous dye ranges with automatic controls, pad mangle design, heating, steaming and washing units (ref. 66).

The Thermosol process, initially developed by DuPont in the 1950's to fix disperse dyes to polyester fabrics, rapidly gained commercial acceptance and is now also used to apply dyes to a variety of fabrics comprised of fiber blends. Figure 3.23 depicts a typical continuous Thermosol dye range where the cotton/polyester blend fabric is passed through a padder containing disperse and vat dyes, the disperse dye fixed by dry heat in the Thermosol unit, then the vat dye fixed by passing the fabric through alkali/hydrosulfite, followed by steaming and soaping to remove excess chemicals and dyestuffs.

There are various types of Thermosol units that have been used to fix dyes to fibers. These include forced convection, conduction-convection, direct contact and infrared radiation units. The forced convection units are either hot flue units [Figure 3.24 (a)] or tenter frames [Figure 3.24 (b)]. Hot flue units have the advantage that large amounts of fabrics may be processed per unit length of equipment relative to that processed in a tenter frame, but the tenter frame has about ten times the heating efficiency of the hot flue unit. Direct contact units are basically different styles of "heated cans" [Figure 3.25 (a)] while conduction-convection units [Figure 3.25 (b)] are those in which hot air is sucked up from the interior of the drum that causes heating of the fabric by conductive and convective heat transfer (ref. 31). The other type of

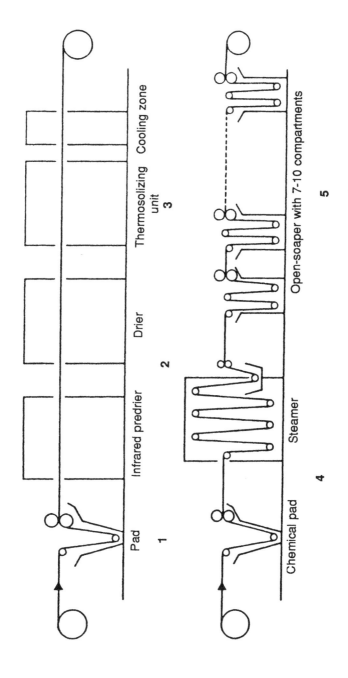

Figure 3.23. Typical continuous plant for dyeing polyester/cotton with disperse/vat dyes (ref. 45). Courtesy of Prof. M. L. Gulrajani, Indian Institute of Technology.

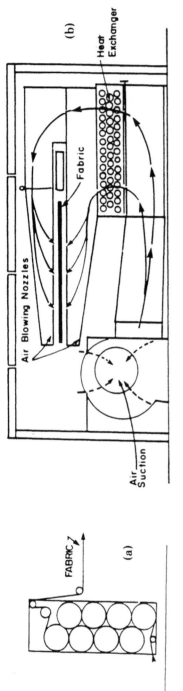

Figure 3.24. (a) Hot flue convection unit and (b) air path in tenter frame convection unit used in Thermosol process (ref. 31). Courtesy of John Wiley and Sons.

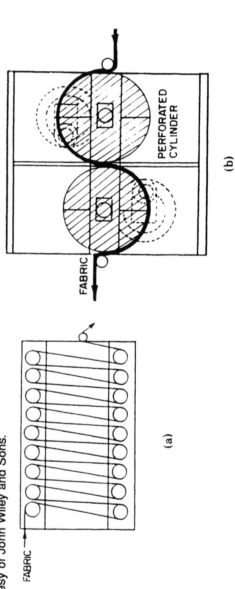

Figure 3.25. Schematic of (a) direct contact and (b) conduction-convection Thermosol units (ref. 31). Courtesy of John Wiley and Sons.

continuous dyeing process uses high temperature steamers that fix the dye after it is padded onto a fabric; these are particularly suitable for continuous processing of cotton/polyester blends.

3.6.4 Machines in which both dye liquor and textile are mobile

Jet dyeing machines are designed to simultaneously move or agitate fabrics and dye liquor. They are practical and economical because of their low fabric to liquor ratios (1:10 or less). These dyeing machines use a venturi jet or tube, whose narrow passage causes suction which imparts movement to the fabric while a standard centrifugal pump imparts movement to the dye liquor. Since their introduction in the late 1960's, jet dyeing machines have replaced many other types of batch dyeing machines (such as winches and becks), and is in many instances the method of choice for commercial batch dyeing processes.

Improvements and modifications of the earlier jet dyeing machines have been critically reviewed and described (refs. 67-69). The first generation of jet dyeing machines were fully flooded with dye liquor and relied only on jet action to move the fabric to minimize problems that occurred in the dyeing of lightweight and certain types of knit fabrics. The jet pressure was not sufficient to prevent crush marks on these types of fabrics and increasing the jet pressure led to undesirable twists and different types of crease mark defects. Second generation jet dyeing machines were based on the "soft-flow" principle in which machines were designed horizontally to make the fabric follow a path similar to that taken by it in a conventional winch. There are basically two types of soft-flow jet dyeing machines: (a) those with a transportation tube outside the machine having a considerable lift from the liquor level to the reel and (b) those with the transportation tube within the dye autoclave, parallel to the dye compartment with a short lift from the dye liquor to the reel. The Platt-Ventura jet dyeing machine (Figure 3.26) is typical of a soft-flow system that has the transportation tube within the autoclave and that is also equipped with a liquor overflow system (refs. 67-69).

The next generation of jet dyeing machines may be characterized as those that use low liquor to fabric ratios (5:1 or less) to reduce energy consumption. Both atmospheric and high pressure dyeing machines of this type have been designed,

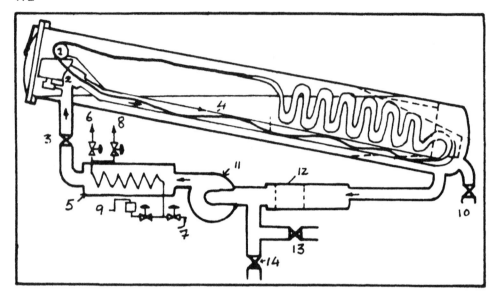

Figure 3.26. Platt Ventura Machine: (1) variable speed-driven reel, (2) fully-flooded nozzle, (3) flow control, (4) fabric delivery tube, (5) heat exchanger, (6) steam in, (7) cooling water in, (8) cooling water out, (9) condensate out, (10) rinse drain, (11) main pump, (12) filter, (13) water supply and (14) drain (ref. 68). Courtesy of Prof. M. L. Gulrajani, Indian Institute of Technology.

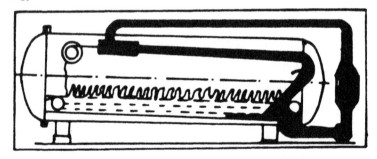

Figure 3.27. General view of the EPSA Miniflott jet system (ref. 68). Courtesy of Prof. M. L. Gulrajani, Indian Institute of Technology.

with the former particular suitable for dyeing cotton knit fabrics with reactive dyes. Initial designs (such as the Thies R-140 and the high-temperature Texpro HI Jet) included a system of circular movement of the fabric with a variable speed reel and powerful jet action. Yet at very low liquor/fabric ratios, some types of fabric became distorted, had crush marks or similar defects in these types of machines (refs. 67-69). Later jet dyeing machines, such as the Thies Roto-Stream and the Scholl Compact, had markedly different designs from the initial low-liquor, jet dyeing machines. The Roto-Stream was designed to avoid crush marks by using a specially designed jet and transportation tube that allows the fabric to pass through the jet with minimum pressure and plaiting it across and inside two perforated circular cages. The Scholl Compact has an overflow chamber above and before the jet with a very long transportation tube that makes the opening of the fabric in rope form more facile (refs. 67-69).

Fully-flooded, soft-flow, and low liquor ratio jet dyeing machines have been further modified to produce a class of rapid jet dyeing machines. Such machines have the advantages of increased productivity, versatility and reduced dyeing time. The reduced dyeing time also gives fewer fabric defects due to an increased liquor circulation rate. Comparative features of conventional and rapid jet dyeing machines are shown in Table 3.7. The ESPA Miniflott machine (Figure 3.27) is a prototype rapid jet dyeing machine that uses the principle of a very long dyetank that has a transportation tube above it of the same length as the dyetank and a pick-up reel and jet at the front of the machine.

TABLE 3.7

Comparative features of conventional and rapid jet dyeing machines (adapted from ref. 68). Courtesy of Prof. M. L. Gulrajani, Indian Institute of Technology.

Conventional jet	Rapid jet
Controlled heating rate due to lower dyestuff absorption and circulation	Fast heating rate due to increased relative contact between fabric and dye liquor.
Comparatively much lower rate of dye	Much increased liquor circulation rate
Because of longer time required for dye liquor circulation, the dyeing cycle takes ca. 130-190 min.	Rapid heating and quicker dye circulation, dyeing cycle takes only 60-70 min.

Later jet dyeing machines such as the Jawetex Jiget in Switzerland, have features such as a J-shaped dye box that make the unit a combination of a jig and jet system. Newer concepts include the development of a continuous jet dyeing machine for treatment of fabric in rope form and the adaptation of ink jet dyeing techniques for paper to the coloration of textiles (refs. 68, 70).

3.7 TEXTILE PRINTING

3.7.1 Introductory remarks

Textile printing is essentially a localized form of dyeing in which colored patterns or designs are confined to a restricted area and the dyebath to fabric ratios are very low. By one of several printing methods, printing ink or paste or dispersions are applied to the fabric, then fixed to the fiber by heat or heat and steam. The excess dye is removed from the fabric by washing and detergent scouring and the fabric is subsequently dried. The types of dyes used in printing textiles are usually the same classes that confer substantivity to given fiber types by solution dyeing processes, e.g., disperse dyes for polyester and disperse/vat dye combinations However, with the use of a suitable binder, pigments may be applied to all types of fibers. Thus, coloration of textiles by pigment printing is a commonly used technique.

A distinction is usually made between printing methods and printing styles. Printing styles are either: (a) direct--all pastes or printing materials contact the fabric surface with no subsequent processing alterations; (b) discharge--subsequent processing of a printed fabric to bleach or destroy certain colors in the pattern; and (c) resist--certain parts of the fabric are originally protected by waxes or other materials to resist coloration by printing pastes or inks.

Printing methods include: (a) block (wooden blocks with printing paste imprints fabric), (b) roller (fabric passes over engraved copper rollers, each containing a different color), (c) transfer (a variation of roller printing in which color is transferred from the surface of paper to the surface of a fabric), (d) screen (silk fabric in a wooden frame serves as a stencil through which the colorant is selectively transferred to a fabric in contact with the screen), (e) spray (the print paste is electrostatically or mechanically sprayed onto the textile), and (f) foam (low wet pickup method in which the dye is part of foam and applied to fabric). There are also some newer printing methods that are being evaluated for their commercial potential such as ink jet printing, direct electrostatic imaging and xerography.

3.7.2 Special coloration techniques and auxiliaries

Application classes for dyes for textile printing and mechanisms of dyeing and obtaining substantivity are basically the same as those previously discussed for solution dyeing of textiles (ref. 71). The major difference with regard to dye application classes for textile printing compared to textile dyeing is the use of various thickeners to localize the print and produce clear and well-defined designs on the fabric and the much more frequent use of pigments in textile printing and binders to affix the pigments onto fabrics.

The desirable properties of thickeners and their classification into natural, modified natural and totally synthetic substances have been critically reviewed (ref. 72). The primary functions of the thickeners are to localize the dye in the desired area of the fabric and promote its adhesion to the fabric surface until the dye is fixed or made substantive. There are three criteria required for a suitable thickener: (a) the properties that it imparts to the moist dye paste, (b) its contribution to an acceptable dried, printed film and (c) its contribution to producing an acceptable printed product. As noted in Table 3.8, there are a wide variety of thickeners available and the choice of which to use is usually dictated by cost and properties such as viscosity of the paste and the durability and stiffness/softness of the resultant printed fabric.

The development of pigment printing over the last 50 years has been reviewed with regard to the types of products used (pigments, binders, thickeners and various auxiliaries such as dispersing agents, protective colloids and hand modifiers) and the advantages and disadvantages of pigments versus the use of substantive dyes to obtain textile prints (ref. 73). It has been estimated that in the mid-1980's, more than half of the textile prints in the world are made by pigment printing. The major advantages of pigment printing are its simplicity, elimination of inexpensive wet aftertreatments, applicability to all fiber types, printing styles and printing methods, and choice of a wide range of pigments with regard to color and lightfastness. Conversely, pigment printing may effect on fabric hand, cause undesirable buildup on rollers and is environmentally undesirable from a health and safety aspect when petroleum-based solvents are used in emulsion thickener formulations.

TABLE 3.8

Classes of thickeners for textile printing (adapted from ref. 72). Courtesy of Marcel Dekker Publishers.

Natural	Modified Natural	Wholly Synthetic
Plant exudates	Starch derivatives	Vinyl polymers
Gum karya, gum arabic and gum tragacanth	British gum; carboxy-methyl and saponified starch-g-acrylic copolymers	Poly(vinyl alcohol); poly(vinyl pyrrolidone)
From seeds and roots	Cellulose derivatives	Acrylic polymers
Locust bean gum; tamarind seed gum; guar gum	Carboxymethyl cellulose; methyl cellulose; hydroxy-ethyl cellulose	Polyacrylamide; polyacrylic acid
Seaweed extracts; alginates	Gum derivatives	Other synthetic copolymers
Starch, crystal gum and pectin	Modified alginates; meyopro gum	Styrene/maleic anhydride; isobutene/maleic anhydride

The most important properties of the pigments are their insolubility or low solubility in water and dry cleaning solvents, their physical form (powder, paste), their particle size, color and brilliance, fastness properties and cost. Pigment finishes have been developed over the years to ensure the highest pigment concentration with good flow properties. The importance of using suitable binders [usually film-formers based on poly(acrylic acid), polyacrylate esters, butadiene or vinyl acetate polymers] cannot be overemphasized. The binder provides the durability of the pigment to the fiber by adhesion. It must be clear and colorless as a dispersion or in solution, be flexible without being tacky, promote adhesion of the pigment to the substrate and be able to withstand subsequent mechanical and thermal stresses. Many binders are used in the form of aqueous polymer dispersions (solid contents as high as 50%) or are prepared directly by emulsion polymerization. The preparation and utilization of binders in this manner give one much flexibility in controlling particle size, morphology, adhesion and stiffness characteristics and durability of the polymers to aging, and fastness to rubbing, washing, light and other important functional characteristics.

In addition to appropriate choice of pigment and binders, there are several different types of auxiliaries used in pigment printing to insure a good and reproducible print (ref. 73). These include the use of dispersing agents and protective colloids to maintain stability of the different printing components, agents to retain water to avoid rapid evaporation, crosslinking agents such as urea-formaldehyde condensation products to improve fastness and binding characteristics of the pigment, and other auxiliaries such as adhesion promoters, hand modifiers, catalysts and defoamers.

Methods of printing cellulosic/polyester blends that use dyes rather than pigments have been critically reviewed (ref. 74). This has been accomplished by several strategies. One technique is to employ two-phase printing (disperse dye for polyester component and reactive component) with an alginate binder followed by a thermosol treatment to fix the disperse dye and alkaline conditions to fix the reactive dye. A new approach is to use a novel catalyst at the print paste stage (various tertiary amines) to quaternize the labile chlorine in representative reactive dyes. This modification greatly improves fixation compared to conventional alkali fixation (ref.74).

3.7.3 Printing styles

As noted earlier, there are three printing styles: direct, discharge and resist. Direct printing is used more frequently than either discharge or resist, but all three are important in commercial applications. Direct printing either by pigments or substantive dyes in the printing paste involves no processing alterations. Fixation of the dye or colorant is achieved by heat and/or steam. Cellulosic and cellulosic/ polyester blend fabrics account for almost 80% of the type of fibers that are printed, irrespective of the printing style.

Discharge and resist printing styles have been critically reviewed with regard to differentiating discharge, resist, and discharge-resist styles, to techniques used in different parts of the world, to adaptability to different types of fibers and fiber blends, and to use of typical formulations and agents (ref. 75). Realistic definitions of discharge, resist and discharge-resist print styles are as follows: (a) discharge printing consists of using a dyed ground with a printing paste containing reducing agents to destroy the dye during its fixation stage; (b) resist printing consists of applying a resist agent (one that blocks dyeing of the fibrous substrate) by chemical or mechanical ground prior to printing,methods, then subsequently applying the dye to all the fabric by padding or printing methods; and (c) discharge-resist printing consists primarily of not fully fixing the dyed then producing a discharge effect by destroying the dye chromophore with a reducing agent, by changing the pH or by a chemical reaction that changes affinity of the dye for the fiber.

Current discharge, resist, and discharge-resist methods for major fiber types and fiber blends are shown in Table 3.9. The most important types of fibers and blends are cellulosic, cellulosic/polyester and polyester. For discharge methods, the appro-

TABLE 3.9

Current discharge and resist methods used to process the major substrates (ref. 75). Courtesy of The Society of Dyers and Colourists.

Substrate	Method
(a) Cellulose	(i) Discharge styles-vat or pigment illuminants on direct, reactive or azoic dyed grounds (reducing agent methods).
	(ii) Chemical resist styles-pigment resists under reactive dyes; reactive/reactive resists using proprietary agents (normally glyoxal-bisulfite adducts.
	(iii) Physical resist styles-ethnic styles using rosin as the resist agent normally referred to as "real wax" prints. An important area for "African" prints and Indonesian "Batiks."
(b) Polyester/ cellulose blends	(i) Technically difficult-small amount processed using a discharge-resist method.
(c) Polyester	(i) Discharge styles-disperse dye illuminants on disperse dye grounds (using reducing agent methods).
	(ii) Discharge-resist--disperse dye illuminants on disperse dye ground shades (agents used include alkali, stannous chloride, sodium and zinc formaldehyde sulphoxylates).
	(iii) Resist-styles-using same approach as (ii).
(d) Wool/silk	(i) Discharge styles-acid/direct/basic dye illuminants on acid or metal complex dyed grounds (reducing agent methods).
(e) Polyamide blends	(i) Discharge styles-using acid/direct/basic dye illuminants on acid or metal complex dye grounds (reducing agent methods).
(f) Acetate/ polyamide blends	(i) Discharge styles-using disperse dyes for both illuminants and ground shades (reducing agent methods).

priate dye class (such as vat or reactive for cellulosics or disperse for polyester) and a reducing agent are essential components of the printing paste. A variety of auxiliary agents are used in discharge printing to improve the performance of reducing agents, to promote print fixation and to control the amount of water or moisture in the printing paste. The use of excimer lasers to pretreat polyester fabrics and change their surface characteristics are claimed to improve the definition and color of discharge prints (ref. 76). Resist styles are divided into chemical (those that employ agents such as glyoxal-bisulfite adducts or stannous chloride) and physical (those that use wax to block the fiber from being dyed). Many of the special ethnic printing styles (African and Batik) are based on the use of wax to impart the desired resist effects.

Reactive resist printing affords a new, versatile technique for achieving a variety of special color effects such as half-tones and double face. Several formulations are described for achieving these effects that are based on the use of two different classes of reactive dyes. Monochlorotriazines are allowed to react with the cellulosic fiber while vinyl sulfone groups can be readily prevented from dyeing by application of an appropriate resist agent (ref. 77).

3.7.4 Printing methods and machinery

Printing methods and machinery have been reviewed from the mid-1930's to the mid-1980's (ref. 78). From a historical perspective, block, conventional roller printing and hand-screen printing were the three earliest methods used. Block printing consists of applying the printing paste to wooden blocks to make an impression on the fabrics and is rarely used today in commercial textile printing operations. Conventional roller printing was the most popular method until it was basically made obsolete by the introduction of transfer printing (a variation of roller printing), rotary screen printing and foam printing. Conventional roller printing is more suitable for very fine or precise designs and rollers last longer than screens under comparable conditions of use. Roller print machines have not changed very much since the introduction of the first roller printing machine in the 1780's. They are extremely durable because they consist of cylindrical copper print rollers with an engraved design on the surface of the rollers inside a larger cast iron cylinder. The printing paste is transferred from a reservoir or trough onto the surface of the engraved rolls, and the colorant completely removed from smooth areas of the roller by a steel doctor blade. Color is transferred from the engraved rolls to the fabric surface under tremendous pressure. Passing the fabrics through a second set of rollers cause the printing paste to adhere to the unengraved surface as well as create lint; the use of a lint doctor blade to minimize contamination in the second color trough is essential for obtaining a good print. Because of the high degree of skill required to obtain good prints from engraved rolls and any faults in doctor blades, more printing faults result from rollers than result from screen printing or heat transfer printing machines.

Heat transfer printing, a new variation of roller printing, has gained commercial acceptance in a short time. It consists in the transfer of color from one surface to another, usually from paper to a textile. Heat transfer printing applications for textiles have been reviewed (refs. 79, 80). There are three major types of heat transfer printing processes: physical or melt, migration and vapor phase. The dyes or pigments on paper are transferred onto fabric by pressure and heat in the melt type. In the migration-type process, water-soluble dyes are printed onto paper with chemicals necessary to cause fixation to the fibers, then transferred to the textile by heat and pressure or by radiofrequency techniques. In the vapor phase method, disperse dyes are printed onto paper and applied to textiles by subliming them at ca. 180-220°C for 0.5 minutes. Migration methods are suitable for wool and cotton, while vapor phase methods are only suitable for polyester fibers. Figure 3.28 is a schematic of a dry heat transfer printing process that sublimes disperse dyes from printing paper onto the fabric by continuous or batch techniques and under pressure or vacuum. There are several advantages to sublimation or dry heat transfer printing: very low capital investments for machinery, use of low skill labor, a pollution-free, dry system and greater design capabilities. Disadvantages of dry heat transfer printing are: processing speeds are slow, cost of transfer paper is high for short runs and the sublimation technique has only proven commercially useful for disperse dyes onto polyester substrates. The mechanism by which dye is transferred from the paper to the fiber is depicted in Figure 3.29. As heat is supplied to the back side of the paper, dye crystals sublime into the vapor phase until they reach the surface of the fiber, and supersaturation of the dye is relieved by condensation onto the fiber surface, followed

HTP - PROCESS DESCRIPTION

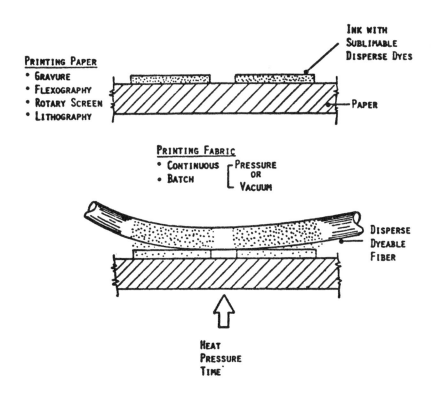

Figure 3.28. Heat transfer printing: process description (ref. 79). Courtesy of American Association of Textile Chemists and Colorists.

by diffusion into the fiber. This transport mechanism continues until final equilibrium is reached between the amount of dye on the paper and in the fabric (ref. 79). Permeable drums with a vacuum have been developed to increase the sublimation rate of the dye from the paper to the fabric and to improve penetration into the fibrous substrate. This is particularly useful for dry heat transfer printing of heavy fabrics such as carpets.

Screen prints have a brighter, well-defined appearance and allow more variation of color applied compared to those obtained by roller printing; screen prints can also be hand-produced to evaluate new designs . Screen printing has evolved from a

182

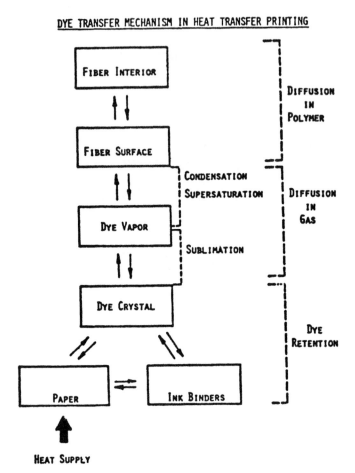

Figure 3.29.   Dye transfer mechanism in heat transfer printing (ref. 79).   Courtesy of American Association of Textile Chemists and Colorists.

specialized, labor-intensive art (hand printing or silk screening over long tables) to a highly mechanized process using flat and rotary screen printing machines. Screen printing is a process that uses a stencil or screen made of silk, polyester, polyamide or nickel mesh through which the printing paste is transferred to the fabric. The screen is coated in certain areas to block the total transfer of the paste and it is stretched over dimensionally stable frames (usually made of light metal alloys). Screens come in various mesh sizes to match the types of fabrics used and desired print designs. Outlines are traced for each color on a dimensionally stable film in which the outlined areas are filled with opaque ink. Films contain a photosensitive additive and a lacquer to promote adhesion of films to the screen; on exposure to light, only unblocked areas of the screen retain the colorant after washing.

For hand screen printing, the fabric is rolled out and fixed in place on long tables (up to 100 yards in length) with waterproof covers. The screen is placed on top of the fabric and the printing paste drawn across the screen in a direction transverse to the fabric length with a rubber squeegee blade. Only one screen is used to apply each color. As one might expect, printing faults and variability could easily be caused by inexperienced operators moving the blade at different angles and speeds. Mechanized tables were later developed for more reproducible print designs, but it was the introduction of flat bed printing that made the screen printing process amenable to large scale, commercial production. In flat bed printing, the fabric is fixed on an endless blanket and moved in an intermittent manner underneath each screen. Screens can be lowered and raised automatically and the speed and angle of the squeegee blades reproducibly controlled.

The advent of rotary screen printing machines for textiles in the last twenty-five years has made this method of printing commercially viable and competitive with heat transfer printing. This is particularly true since computer-aided design (CAD) systems can now be used to make perforated cylindrical screens of metal of precise mesh size as well as modern techniques to directly transfer the design to a lacquer-coated screen by a laser beam or laser engraving method (ref. 81).

A schematic of the fundamental aspects of a modern rotary screen printing machine is depicted in Figure 3.30. The printing paste is applied to the fabric by the combined action of: the screen rotating around its own longitudinal axis, the printing paste, the continuously moving fabric and the squeegee. The printing paste is fed from a pipe onto the inside of the screen that forms, with the aid of the squeegee, a printing paste wedge. Movement of the screen creates the dynamic pressure required to force the printing paste through the holes; this dynamic pressure is counteracted by the penetration resistance of the fabric and flow resistance of the screen. The amount of paste that is transferred to the fabric depends on the rate of discharge, paste viscosity, nature of the substrate and the contact time (ref. 82).

An important alternative method of printing fabrics using roller or screen printing machines involves the use of foam rather than printing paste. The same considerations apply to foam printing that are of importance in foam dyeing:

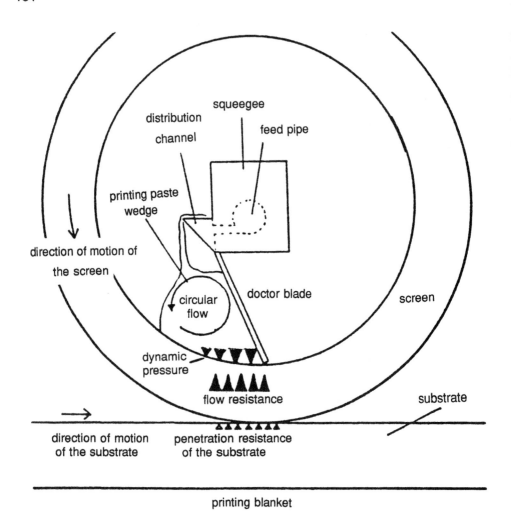

Figure 3.30. Principle of printing paste application (ref. 82). Courtesy of Melliand Textilberichte.

generation of foam, blow ratio (ratio (ratio of air to liquid) and stability of the foam as well as its rheological properties. Several advantages of foam printing as contrasted with use of printing pastes are: hardly any need for thickeners since rheology of print paste is controlled by blow ratio of foam, suitable for wet-on-wet applications because method not dependent on moisture content of fabrics, reduced energy consumption. and use of dyes and chemicals and drying time, and increased rates of printing production (ref. 83).

New and emerging methods for printing textiles include ink jet printing, spraying, electrophotography (xerography and direct electrostatic imaging). Ink jet printing has been used for some time for coloration of paper and documents, and these types of machines can be adapted to print on textiles with little or no modification (refs. 70, 84, 85). This type of printing is characterized by a noncontact method that projects drops of ink on the surface of the substrate that is printed and affords high print quality at rapid speeds (printing rates of up to 100 characters/second). The two most important types of ink jet printers are the continuous type characterized by high speed and cost, and the drop-on-demand (DOD) or impulse jet printer. The latter is more rapidly gaining commercial acceptance than the former because of its capability to be interfaced with standard CAD systems and because of its versatility for prototype/ design use. The impulse ink jet printer generates bubbles either by thermal or piezoelectric effects that are rapidly driven towards the substrate. Prints produced by this method are generally of equal high quality for paper and textile substrates.

Spray printing methods have had limited use for short production runs of upholstery fabrics and carpets, but have yet to be adapted for printing a diverse group of fabrics. Electrophotography methods are still in the experimental stage, but xerographic techniques used on paper have been adapted to textiles. The use of direct electrostatic imaging on textiles is even less developed, but also has some potential as an new method of textile printing (ref. 86). It is likely that refinements will make some of these more progressive methods of textile printing competitive with rotary screen and transfer printing methods. Discovery of new printing methods and/or adaptation of printing methods from other technologies is inevitable.

GENERAL REFERENCES FOR METHODS OF APPLYING DYES TO TEXTILES

AATCC, **Dyeing Primer**, Am. Assoc. Text. Chem. Color., Raleigh, N. C., 1981, 40 pp.

J. R. Aspland, **A Series on Dyeing**, Text. Chem. Color. 23 (10) 13-20 (1991) through 25 (9) (1993) 79-85. 21 articles occurring monthly on dye application classes.

B. Glover, **Dyes, application and evaluation**, in: J. I. Kroschwitz (Exec. ed.), Kirk-Othmer Encyclopedia of Chemical Technology, Vol. 8, 4th Ed., Wiley-Interscience, New York, 1993, pp. 672-753.

M. L. Gulrajani (Ed.), **Dyeing of Polyester and Its Blends**, Indian Inst. Tech., Hauz Khas, New Delhi, India, 1987, 378 pp.

S. V. Kulkarni, **Textile Dyeing Operations**, Noyes Publishers, Park Ridge, N. J., 1986, 387 pp.

L. W. C. Miles (Ed.), **Textile Printing**, 2d Ed., Merrow Publishers, London, 1981, 293 pp.

H. L. Needles, **Textile Fibers, Dyes, Finishes and Processes: A Concise Guide,** 2d Ed., Noyes Publishers, Park Ridge, N. J., 1986, 227 pp.

C. Preston (Ed.), **Dyeing of Cellulosic Fibers**, J. Soc. Dyer. Color. Publishers, Manchester, England, 1986, 408 pp.

J. S. Schofield, **Textile printing 1934-1984**, Rev. Prog. Color. Rel. Topics 14 (1984) 69-77.

E. R. Trotman, **Dyeing and Chemical Technology of Textile Fibers**, 6th Ed., Hodder and Stoughton Ltd., London, 1991, 587 pp.

K. Venkataraman (Ed.), **The Chemistry of Synthetic Dyes, Vols. I-VIII**, Academic Press, New York, 1952-1978.

D. R. Waring and G. Hallas (Eds.), **Chemistry and Application of Dyes**, Plenum Press, New York, 1990, 430 pp.

H. Zollinger, **Color Chemistry: Synthesis, Properties and Applications of Organic Dyes and Pigments**, 2nd Rev. ed., VCH Publishers, New York, 1991, 496 pp.

TEXT REFERENCES FOR METHODS OF APPLYING DYES TO TEXTILES

1 H. L. Needles, **Handbook of Dyes, Finishes and Textile Auxiliaries**, Noyes Publishers, Park Ridge, N. J., 1986, pp. 164-165.

2 E. N. Abrahart, **Dyes and Their Intermediates**, Chapter 4: Classification of dyes, Edward Arnold Ltd., London, 1977.

3 R. H. Peters, **Textile Chemistry, Vol. III: The Physical Chemistry of Dyeing,** Elsevier, Amsterdam, 1975, pp. 15-16, 204, 421, 550-553, 649, 723-739.

4 F. Beffa and G. Back, **Metal-complex dyes for wool and nylon - 1930 to date**, Rev. Prog. Color. Rel. Topics 14 (1984) 33-42.

5 P. A. Duffield, R. R. D. Holt and J. R. Smith, **Low chrome effluent dyeing**, Mell. Textilber. 72 (11) (1991) 938-942.

6 A. P. B. Maasdrop, **A review of the chrome mordant dyeing wool with special reference to the Afterchrome process**, SAWTRI Spec. Publ. 61 (1983) 45 pp.

7 J. Shore, **Developments in direct dyes for cellulosic materials**, Rev. Prog. Color. Rel. Topics 21 (1991) 23-41.

8 D. R. Hilderbrand, **Fiber reactive dyestuffs**, Chemtech. 8 (4) (1978) 224-228.

9 I. D. Rattee, **Reactive dyes for cellulose 1953-1983**, Rev. Prog. Color. Rel. Topics 14 (1984) 187-203.

10 M. R. Fox and H. H. Sumner, **Dyeing with reactive dyes,** in: C. Preston (Ed.), The Dyeing of Cellulosic Fibers, Dyers' Co. Publications Trust, Bradford, England, 1986, Chapter 4, pp. 142-195.

11 A. H. M. Renfrew, **Reactive dyes for cellulose: replacement of anthraquinone blues by triphenodioxazines**, Rev. Prog. Color. Rel. Topics 15 (1985) 15-20.

12 A. H. M. Renfrew and J. A. Taylor, **Cellulose reactive dyes: recent development and trends**, Rev. Prog. Color. Rel. Topics 20 (1990) 1-9.

13 K. Imada, N. Harada and T. Yoshida, **Recent developments in optimizing reactive dyeing of cotton**, Text. Chem. Color. 24 (9) (1992) 83-87.

14 R. Raue, **Cationic dyestuffs**, Rev. Prog. Color. Rel. Topics 14 (1984) 187-203.

15 U. Baumgarte, **Dyeing with vat dyes**, in: C. Preston (Ed.), The Dyeing of Cellulosic Fibers, Dyers' Co. Publications Trust, Bradford, England, 1986, Chapter 6, pp. 224-255.

16 U. Baumgarte, **Developments in vat dyes and in their application 1974-1986**, Rev. Prog. Color. Rel. Topics 17 (1987) 29-38.

17 U. Baumgarte, **Reduction and oxidation processes in dyeing with vat dyes**, Mell. Textilber. 68 (3) (1987) 189-195 & E83-86.

18 U. Baumgarte, **Reduction and oxidation processes in dyeing with vat dyes**, Mell. Textilber. 68 (4) (1987) 276-280 & E123-125.

19 C. Senior and D. A. Clarke, **Dyeing with sulfur dyes**, in: C. Preston (Ed.), The Dyeing of Cellulosic Fibers, Dyers' Co. Publications Trust, Bradford, England, 1986, Chapter 7, pp. 256-289.

20 R. A. Guest and W. E. Wood, **Sulfur dyes**, Rev, Prog. Color. Rel. Topics 19 (1989) 63-71.

21 The Colour Index, **Sulfur dyes and condense sulfur dyes,** Vol. III., Society of Dyers and Colorists, Bradford, England, and American Association of Textile Chemists and Colorists, 1971, pp.3649-3650 and 3705.

22 M. L. Gulrajani, **Disperse dyes; mechanism of dyeing**, in: M. L. Gulrajani (Ed.), Dyeing of Polyester and Its Blends, Indian Inst. Tech., Hauz Khas, New Delhi, India, 1987, pp. 21-41 and 47-71.

23 O. Annen, R. Egli, R. Hasler, B. Henzi, H. Jakob and P. Matzinger, **Replacement of disperse anthraquinone dyes**, Rev. Prog. Color. Rel. Topics 17 (1985) 72-87.

24 A. T. Leaver, B. Glover and P. W. Leadbetter, **Recent advances in disperse dye development and applications**, Text. Chem. Color. 24 (1) (1992) 18-21.

25 Imperial Chemicals Ltd. Technical Information Bulletin D1389, Manchester, England, 1973, 12 pp.

26 J. F. Dawson, **Fifty years of disperse dyes (1934-1984)**, Rev. Prog. Color. Rel. Topics 14 (1984) 90-97.

27 E. De Guzman and B. Sutton, Jr., **Dye carriers**, in: J. I. Kroschwitz (Exec. ed.), Kirk-Othmer Encyclopedia of Chemical Technology, Vol. 8, 4th Ed., Wiley-Interscience, New York, 1993, pp. 533-541.

28 V. A. Shenai, R. K. Panadikar and D. N. Panda, **Carrier dyeing of polyester fibers. A review**, Text. Dyer & Printer 19 (2) (Jan. 15, 1986) 13-18; 19 (3) (Jan. 29, 1986) 17-20.

29 W. Ingamells and M. N. Thomas, **The mechanism of carrier dyeing**, Text. Chem. Color. 16(3) (1984) 55-60.

30 V. Ravichandran, M. Wilde, R. Gadoury, A. T. Lemley and S. K. Obendorf, **Some observations on the effects of selected dye carriers on poly(ethylene tere-phthalate)**, Text. Chem. Color. 19 (11) (1987) 35-37.

31 S. E. Akaoui, **A heat and mass transfer study - analysis of the continuous processes for the coloration of polyester fabric. I. Heat transmission to the fabric system - fabric, fiber and dye particles**, J. Appl. Poly. Sci. 27 (12) (1982) 4693-4711.

32 S. E. Akaoui, **A heat and mass transfer study - analysis of the continuous processes for the coloration of polyester fabric. II. Modelling and characteristics of the (polyester) fabric heating**, J. Appl. Poly. Sci. 27 (12) (1982) 4713-4733.

33 H. Herlinger, W. Koch and R. Lim, **Dye transfer via the vapour phase as a basis for a new dyeing method**, Mell. Textilber. 68 (3) (1987) 341-346 & E149-152.

34 K. V. Datye and A. A. Vaidya, **Mass coloration of nylon-6**, Colourage (April 7, 1983) 7-11.

35 H. J. Sohn, **Mass coloration of PP fibres and filaments**, Chemiefasern/Textileindustrie 32/84 (10) (1982) 712-717.

36 M. Wishman, J. Leininger and D. Fenton, **The coloration of polypropylene fibers**, in: R. B. Seymour and T. Cheng (Eds.), Advance in Polyolefins, Plenum Press, New York, 1987, pp. 523-535.

37 U. Einsele, **New methods for the dyeing of polypropylene. First report: dyeing with the aid of binders**, Mell. Textilber. 72 (10) (1991) 847-851.

38   H. Herlinger, G. Augenstein and U. Einsele, **Innovative methods for the dyeing of polypropylene. 2nd report: the influence of the dyestuff constitution and auxiliaries**, Mell. Textilber. 73 (9) (1992) 737-742.

39   R. H. Peters and L. W. C. Miles, **New developments in textile coloration**, in: K. Venkataraman (Ed.), The Chemistry of Synthetic Dyes, Vol. VIII, Academic Press, New York, 1978, pp. 136-148.

40   T. Takagishi , **The role of solvents in dyeing of synthetic fibers**, Sen-i Gakkaishi 36 (10) (1980) P401-410.

41   K. Solanki, **Solvent dyeing of polyester fiber**, Text. Dyer Printer 21 (20) (1988) 17-24b.

42   F. O. Harris and T. H. Guion, **A new approach to solvent dyeing with non-ionic dyes**, Text. Res. J. 42 (1972) 626-627.

43   A. J. Cofrancesco, **Dyes, natural**, in: J. I. Kroschwitz (Exec. ed.), Kirk-Othmer Encyclopedia of Chemical Technology, Vol. 8, 4th Ed., Wiley-Interscience, New York, 1993, pp. 784-809.

44   W. J. Marshall, **Dyeing of cellulosic/polyester blends**, in: C. Preston (Ed.), The Dyeing of Cellulosic Fibers, Dyers' Co. Publications Trust, Bradford, England, 1986, Ch. 9, pp. 320-358.

45   R. B. Chavan and A. K. Jain, **Dyeing of polyester/cotton blends**, in: M. L. Gulrajani (Ed.), Dyeing of Polyester and Its Blends, Indian Inst. Tech., Hauz Khas, New Delhi, India, 1987, pp. 205-236.

46   P. W. Leadbetter and A. T. Leaver, **Disperse dyes - the challenge of the 1990s**, Rev. Prog. Color. Rel. Topics 19 (1989) 33-39.

47   A. I. Wasif, S. K. Chinta, S. K. Kaul and S. Chandra, **A novel method of dyeing polyester/cellulosic blended fabrics - Part II**, Colourage 38 (12) (1991) 15-21.

48   N. E. Hauser, **Dyeing cotton/wool blends**, Text. Chem. Color. 18 (3) (1986) 11-18.

49   P. G. Cookson, **The dyeing of wool/cotton blends**, Wool Sci. Rev. 62 (1986) 3-119.

50   A. Laepple and R. Jenny, **Färbenvarianten für sport- und freizert-artikel aus PAC/CO**, Textilveredlung 23 (7/8) (1988) 248-252.

51   W. Haertl, **Dyeing of acrylic/wool and acrylic/cotton yarns and its influence on the processing of the yarns**, Mell. Textilber. 70 (1989) 354-357 & E150-151.

52   R. B. Chavan and A. K. Jain, **Dyeing of polyester/wool blends**, in: M. L. Gulrajani (Ed.), Dyeing of Polyester and Its Blends, Ind. Inst. Tech., Hauz Khas, New Delhi, India, 1987, pp. 237-258.

53   S. M. Doughtery, **The dyeing of polyester/wool blends**, Rev. Prog. Color. Rel. Topics 16 (1986) 25-38.

54 G. B. Guise and I. W. Stapleton, **Washfast dyes for wool: II. Synthesis and dyeing properties of model isothiouronium dyes**, J. Soc. Dyer. Color. 91 (1975) 259-264.

55 C. C. Cook, **Aftertreatments for improving the fastness of dyes on textile fibres**, Rev. Prog. Color. Rel. Topics 12 (1982) 73-89.

56 J. A. Bone, P. S. Collingshaw and T. D. Kelly, **Garment dyeing**, Rev. Prog. Color. Rel. Topics 18 (1988) 37-46.

57 G. R. Turner, **What's new in carpet dyeing**, Text. Chem. Color. 19 (8) (1987) 16-18; 19 (9) (1987) 69-72.

58 A. Hoehner, **Carpet dyeing: Continuous versus exhaust. Can there be the same quality?**, Text. Chem. Color. 22 (5) (1990) 29-33.

59 T. L. Dawson, **The dyeing of fibre blends for carpets**, Rev. Prog. Color. Rel. Topics 15 (1985) 29-36.

60 C. Smith, **Coloration and dyeing of nonwovens**, Text. Chem. & Color. 22 (2) (1988) 25-29.

61 G. H. J. van der Walt and N. J. J. van Rensburg, **Low-liquor dyeing and finishing**, Text. Prog. 14 (2) (1986) 60 pp.

62 M. Capponi, A. Flister, R. Hasler, C. Oschatz, G. Robert, T. Robinson, H. P. Stabelbeck, P. Tschudin and J. P. Vierling, **Foam technology in textile processing**, Rev. Prog. Color. Rel. Topics 12 (1982) 48-57.

63 R. G. Kuehni, J. C. King, R. E. Phillips, W. Heise, F. Tweedle, P. Bass, C. B. Anderson and H. Teich, **Dyes, application and evaluation**, in: M. Grayson (Exec. ed.), Kirk-Othmer Encyclopedia of Chemical Technology, Vol. 8, 3rd Ed., Wiley-Interscience, New York, 1979, pp. 280-350.

64 A. A. Vaidya, **Dyeing of polyester in loose fibre and yarn forms**, in: M. L. Gulrajani (Ed.), Dyeing of Polyester and Its Blends, Indian Inst. Tech., Hauz Khas, New Delhi, India, 1987, pp. 132-152.

65 R. S. Bhagwat, **Developments in beam dyeing machines**, in: M. L. Gulrajani (Ed.), Dyeing of Polyester and Its Blends, Indian Inst. Tech., Hauz Khas, New Delhi, India, 1987, pp. 153-166.

66 H. Ellis, **Changes in machinery design for continuous dyeing and printing: 1930 to 1983**, Rev. Prog. Color. Rel. Topics 14 (1984) 145-156.

67 J. D. Ratcliffe, **Developments in jet dyeing**, Colourage 28 (14) (July 2, 1981 Supplement) 1-15.

68 R. S. Bhagwat, **Developments in jet dyeing machines**, in: M. L. Gulrajani (Ed.), Dyeing of Polyester and Its Blends, Indian Inst. Tech., Hauz Khas, New Delhi, India, 1987, pp. 167-185.

69 R. S. Bhagwat, **Jet dyeing machines**, Colourage 39 (2) (1992) 71-83.

70 L. A. Graham, **Ink jet systems for dyeing and printing of textiles**, Text. Chem. Color. 21 (6) (1989) 27-32.

71 R. J. Thomas, **Dyeing mechanisms in textile printing**, Colourage 23 (22A) (Oct. 22, 1981) 3-15.

72 P. Bajaj, R. B. Chavan and B. Manjeet, **A critique of literature on thickeners in textile printing**, J. Macro. Sci. Rev. Macro. Chem. Phys. C24 (3) (1984) 387-417.

73 W. Schwindt and G. Faulhaber, **The development of pigment printing for the last 50 years**, Rev. Prog. Color. Rel. Topics 14 (1984) 166-175.

74 J. R. Provost and H. Burchardi, **Printing on polyester/cellulose blends - a review**, Mell. Textilber. 73 (1) (1992) 55-59.

75 J. R. Provost, **Discharge and resist printing - a review**, Rev. Prog. Color. Rel. Topics 18 (1988) 29-36.

76 D. Knittel, W. Günther and E. Schollmeyer, **Discharge printing on laser-irradiated polyester fabrics,** Text. Praxis Intl. 45(1) (1990) 46-47.

77 H. D. Opitz, **Reactive resists - new techniques for textile printing**, Mell. Textilber. 71 (10) (1990) 775-781.

78 J. S. Schofield, **Textile printing 1934-1984**, Rev. Prog. Color. Rel. Topics 14 (1989) 69-77.

79 E. J. Gorodony, **Analysis of what the textile printer should know about heat transfer printing**, in: Symposium Proceedings AATCC: Textile Printing: An Ancient Art and Yet So New, AATCC, Research Triangle Park, N. C., 1975, pp. 83-97.

80 C. E. Vellins, **Transfer printing**, in: K. Venkataraman (Ed.), The Chemistry of Synthetic Dyes, Vol. VIII, Academic Press, New York, 1978, pp. 191-220.

81 J. Hilden, **Engraved rotary printing screens**, Intl. Text. Bull. (Dyeing/Printing/Finishing Ed.), 1st Quarter 1989, 35-36.

82 U. Ferber and J. Hilden, **Application of printing paste in rotary screen printing: a comparative analysis of the magnetic roller and doctor blade techniques**, Mell. Textilber. 70 (9) (1989) 705-709; E303-305.

83 B. N. Bandopadhyay, **Chemical wet processing of textiles through foam application, Part I**, Colourage 29 (9) (May 6, 1982) 3-8.

84 B. Smith and E. Simonson, **Ink jet printing for textiles**, Text. Chem. Color. 19 (8) (1987) 23-29.

85 T. L. Dawson, **Jet printing**, Rev. Prog. Color. Rel. Topics 22 (1992) 22-31.

86 W. W. Carr, W. C. Tincher, F. L. Cook and W. R. Lanigan, **New techniques for fabric printing**, Proc. AATCC 1986 (1986) 176-177.

Chapter 4

FABRICS WITH IMPROVED AESTHETIC AND FUNCTIONAL PROPERTIES

4.1 INTRODUCTION

Techniques for improving one or more aesthetic and functional properties of textiles encompass several major approaches that are far more comprehensive and diverse than employing only conventional textile finishing methods. Such improvement can be achieved by three major approaches: (a) dimensional stabilization, (b) surface modification and (c) physicochemical and chemical techniques. Dimensional stabilization of fabrics is usually achieved during processing by various mechanical and/or thermal techniques. Modification of the surface of fibers in fabrics by a variety of techniques (chemical, physical, exposure to high energy sources) changes the appearance, aesthetics and surface characteristics of fabrics. Physicochemical and chemical methods include coating, exposure to high energy sources, microencapsulation and application of chemical finishes and polymers to improve one or more textile properties for specific end uses.

4.2 DIMENSIONAL STABILIZATION BY MECHANICAL AND THERMAL METHODS

4.2.1 Mechanisms of stabilization

The purpose of dimensional stabilization is to set the fibers and yarns in various fabric constructions so that the textile will not shrink or alter its dimensions during use and/or refurbishing. Dimensional stabilization of fibers and of fabrics has been accomplished by a basic knowledge of the effects that heat and mechanical forces have on textile structures during processing. Such stabilization occurs by application of techniques such as tenter frame control, heat and/or steam setting and compressive shrinkage. All fibers in fabrics are subject to the constraints imposed by fabric geometry and construction. However, several setting mechanisms can occur between and within fibers that will ultimately influence the dimensions and form of the fabric during processing. The different types of mechanisms of set have been reviewed by Hearle (ref. 1) and are shown in Table 4.1.

Mechanisms that are relevant to obtaining set between fibers by mechanical and thermal methods during processing include frictional resistance and fiber-to-fiber bonding (such as hydrogen bonding).

TABLE 4.1

Mechanisms of set (ref. 1). Courtesy of Merrow Publishing Co.

A. *Between fibres*

    1. Friction, related to normal forces in structure.

                weak (hydrogen bonded).
    2. Bonding
                strong (adhesive).

    3. Impregnation with matrix.

B. *Within fibres*

    1. Chain stiffness.

    2. Temporary crosslinks   sensitive to: temperature,
       (strong dipole; hydrogen bonds).   moisture, stress

    3. Crystallization.

    4. Chemical crosslinks.

    5. Impregnation with matrix.

Mechanisms that are important for obtaining set within fibers during processing include chain stiffness, intermolecular hydrogen bonding and crystallinity. The energy change between fibers (δE_b), derived from frictional resistance, is always positive. The relationship of δE_b to the frictional resistance is given by the equation: $\delta E_b = (\mu N).|x_b|$, where μ is the coefficient of friction, N is the normal load between fibers and $|x_b|$ is the relative fiber displacement. If the normal load for a system is known (as in the forces at crossover points for woven or for knitted fabrics), then the displacement in the textile structure may be measured or predicted (ref. 1).

Hydrogen bonding is perhaps the most important type of bond between and within fibers that influences setting and stabilization. Cellulosics (hydroxyl groups) and polyamides and wool (amide groups) have strong hydrogen bonding networks. However, these types of bonds may be broken or disrupted at elevated temperatures with hot water or with steam. Thus, desetting or relaxation of these types of fibers may occur in the presence of sufficient moisture near the boiling point of water. Conversely, setting occurs when these fibers or fabrics derived from them are dried to remove water, and new hydrogen bonds are formed.

Chain stiffness and changes in crystallinity occur in synthetic fibers such as polyester and polypropylene and in natural fibers such as cellulosics and wool and thus may be advantageously used to set or deset these types of fibers. Heating thermoplastic or elastomeric fibers increases their flexibility while subsequent cooling of these fibers causes chain stiffness or set to the desired dimensions. Similarly, crystallization of many fibrous polymers occurs on heating with an accompanying increase in irreversible set. Depending on the end use of the textile, wet or dry heat is employed to effect various degrees of temporary or permanent set.

A useful definition of setting was suggested by Lindberg (ref. 2):

> Setting is a change in the stress-free form of a fiber resulting from a deformation and subsequent chemical or physicochemical treatments prior to release of the fibre.

The degree of set may also be described in mathematical terms after a combined setting and desetting process by $G \cdot (1 - H)$. G = the permanent deformation or degree of set that may also be represented by $(l_2 - l_o)/(l_1 - l_o)$ (l_o = the original fiber length; l_1 = the fiber length at a deformation energy of ε_o and l_2 = the fiber length at a deformation energy of ε_1. H = the degree of deset that is given by the expression $1-(l_3 - l_o)/(l_2 - l_o)$, where l_3 = the fiber length at a deformation energy of ε_2 (ref. 2).

The states at which the fibers are set have a profound influence on the mechanical properties and dimensional stability of the fabric. The non-swollen fiber state is important in the dry heat setting of synthetic fibers. Cohesive setting occurs in the drying of hygroscopic fibers such as cotton and wool. The swollen fiber state is important in the bisulfite setting of wool, the wet setting of polyamides and in the mercerization of cotton (ref. 2).

4.2.2 Heat setting for stabilization

Since natural and synthetic fibers undergo different responses to heat and to moisture, procedures for the dimensional stabilization differ from one fiber type or fiber blend to another. This is particularly true concerning desirable setting temperatures and the order in which setting is conducted in a textile processing sequence. This is readily observed by comparing the sequences for different types of fibers and even for the same types of fibers with different constructions (Table 4.2).

The tenter frame is the apparatus most frequently used to heat set all types of fabrics. It consists of three parts: (a) an entry frame and chain whereby the fabrics

TABLE 4.2

Position of heat setting in the finishing sequence of fabrics (adapted from ref. 3). Courtesy of Merrow Publishing Co.

Filament fabrics			Polyester/wool fabrics		Polyester/cellulosic fabrics	Textured polyester fabrics		
Woven	Warp-knit		Clear-cut finish	Milled finish		Warp-knit		Weft-knit
scour	(a)	(b)	pre-set	pre-set	desize	(a)	(b)	
heat set	heat set	scour	scour dye	scour dye	scour mercer.	heat set	dye	dye
dye stenter	scour dye	heat set dye	brush & crop	brush & crop	optional bleach	dye	dry	dry
finish	stenter	finish	heat set condition	singe soap mill	heat dye set. heat dye set	dry	heat set	heat set
			decatize press	heat set crop condition	crop singe			
				decatize press	compr. shrink.			

are dimensionally stabilized by clips or pins before heating, (b) a heating zone that varies in temperature from 140-230°C depending on the fiber type and fabric construction, and (c) a delivery system that holds the fabric under minimum tension while cooling to retain the desired dimensions (ref. 3). Because of the importance of the tenter frame, there have been several studies to measure and predict the appropriate or required amounts of moisture and heat necessary to stabilize fabrics dimensionally in it as well as to conserve energy during processing. The latter topic will be discussed in the section on **energy conservation**.

The importance of the relationship between moisture regains for fibers and the dewpoint or derived wet-bulb temperature at a given air temperature in an automatic control system was demonstrated with natural and synthetic fibers in an experimental drying apparatus. The suggestion was made that the desired moisture regain of a fabric exiting a tenter frame could be obtained by measuring only the air temperature and moisture content inside the tenter frame (ref. 4). However, this assumption was disputed on the basis that tenter frame air is rarely in equilibrium with fabric (ref. 5).

Dry heat setting is of predominant importance for thermoplastic synthetic fibers such as polyester and polyamide. Figure 4.1 schematically illustrates how these fibers are set by a heat process (ref. 6). Many fundamental studies have attempted to elucidate the important factors that influence heat setting of polyester fibers and fabrics. In a series of papers by Gupta and co-workers (refs. 7-10), it was demonstrated that free-annealed fibers have more perfect crystals and a more distinct phase separation between crystalline and amorphous regions to cause morphological reorganization than fibers held at constant length during heating. Moreover, it was observed that recovery of free-annealed fibers is dominated by their amorphous phase while both crystalline and amorphous phases affected recovery behavior of fibers held at constant length. An experimental apparatus was devised to study the kinetics of crystallization of polyester fibers in microseconds to simulate the transformations occurring under commercial processing conditions (ref. 11). It is likely that more sophisticated theoretical models and processing simulations will be discovered to predict and optimize desirable heat setting conditions for thermoplastic fibers.

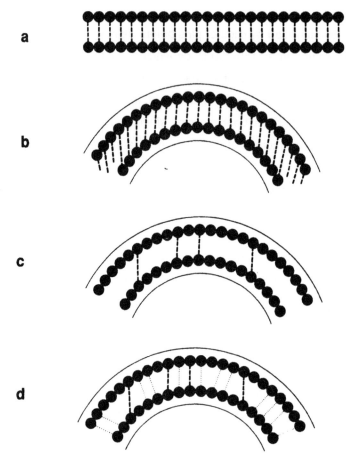

Figure 4.1. Diagrammatic representation of the heat-setting process in thermoplastic polymers: (a) bonds between polymer molecules in a fibre; (b) distorted fibre;intermolecular bonds are strained and act to restore the original molecular (and hence fibre) configuration; (c) heat treatment destroys many of the original intermolecular bonds; residual bonds provide weak restoring forces; (d) new (dotted) bonds form on cooling; these act to 'set' the molecular structure in a new configuration and overcome the weak restoring effect of residual original bonds (ref. 6). Courtesy of The Textile Institute.

4.2.3 Steam and chemical setting

Because of their hygroscopic nature and unique structural features, cellulosic and particularly proteinaceous fibers, require additional mechanical finishing steps and careful shrinkage control in their drying and heat setting. Shrinkage of hygroscopic fibers may occur by three mechanisms: hygral expansion, relaxation shrinkage and felting. Hygral expansion is a reversible change in fibers, yarns and/or fabrics that occurs due to changes in relative humidity. Relaxation shrinkage is an irreversible process that occurs due to longitudinal contraction of fibers in yarns when they are immersed in water or when yarns or fabrics undergo extension during processing. Felting is a progressive type of fiber shrinkage that occurs on repeated agitation and immersion of textiles in water (ref. 12). The first two types of shrinkage are known to occur in both cellulosic and proteinaceous fibers, while felting shrinkage is unique to the wool fiber. Thus, careful control is required when fabrics composed of natural fibers are exposed to heat and moisture to achieve the desired degree of dimension stabilization and set. It is not surprising then that stabilization of natural fibers and occasionally cellulosic/polyester blends are achieved by steam and/or chemical methods.

Over the last fifty years, there have been many studies to determine the mechanisms by which wool fibers undergo setting. The amounts of time that extended wool fibers are exposed to steam determines whether they contract (supercontraction) or expand (permanent set) relative to their original length. The nature of the original disulfide crosslinks and their rearrangement to more stable crosslinks has continually been a subject of disagreement and controversy. The formation of new lanthionine and lysinoalanine crosslinks, thiol-disulfide interchange reactions, and hydrogen bonding rearrangement without reformation of disulfide crosslinks are three of the more current proposals advanced to explain the mechanisms of setting of wool. Critical evaluations of the mechanisms as they relate to actual processing conditions, indicate the importance of both hydrogen-bonding rearrangement and thiol-disulfide interchange reactions (latter mechanism shown in Figure 4.2). However, the formation of lanthionine and lysinoalanine crosslinks is currently not considered important in the setting mechanism for wool (ref. 13).

A formula has been derived that accurately describes and predicts the degree of set or contraction of wool fibers after they are extended and treated with either hot water or aqueous urea/bisulfite solutions. The set in the fibers S may be defined in linear viscoelastic terms by:

$$S = 1 - (F_R/\varepsilon_o)/Y_i \phi(t, 0)) \tag{4.1}$$

where F_R = the force of fiber retraction, ε_o = the setting strain, $\phi(t, 0)$ = the relative force decay and Y_I = the initial stiffness of the fiber. Although all three of these variables influence the degree of set, it is important to measure the stiffness value carefully since it usually increases with the time of release of fiber tension (ref. 14).

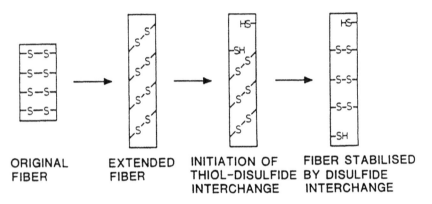

| ORIGINAL
FIBER | EXTENDED
FIBER | INITIATION OF
THIOL–DISULFIDE
INTERCHANGE | FIBER STABILISED
BY DISULFIDE
INTERCHANGE |

Figure 4.2. The thiol-disulfide interchange mechanism for wool setting (ref. 13). Courtesy of Marcel Dekker Publishers.

There are several wet and dry techniques used for dimensional stabilization of wool fabrics at various stages of processing. These include crabbing, milling or fulling, wet flat setting and pressing. Crabbing is usually the first setting process to which the wool fabrics are subjected. In this process, the fabric is wound under tension into boiling water onto a roller to prevent uneven shrinkage from occurring in later finishing steps. Milling is also performed at earlier stages of wool processing; it is also described by the terms fulling or felting. This process is used to entangle wool fibers in fabrics progressively by the combined application of heat, friction and moisture. There are three types of liquor that are used in milling processes: grease (solutions of sodium carbonate or sodium carbonate/bicarbonate to form soaps from natural wool grease), soaps or synthetic detergents, and acid (pH 2-4) to increase the milling rate of heavy fabrics (ref. 13).

Wet flat setting and pressure methods (such as decatizing in the presence of steam) are setting processes used in the later stages of wool processing and finishing. Wet flat setting is usually conducted with reducing agents (sodium bisulfite or monoethanolamine sulfite) and hydrogen-bonded breaking agents (urea). Although the setting may be conducted at room temperature, it requires an extra wet processing step, and thus increases the cost of finishing the fabric. Gaseous treatment of wool fabrics with sulfur dioxide has been demonstrated to be comparable to wet flat setting, provided the wool fabrics have high water content (at least 30%). This SO_2 treatment is most effective at pH 4 and 5.5, the same acidity maxima for wet flat setting with bisulfite or similar agents (ref. 15).

Pressure or mechanical methods that do not use chemical treatment are generally more economical for setting of wool fabrics. Wet flat setting has been essentially replaced by pressure decatizing and continuous decatizing processes (ref. 12). Decatizing is a process for improving the hand and appearance of fabrics (primarily wool) by winding them tightly around a perforated roller that has steam passed through it. In pressure decatizing, penetration of steam into the fabric is achieved by creating a vacuum in the apparatus before introduction of steam; the pressure and time of treatment can be varied to achieve different degrees of permanent set. Continuous pressure decatizing machines now allow fabrics to be stabilized by short contact times by engineering modifications of the decatizing apparatus.

In addition to the many stabilization processes for wool fabrics, sanforization™ is a very important process for dimensional stabilization of fabrics composed of cotton and/or cotton/synthetic fiber blends. It also uses pressure equipment that causes compressive shrinkage in the fabric by feeding it, along with a rubberized backing, through a roller under tension. The fabric is fed in such a manner that when the tension is released from the rubber backing, the fabric undergoes the desired amount of shrinkage and is thus dimensionally stabilized.

4.3 PRODUCTION OF SURFACE EFFECTS IN FABRICS

There are a variety of techniques that are employed to cause changes in the surface characteristics of fabrics. Although such modifications were until recently used for altering the smoothness, roughness, luster and other aesthetic fabric qualities, they are now also used to change adhesion, dyeability and wettability of fibrous surfaces. Modification of textile surfaces is accomplished by the application of heat, pressure, moisture, electrostatic, mechanical forces, or exposure to high energy sources or a combination of these techniques.

4.3.1. Calendering

Luster, smoothness and other surface effects on fabrics are usually achieved by calendering. The fabric is inserted between heavy rollers varying in hardness, surface speed and temperature to achieve the desired fabric surface effects. Variations of calendering include embossing, schreinerizing and frictional calendering. Embossing is the production of a fabric pattern in relief by passing the fabric through a calender containing a rigid, heated and engraved metal roller and a soft or elastic roller. Schreinerizing is the technique for producing a lustrous fabric by passing it through rollers, one of which is engraved with very fine lines at an angle to the fabric being processed. In frictional calendering, fabrics are passed through a calender in which a rigid, highly polished and heated metal roller rotates at a higher surface speed than a soft roller to produce glazed or lustrous surface effects.

The latest developments in calendering have been reviewed from several perspectives (refs. 16-18). Variables that produce different surface effects may be achieved by using metal or filled, soft rollers, by varying the surface temperature, nip pressure, web speed and fabric humidity and different bowl combinations. In addition to high luster finishes and embossing effects, dull or matte finishes can be produced by using filled rollers (such as cotton or wool/paper) (ref. 16). The Nipco roll system shown in Fig. 4.3 is considered a modern prototype calendering system with much versatility for producing a variety of desired surface effects in fabrics.

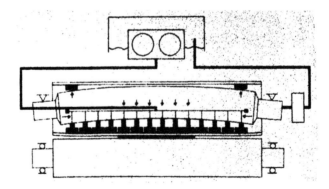

Figure 4.3. The Nipco roll system features individual hydraulic pistons mounted across the width and facing the nip roll (ref. 16).

With the exception of calendering, many of the processes to alter the surface characteristics of fabrics were initially developed for wool, then later adapted for use with cellulosic and synthetic fibers. An excellent example is the process of raising, in which open width fabric is passed between rotating rollers containing teasels (called teaseling or gigging) or containing fine wires (called napping). Raising produces a layer of protruding fibers on the fabric surface. This protrusion forms a pile, nap, or cover on the fabric surface, changes the softness or hand of the fabric and its bulk, and/or changes the appearance of the weave, design or color of the fabric. There are four different raising procedures (Fig. 4.4) that may be achieved by varying the direction of rotation of the raising roller and the drum. In addition, raising effects can be controlled or varied by using different geometries of wires, speed of the fabric, drum speed, raising energy of the pile/counterpile, fabric tension and associated warp angle and center-driven fabric lifting devices (ref. 19).

After the fibers are raised, the fabric may be subjected to other processes to alter its surface characteristics further. Shearing generally follows raising, and produces fibers of uniform length by application of a helical blade to remove long fibers from

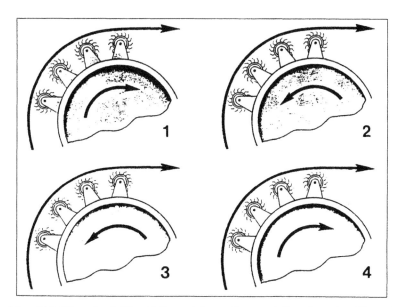

Figure 4.4. Direction of rotation of the raising rollers and the drum in different raising routines: (1) pile/counterpile/normal; (2) pile/counterpile/contra; (3) full felting, normal; (4) full felting, contra (ref. 19). Courtesy of ITS Publishing Co.

the fabric surface. Sanding may also be employed after raising, and consists of breaking protruding surface fibers by contact with abrasive paper containing carborundum or other suitable abrasives. After shearing or sanding, brushing may be employed to remove loose, short fibers that are present on the fiber surface.

4.3.2 Flocking and other surface modifications

Flocking is a surface modification used for imparting a velvet, chamois or other special appearance to a smooth textile substrate by various techniques. The basic process consists of affixing short fibers to a fibrous substrate that has been previously coated with an adhesive. The earliest flocking techniques employed mechanical or electrostatic methodologies; however, other techniques such as pneumatic and transfer flocking or combinations of mechanical/electrostatic or pneumatic/electrostatic are also currently practiced. The loose flock may be any major fiber type, but cotton, rayon, polyamide, polyester, acrylic and modacrylic are the most frequently used fibers. Figure 4.5 depicts the important steps and options in a typical

Figure 4.5. Schematic of a flocking process: (a) adhesive application, full surface; (b) adhesive application, partial for pattern flocking; (c) flocking; (d) drying - setting; (e) cleaning; (f) finishing, e.g., printing or embossing (ref. 20). Courtesy of ITS Publishing Co.

flocking process (ref. 20). In this schematic, the adhesive may be applied to the full surface or partially applied to produce a surface pattern. Flocking is normally conducted by a combination of electrostatic and mechanical processes whereby the fibers are brush-metered through a screen for uniformity. The adhesive is usually set by hot air drying, then unbound fibers removed to clean the surface. The flocked fabric can then be finished, dyed or printed.

The general requirements to produced acceptable flocked fabrics have been critically reviewed and discussed (refs. 20-22). Loose fibers that serve as the source for the flocked surface should be uniform in length, have good resiliency and possess acceptable separating power and flow capacity. Adhesives used to attach the flock to the substrate should have high electrical conductivity, long pot life, certain rheological characteristics, dry quickly for fixation, and be resistant and fast to rubbing, washing, ageing and ultraviolet light. Maximum fiber-packing density in electrostatic flocking processes has been calculated using a Monte Carlo technique. It was determined that the maximum fiber-packing density (expressed as fraction of substrate surface area occupied) was 53%, but far less than the value of 91% expected for close-packed circles (ref. 23).

The exposure of fiber surfaces to high energy sources has progressed from initial demonstrations of altered surfaces with enhanced aesthetic and functional properties to current commercial interest to replace conventional wet chemical and mechanical finishing techniques with this high-tech methodology. The pioneering work in this area was conducted only with natural fibers (wool, mohair and cotton) from the early 1960's to the mid 1970's. Relevant surface properties that were improved included enhanced dyeability, dimensional stability, adhesion and wettability. Kassenbeck was the first to recognize many of these improvements in wool (tensile strength, dimensional stability and dyeability) after it was exposed to glow discharge (ref. 24). Thorsen later showed that cotton yarns had improved strength and related properties when exposed to a corona discharge (ref. 25). Pavlath critically reviewed the early work in this area, including his own studies on the shrinkproofing on wool after exposure to low temperature plasma (ref. 26).

In the late 1970's, several investigators demonstrated that synthetic as well as natural fibers had improved dyeability and wettability when exposed to laser irradiation (ref. 27) or to plasma treatments in different atmospheres (ref. 28). More detailed and current studies of textiles exposed to high energy sources are discussed more comprehensively in the next section with regard to enhancement of functional as well as aesthetic properties by physicochemical methods. Other processes that modify the surface characteristics of fabrics to change their appearance and aesthetics include crushing, creasing, burn out and pleating. These however are unimportant and will not be discussed further.

4.4 METHODS FOR IMPROVING FUNCTIONAL PROPERTIES

4.4.1 Overview

Numerous functional fabric properties may be improved by using suitable chemical and/or physicochemical techniques. Application of textile finishing agents is still the primary approach for achieving such property improvement. However, physico-chemical methods (such as coating and exposure to high energy sources) are becoming commercially attractive and have begun to supersede conventional wet chemical methods. Also, there has been an increasing use of polymers to improve several functional fabric properties rather than use of a simple chemical agent to improve only one functional property. Properties of fabrics or fibrous substrates are altered to improve their performance with regard to various physical, chemical and/or biological agents and influences. Thus, they either enhance or change the ability of the textile to withstand such influences. In doing so, this imparts health, safety, and/or aesthetic and practical attributes to the fabric for the benefit of its wearer and/or user. Such property modifications include those that provide resistance to wrinkling, fire, soils and stains, water, microorganisms and insects, light, heat and cold, shrinkage, air pollutants and chemical agents, weathering, mechanical changes caused by abrasion, pilling and various types of deformation and buildup of static charge. Each of these modifications is briefly described in Table 4.3 with regard to its purpose, the property or properties that it imparts and its general applicability or suitability to different fiber types.

TABLE 4.3

Improvement of functional textile properties

Property imparted to fiber	Purpose	Application/suitability to fiber type
Wrinkle resistance or resiliency	Aesthetic for user	Cellulosics and their synthetic fiber blends
Flame retardancy	Safety of user	Most natural and synthetics
Absorbency	Aesthetic and functional for user	Usually hydrophobic synthetics to impart hydrophilicity
Soil release	Aesthetic for user	Primarily synthetics and their blends
Repellency (soil, stain and water)	Primarily aesthetic for user	Primarily synthetics, sometimes natural fibers
Resistance to microorganisms and insects	Functional protection of fiber and/or health of user	Primarily cellulosics for fiber protection, wool for insect protection, all fibers for health and medical aspects
Resistance to UV light, heat and pollutants	Aesthetic for colorfastness, functional to prevent fiber degradation	Most natural and synthetics, especially polyamides
Heat and cold (thermal comfort)	Aesthetic and functional to user	All natural and synthetic fibers
Shrinkproofing	Aesthetic and functional to user	Primarily wool, but cellulosics and synthetic also treated
Abrasion and wear resistance	Functional protection of fibers	Primarily cellulosics and their blends
Resistance to pilling	Aesthetic for user	High-tenacity synthetic fibers and their fiber blends
Resistance to static charge	Functionality and safety to user	Primarily synthetic fibers

There are various methods of applying these functional finishes to textiles that are either primarily physicochemical or chemical in their approach. Physicochemical methods include alteration of the fiber surface by high energy, coating, insolubilization or deposition and microencapsulation. Chemical methods include graft and/or homo-polymerization, crosslinking and resin treatment, covalent bond formation and ion-exchange/ chelation. It is also possible to combine a physicochemical technique (such as laser surface treatment) with chemical methods (such as graft poly-merization) to improve functional textile properties. Table 4.4 contains examples of each type of technique, the property or properties imparted and representative chemical species (when applicable) used for such property enhancement.

4.4.2 Physicochemical methods

Insolubilization by simple deposition of active chemical agents on or in fabrics is the oldest method to alter or enhance functional textile properties. This insolubilization is usually achieved by applying the finishing agent to the textile from non-aqueous solvents or from emulsions. Excess solution is then removed from the fabric by passing it through pressure rollers, then allowing it to either air dry or be dried and/or cured by the application of heat. Early flame retardants such as ammonium salts and borax/boric acid are in this category. Quaternary ammonium salts are excellent antistatic agents in low quantities when they are deposited on fibrous surfaces after evaporation from aqueous solution. Aluminum salts were used with soaps to impart water repellency to fabrics. However, most fabrics modified by this method have little or no durability to laundering or even leaching. With the advent of regenerated and synthetic fibers, it became possible to incorporate the desired finishing agent into wet or melt spinning baths before the fibers are transformed into yarns and fabrics. Heat and ultraviolet stabilizers and antimicrobial agents are frequently incorporated into fiber spinning baths. Although this method of insolubilization confers better durability than simple surface deposition, other methods are usually used to confer moderate to long term durability of the active agent on fabrics.

Another physicochemical technique, fabric coating, has evolved from a relatively crude and minor method using drying oils, natural rubber and cellulose derivatives as coating materials to a versatile and important method to improve functional textile properties by employing a variety of synthetic and natural elastomers and thermoplastics as coating materials. A conventional and useful definition of a coated fabric is given by Teumac (ref. 29):

> A coated fabric is a construction that combines the beneficial properties of a textile and a polymer. The textile (fabric) provides tensile strength, tear strength and elongation control. The coating is chosen to provide protection against the environment in the intended use. A polyurethane might be chosen to protect against abrasion or a polychloroprene (Neoprene) to protect against oil.

A broader and more current definition of a coated fabric considers it to be a composite consisting of fiber and matrix phases (coating containing fillers and acting as a binder) and the interface between the fiber and polymeric coating (ref. 30).

TABLE 4.4

Techniques for improving functional textile properties [a]

Method or technique	Example of property imparted	Representative agent or scientific principle
Insolublization or deposition	Resistance to UV or sunlight	Cr^{3+}
Coating	Water repellency	Poly(vinyl chloride)
Plasma treatment	Wettability and soil release	Modification of polyester
Microencapsulation	Antimicrobial activity	Quaternary ammonium salts
Graft and/or homo-polymerization	Soil release	Poly(acrylic acid)
Crosslinking agents or resins	Wrinkle resistance or resiliency	Cyclic urea-formaldehyde resins
Covalent bond formation	Flame retardancy	Phosphorylation with phosphoric acid-urea
Ion-exchange/chelation	Antistatic	Amphoteric surfactants (N-oleyl-N,N-dimethylglycine)

[a] The first four methods listed are primarily physicochemical while the latter four methods are chemical.

Coated fabrics may be classified according to the fiber types or fibrous substrate methods by which the coatings are applied, types of coatings (conventional impermeable or breathable/microporous) and specific polymers used to coat the substrate. The major types of fibers that are frequently used for producing coated fabrics include cotton, cotton/polyester blends, polyester, wool and nylon. However, advances in surface modification of fibers and coating science and technology now make it possible to coat any type of fiber or fibrous substrate, such as glass, polypropylene and even polyethylene. All major types of fabric constructions (woven, knit and nonwoven) are amenable to coating processes. Coated, woven fabrics are primarily used in applications that require high strength, while coated, knitted fabrics are used in applications that normally require moderate strength but good elongation. Coated, nonwoven fabrics were initially used only in specialty applications such as the production of shoe liners. However, they are now employed in a variety of diverse applications.

Coating techniques and processes are numerous and have been comprehensively reviewed and described (refs. 31-35). Major types of coating methods are: direct, indirect or transfer coating and those in which the polymer is melted onto the fabric surface. The simplest of the direct coating processes on fabrics or fibrous webs involve use of a doctor blade, a device that controls the thickness of the coating on the fabric surface. Some of the more commonly used doctor blade configurations are shown in Fig. 4.6. A gap is set between the doctor blade or knife and the roll (steel or rubber-coated) in a typical knife-over-roll configuration [Fig. 4.6 (a)]. Non-uniform surfaces and high speed coating are not amenable to this method due to non-uniform coating thickness and high shear rates. A floating knife arrangement [Fig. 4.6 (b)] is used on porous materials to avoid "strikethrough" of the coating; the degree of coating penetration is thus controlled by tension of the web or substrate. Knife-over-blanket configurations [Fig. 4.6 (c)] are usually employed when the substrate cannot be subjected to much tension. In knife-over-plate configurations [Fig. 4. 6 (d)], the fabric should be fairly uniform to pass over a steel plate; otherwise, the fabric may tear if it caught between the support plate and the doctor blade.

Roll coating techniques are more complex and expensive than knife coating methods and normally make use of engraved rolls. Engraved-roll or gravure coating methods are more precise because metered amounts of a coating may be applied to the fibrous substrate. Engraved rolls can be employed in direct, offset, reverse roll coating and many other coating techniques. Figure 4.7 is a schematic of one popular type of reverse roll coating configuration. This type is a nip-fed drive that allows the coating to be uniformly applied to the fabric on a continuous conveyer belt or applied directly onto the fabric in a conventional manner.

Rotary screen coating is a process that is analogous to commercially practiced dyeing processes. Figures 4.8 (a) and 4.8 (b) show the two most frequently used processes of this type. In both processes, doctor blades press the coating medium

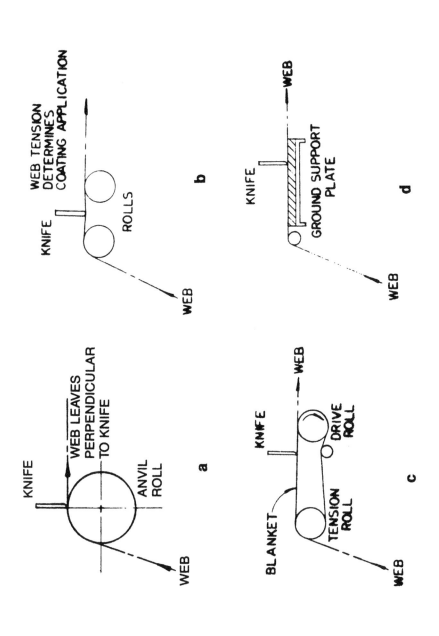

Figure 4.6. Doctor blade or knife configurations for coating fabrics: (a) knife-over-roll; (b) floating knife; (c) knife-over blanket; and (d) knife-over-plate (adapted from ref. 34). Courtesy of Technomic Publishing Co.

212

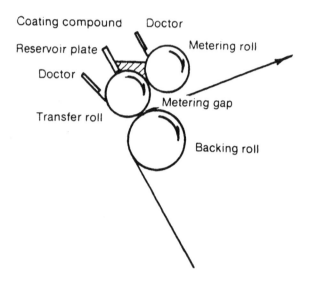

Figure 4.7. Nip-fed reverse-roller coater (ref. 33). Courtesy of Technomic Publishing Co.

I.

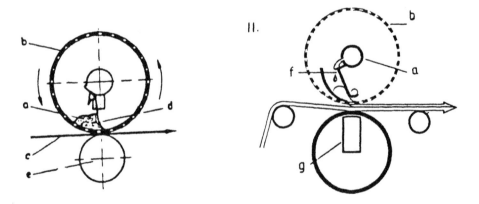

Figure 4.8. The rotary screen coating principle. I. Stork rotary squeegee principle. II. Zimmer rotary screen magnetic squeegee principle. (a) paste supply; (b) rotary screen; (c) textile material; (d) squeegee; (e) counter-roller; (f) magnetic squeegee and (g) magnet system (ref. 35). Courtesy of Stork Brabant B.V. for 4.8 (a) and of Johannes Zimmer Co. for 4.8 (b).

through perforations in the screen wall onto the fibrous substrate. With the Stork machine [Fig. 4.8 (a)], this is achieved mechanically with a squeegee, while with the Zimmer machine [Fig. 4.8 (b)], pressure is applied by an electromagnetic squeegee device. Transfer coating is analogous to transfer printing and has become an increasingly popular coating process. This type of process normally involves the use of release paper (as shown in the schematic in Fig. 4.9) in which the coating is transferred from the paper to the textile and bonded by lamination.

1 Release Paper   3 Drying Units   5 Knife Over Roller 7 Laminating Unit
2 Reverse Roll Coater 4 Cooling Rollers 6 Textile Let Off   8 Coated Fabric Take Up

Figure 4.9. Transfer coating line with two coating heads (ref. 36). Courtesy of American Association of Textile Chemists and Colorists.

All polymers used for coatings are not amenable for use in solutions or dispersions, and thus are applied by several melt coating methods that take advantage of the thermoplastic nature of such polymers. The two most common methods are extrusion coating and melt calender coating. In extrusion coating, the thermoplastic polymer is extruded onto the fabric, cooled on a subsequent roller, than bonded by pressure. In melt calendering, the thermoplastic polymer is oriented into a film that is transferred to the fibrous substrate (sometimes with the use of release paper shown in Fig. 4.10). After embossing, the coated fabric is cooled in a manner similar to that employed in extrusion coating. Another common technique for applying thermoplastic polymers to fabrics is powder coating, which is applied to textiles and to other substrates by fluidized bed, electrostatic, spraying or spreading methods. A schematic of a powder coating technique for textiles is shown in Figure 4.11.

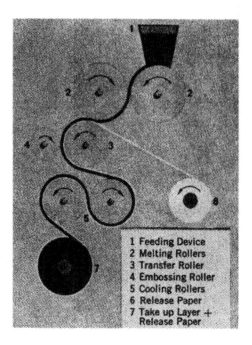

Figure 4.10. Melt calendering (ref. 36). Courtesy of American Association of Textile Chemists and Colorists.

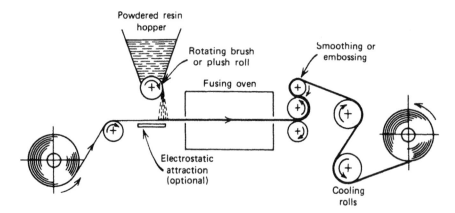

Figure 4.11. Powdered-resin coating of textile substrate (ref. 31). Courtesy of John Wiley and Sons Publishers.

Some of the many other coating techniques and methods used are: cast coating, foam coating, spray coating, curtain coating and rod coating. All of the coating methods described thus far produce fabrics that are significantly impermeable to both liquid and water vapor. Such conventionally coated fabrics are not especially comfortable when they are used in protective clothing and apparel. However, in the last ten to fifteen years, a series of "breathable" coated fabrics have been produced by coagulation coating processes similar to the one shown in Figure 4.12. Typically, polyurethanes are coagulated or precipitated in an aqueous bath containing a water-soluble solvent such as N,N-dimethylformamide (DMF). On removal of the solvent, micropores are formed in the coating that allow water vapor to be transmitted while retaining resistance to liquid water. The first of many microporous, breathable fabrics was Gore-Tex, a thin polytetrafluoroethylene membrane that is laminated to a conventional fabric by use of an adhesive. However, most poromeric or microporous coatings are still based on conventional or modified polyurethanes due to cost, availability and ease of processing.

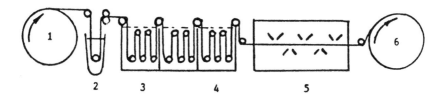

Figure 4.12. Diagram of a coagulation coating line: (1) substrate let-off; (2) impregnation; (3) coagulation bath; (4) washing bath; (5) drying; and (6) batching (ref. 35). Courtesy of ITS Publishing Co.

The types of polymers used to coat fibrous substrates range from elastomers (natural and synthetic rubbers such as polyisoprene, polychloroprene, styrene-butadiene, acrylonitrile-butadiene, isobutylene-isoprene; polysiloxanes; polysulfides and polyurethanes) to thermoplastics (polyvinyl chloride, polyvinylidene chloride, polyethylene, polypropylene, polystyrene, polyester, polyamide and segmented polyurethanes). Each of these polymers or copolymers produces coated fabrics with several improved functional properties suitable for specific applications or end uses. However, polyvinyl chloride and polyurethane are used much more extensively than other classes of polymers for coating fabrics.

Polyvinyl chloride may be applied to fibrous substrates as a fine dispersion or in a solution (**plastisol** or **organosol**) or by various melt coating processes. This polymeric coating gives fabrics with good resistance to weathering, ageing, fire and is water-impermeable. A **plastisol** is a resinous form of a polymer (in this case, polyvinyl chloride) of very fine particle size that is dispersed in a plasticizer such as dimethyl phthalate and which then forms a film when heated so that the powdered polymer is absorbed by the plasticizer with no solvent being evaporated. An **organosol** is a **plastisol** whose viscosity is reduced by diluting a solution of it with polar or non-polar solvents. Plastisols or organosols of polyvinyl chloride are usually applied to fabrics by knife, reverse roll or transfer coating techniques. A typical formulation for plastisol coating contains an emulsified polyvinyl chloride resin with particle-size ranges of 0.5-10 microns, a plasticizer, fillers, pigments, heat and ultraviolet stabilizers and other specialty additives for specific end uses. The desirable properties of plasticizers have been discussed in detail (ref. 37) with regard to their inherent properties (such as flash point, acidity, stability to heat and hydrolysis), compatibility with other ingredients, stability to oxidation and humidity, performance characteristics for desired properties (such as color, mechanical integrity, abrasion, low toxicity) and permanence of the coating (to weathering, leaching, microbes, chemicals).

Polyurethanes are an extremely versatile class of polymers that are frequently applied as coatings onto textiles. They are used in applications that require fabrics that are lightweight and possess good abrasion resistance. Moreover, most of the new microporous coatings that give improved comfort to fabrics are based on some type of polyurethane coating. Figure 4.13 exemplifies the diversity of textile constructions and fabric coating processes available for application of polyurethanes to textiles. However, many of the variations and combinations shown are also applicable to most thermoplastic and elastomeric polymers used for coating fabrics.

From a chemical perspective, polyurethanes may be classified as one or two components. Two-component polyurethanes are formed by reacting a linear OH-terminated polyester of diethylene glycol adipate with toluene diisocyanate (Figure 4.14). These types of polyurethanes are normally coated onto textiles by direct, reverse roll or transfer coating methods (ref. 36). One-component polyurethane systems shown in Figure 4.14 consist of alternating soft (polyester or polyether) and hard polyurethane segments. If excess diisocyanate is employed, then the resultant

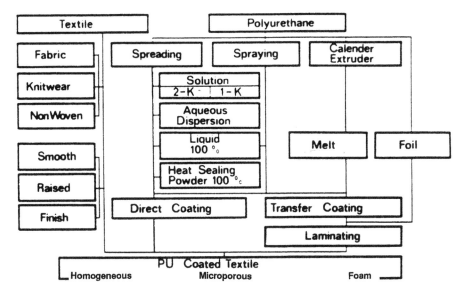

Figure 4.13. Coating of textiles with polyurethanes (ref. 36). Courtesy of American Association of Textile Chemists and Colorists.

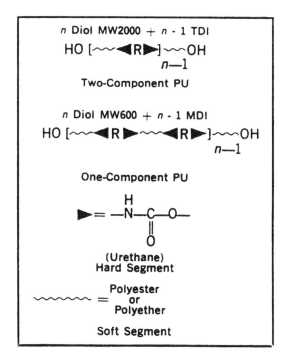

Figure 4.14. One- and two-component polyurethanes (ref. 36). Courtesy of American Association of Textile Chemists and Colorists.

terminal isocyanate groups can be further reacted with amines to form terminal urea end groups. If water is present, some of the isocyanate will automatically be converted to amines (releasing CO_2) that can then subsequently form internal urea hard segments. One-component polyurethane systems may be coated onto fabrics by the same techniques as those described for coating two-component polyurethane systems onto fabrics. However, these one-component systems may also be applied to fabrics by techniques used for applying thermoplastic polymers, namely calendering, extrusion and lamination.

A specialty application that consumes large quantities of polymeric coatings is the backcoating of carpets. A variety of synthetic elastomers and thermoplastics are applied to carpets. While natural polymers such as starches, cellulose derivatives and rubber are still used as coating materials, most current coatings are synthetic latexes such as styrene-butadiene rubber, acrylic, poly(vinyl acetate) or poly(vinyl chloride) (ref. 38). Most coatings are cured onto textile substrates by heat (e.g., convective drying). However, there has been some commercial interest in non-thermal curing methods such as radiation curing or related techniques.

The last and perhaps most promising physicochemical method of applying finishing agents to textiles is microencapsulation of the agent and affixing of the microcapsule to the fibrous substrate. The agent then diffuses out of the microcapsule by a controlled release mechanism to impart the desired properties to the substrate. This technique has been used extensively for the delivery of drugs and of pesticides through plastics and films as substrates. However, it has only recently begun to be used for fabrics and fibrous substrates and is discussed in a critical review of this topic (ref. 39).

An example of its utility for fabrics is the microencapsulation of catalysts and/or durable press resins for fabrics. This approach avoids premature curing of the resins and makes it possible to use unstable resin systems with active catalysts to obtain wrinkle-resistant and smooth-drying fabrics (ref. 40). Another example is the diffusion of antibacterial agents or pesticides from textile materials laminated or sandwiched between two protective plastic barriers (ref. 41). Microencapsulation technology is increasingly being used in Japan to produce modified fabrics having novel functional properties. Microcapsules containing perfume have been attached to fabrics that release a fragrance, and cholesteric liquid crystals have been microencapsulated into coated fabrics to produce swimsuits and skiwear that change color when the temperature rises or falls (ref. 42). Other ways in which microencapsulation techniques have been used to modify fibers are to produce long thin, fibers that effectively act as microcapsules and release fragrances and to produce acrylic fibers with different pore sizes that can release antibacterial agents at different rates (ref. 39). It is probable that future developments and innovations will occur in the attachment of microcapsules to fibrous substrates to produce various improved properties. In some cases, hollow fibers themselves can function as modes of encapsulating fragrances and/or antimicrobials.

4.4.3 Polymerization and chemical methods

Many of the modern functional finishes for textiles use, in some manner, the physical and/or chemical attachment of a polymer to a fibrous substrate. Various types of polymerization processes have been conducted in the presence of fibrous substrates. Many types of polymerization involve the formation of a chemical bond between the polymer and the fiber; however, physicochemical methods such as fabric coatings are also used in conjunction with chemical attachment to improve the durability of a textile finish. One way of classifying the various types of polymerization processes that are employed in textile finishing is: (a) homopolymerization, (b) graft polymerization and (c) network polymerization and/or crosslinking. Schematics for each of these types of polymers are depicted in Figure 4.15.

Homopolymerization generally denotes the formation of a linear polymer from a monomer or an oligomer, and may be accomplished in many ways. This includes addition polymerization (normally using various vinyl monomers), condensation or step-growth polymerization and ring-opening polymerization. Polymerization may be initiated by free radical, cationic, or anionic species or by chain-transfer agents. Chemical classes of polymers used and techniques for affixing these homopolymers to fibers are extremely diverse and depend to a large extent on the suitability of the polymer to impart the desired property to the textile or fibrous substrate. Some examples of homopolymers used in this regard are: polyurethanes, polydimethyl-

220

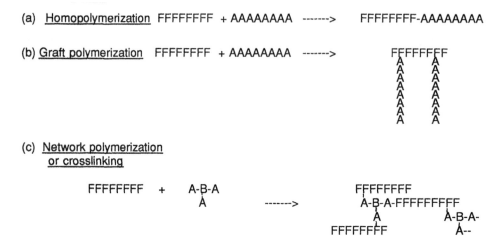

(a) <u>Homopolymerization</u> FFFFFFFF + AAAAAAAA ------->   FFFFFFFF-AAAAAAAA

(b) <u>Graft polymerization</u> FFFFFFFF + AAAAAAAA ------->

(c) <u>Network polymerization</u>
     <u>or crosslinking</u>

Figure 4.15. Representative polymerization processes in the presence of fibrous substrates (F = fiber site, A = monomer and repeating unit of polymer and ABA = trifunctional monomer).

siloxanes and polyacrylic acids. Emulsion polymerization and interfacial polycondensation are only two of the many possible techniques for forming these homopolymers and applying them to fabrics.

Graft polymerization is a special type of copolymerization in which a reactive site is created on the main polymer chain. In this instance, the fiber is the main chain that is available for chemical attachment or grafting of a polymerizable monomer. The formation of a reactive site in the polymer backbone of the fiber is usually achieved by generating free radicals at that site (FFFFFFFFFFFFFFFFFFFFF).

Free radicals can be produced by a variety of techniques. Chain transfer between the monomer (A) and the polymer backbone (F) can be caused by using a free radical initiator such as a hydroperoxide (°OOH). The fiber may be subjected to ionizing radiation before contact with the monomer (pre-irradiation technique) or in the presence of the monomer (mutual irradiation technique) under vacuum in an inert atmosphere such as nitrogen to avoid termination or inhibition of polymerization due to the presence of oxygen. Free radicals can also be produced by exposing the fiber and the monomer to ultraviolet light in the presence of a photosensitizer or by using redox agents such as Ce^{+4}.

Ionic graft polymerization, using strong bases as anionic initiators or carbocations as cationic initiators, is less frequently used to graft monomers onto fibers. Even in the most carefully designed and controlled systems, it is common to have homopolymer formed from the monomer in addition to grafting of the monomer onto the fiber. Homopolymer is difficult to extract from the fiber with various solvents since long chain molecules migrate from the fiber pores only with difficulty, but extensive extraction will indicate the amount of polymer remaining on the fiber that is due to grafting. Monomers commonly grafted onto fibers include acrylonitrile, styrene, acrylic acid and methyl methacrylate.

Crosslinking of fibers can occur when difunctional agents react with adjacent polymer chains in the fiber to form stable bridges or links between the chain. It may also occur when multifunctional agents such as resins form network polymers that are physically occluded in the fiber and/or chemically bound to the fiber at various sites. Cyclic urea-formaldehyde compounds and various resins with appropriate catalysts are agents most frequently used to produce wrinkle-resistant cellulosic fabrics. The smooth drying properties and resiliency are attributable to various types of crosslinks and network polymers in the fibrous matrix. A more detailed discussion follows later in the section on durable press and wrinkle-resistant textile finishes. Other crosslinking agents and substances used in functional finishes include the free radical crosslinking of polysiloxanes and acid- or base-catalyzed crosslinking of epoxides.

In contrast to the various polymerization methods described, covalent bond formation imparts functional properties to textiles by chemical modification of reactive fiber sites to form fibrous derivatives. Since cellulosic and proteinaceous fibers have reactive hydroxyl, amino acid and sulfhydryl groups, these fibers have been more extensively investigated with regard to their chemical alteration than have synthetic fibers. Synthetic fibers such as acrylics or polyesters do not have functional groups capable of undergoing the variety of chemical reactions that are possible with the naturally-occurring and regenerated cellulosic and protein-based fibers.

The conversion of fibrous cotton cellulose (Cell-OH) into fibrous cellulose acetate (Cell-O-C-CH$_3$) to impart rot resistance and into cellulose phosphate (Cell-O-P-OH) to impart flame resistance are two representative examples of chemical modification of textiles to impart improved functional properties. The shrinkproofing of wool by oxidation of its disulfide linkages to sulfonic acid linkages with chlorine water (RS-SR $\xrightarrow{[O]}$ 2RSO$_3$H) is illustrative of how protein fibers have been chemically modified to improve their functional properties.

Formation of sites on the fiber to promote ion-exchange or to effect chelation with various metals is sometimes employed to impart functional properties to fibers. These include: antimicrobial activity, resistance to soils and stains, resistance to static electricity and to heat and/or ultraviolet light. The ion-exchange sites or metal chelates may be chemically attached to fibers by derivatization or by some form of polymerization. Alternatively, ionic substances or metal chelates may be deposited, insolubilized in, or coated onto the textile substrate. The use of copper 8-hydroxy-quinolinate to impart mildew resistance or quaternary ammonium salts ($R_4N^+X^-$) to impart antistatic properties to textiles are two examples that involve respectively, metal chelates and ion-exchange chemistry for affixing functional agents to fibers.

4.5 IMPROVEMENT OF SPECIFIC FUNCTIONAL PROPERTIES

4.5.1 Overview

The improvement of one or more functional properties of textiles has been accomplished by using one or more of the techniques described in Table 4.4. Earlier methods of improving functional properties focused on simple chemical modification of functional groups of fibers and deposition of active agents on fabrics. Finishing techniques evolved later that used the incorporation of a polymer or copolymer in the fibrous material to impart functional property improvement. Current techniques use one or more physicochemical and chemical approach to produce textile materials with improved functionality in one or several properties. The more important functional property improvements for textiles will now be discussed in this context.

4.5.2 Crease resistance or resiliency

Crease-resistant textiles can now be produced by using knit fabrics rather than woven fabrics. Furthermore, fabrics derived from synthetic fibers have excellent crease resistance that can be obtained without application of finishes. However, effective wrinkle-resistant finishes have been developed for natural textile fibers. These developments may be arbitrarily divided into four periods of activity. The first period was that in which the concept of imparting resiliency to fabrics was discovered when they were treated and cured with appropriate resins. In the second period, commercialization of such finishes followed and there were numerous investigations that embodied a variety of chemical and physicochemical approaches for improving wrinkle resistance and associated fabric properties. In the third period, studies were focused on the elucidation, prediction and explanation of the differences in reactivity and efficiency of the various chemical agents and catalysts to achieve smooth drying fabric properties. Structural features of fibers that enhanced or contributed to this property were also extensively investigated. The last period has been concerned with development of crease-resistant fabric finishes that liberate little or no formaldehyde and in identifying conditions under which formaldehyde is liberated.

Foulds, Marsh and Wood demonstrated in the late 1920's that crease resistance could be imparted to cellulosic fabrics with thermosetting resins (such as phenol-formaldehyde and urea-formaldehyde) (ref. 43). Since that time numerous publications and patents on this subject have been issued. These investigations and inventions range from the description of novel reagents to impart crease resistance to the mechanistic aspects of optimum performance of fabrics to recover from deformations.

As shown in Table 4.5, a variety of finishing techniques and crosslinking agents and resins has been employed to impart dry and/or wet crease recovery to cellulosic fabrics. For imparting dry crease recovery, the reaction is usually acid-catalyzed, and involves application of formaldehyde or di- or polyfunctional linear and cyclic ureas, carbamates, amides or melamines (containing $-N-CH_2OH$ or $-N-CH_2OR$ groups) to the fabric by a pad-dry-cure process. This produces smooth drying fabrics with crease recovery angles of 260-300° (360° is the optimum). Catalysts include ammonium salts, zinc chloride, zinc nitrate and magnesium chloride as well as organic acids.

The acid-catalyzed crosslinking of cellulosics under moist and wet conditions is achieved at lower pH values than under dry conditions by using mixtures of inorganic and organic acids or strong inorganic acids (such as HCl). When cyclic N-methylol

TABLE 4.5

Resin-finishing processes and the basic crosslinking agent used (adapted from ref. 44). Courtesy of BASF.

Finishing Process	State of Cellulosic Fibre: Reaction	Constitution of Basic Crosslinking Agent
Dry crosslinking (elevated temp.) Conventional finish Flash curing Permanent Press Combined finishing and dyeing	De-swollen; Acid catalysis	$(X = CH_2OR: R = H, Alkyl)$ $X–HN–CO–NH–X$ $X–HNCO–(CH_2)_n–CONH–X$ $X–HNC(O)O(CH_2)_nO(O)CNH–X$ $X–O(CH_2)_nO–X$ $ROCONX_2 \quad RCONX_2$ triazine and imidazolidinone structures
Moist crosslinking (various temps.) Moist-batch W 111 Speed-dry	Partially swollen; Special acid catalysts	Special types of imidazolidinone structures
Wet crosslinking (low temps.) Alkaline (a) Acid (b)	Swollen; Alkaline or acid catalysis	$S^+(CH_2–CH_2–OSO_3^-)_3 \, 2Na^-$ (a) $CH_2–CH–CH_2Cl$ (epoxide) (a) imidazolidinone structure (b)
Multi-stage crosslinking (various temps.) Bel-O-Fast Teb-X-Cel Ambivalent crosslinking	De-swollen and swollen; Acid and alkaline catalysis	1st Stage 2nd Stage $CH_2–CH–CH_2Cl$ (epoxide) $\longrightarrow$ imidazolidinone $S^+(CH_2–CH_2–OSO_3)_3 \, 2Na \longrightarrow$ $CH_2=CH–C(O)–NH–X \longrightarrow$

ureas are used, this produces fabrics with good wet and moderate dry crease recovery (moist conditions) or with only good wet crease recovery (wet conditions). In multi-stage crosslinking, alkaline catalysts are used in the first step (swollen state to impart wet recovery) with compounds such as inner salts of sulfonium betaines, epoxides, and N-methylolacrylamide to cause nucleophilic attack by cellulosic hydroxyl groups. Cyclic ureas with acid catalysts are then employed in the second step (deswollen state) to impart dry and wet crease recovery, respectively to the fabrics. The model proposed by Marsh (ref. 43) and verified by other investigators satisfactorily explained how fibers recover from deformations in the collapsed state to impart dry crease recovery and in the swollen state to impart wet crease recovery (Figure 4.16). Solid lines in the diagram represent crosslinks between cellulosic chains and the dotted lines represent hydrogen bonds between chains. When the fiber is crosslinked in the collapsed state the hydrogen bonding network is retained and contributes to dry recovery. When the fiber is crosslinked in the wet state the hydrogen bonding network is disordered and only wet recovery is afforded by covalent crosslinks between polymer chains.

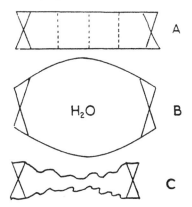

Figure 4.16. Diagrammatic representation of A, the dry crease recovery of the collapsed fibre; B, wet crease recovery of the swollen fibre; C, the absence of crease recovery in B when dried (ref. 43). Courtesy of Chapman and Hall Publishing Co.

Numerous methods for imparting crease recovery to fabrics and explaining how this may be accomplished have contributed to commercialization of these types of finishes. However, there are also several adverse effects produced on application of these agents to the fiber, namely increased stiffness, loss of tensile strength and abrasion resistance, and in some instances, retention of chlorine bleach that produces yellowing when the fabrics are subsequently ironed or dried at high temperatures.

Losses in tensile strength and abrasion resistance led to many theoretical and practical studies to improve these deficiencies. Such improvements were generally achieved by using polyester/cellulosic blends to offset the loss of mechanical properties in 100% cellulosics, by resin-treating fabrics woven from tension-mercerized yarns possessing a high degree of crystallinity and crystallite orientation, and by using finishing techniques such as core crosslinking (ref. 45). This latter technique allows for a more uniform distribution of crosslinks in the fibers and thus, a more uniform distribution of stresses and deformations in the treated fibers. Although it was originally believed that chlorine retention in crease-resistant fabrics was primarily due to the conversion of free N-H groups into N-Cl groups (chloramines), it was later discovered that the rate at which this conversion occurred was the most important factor in selecting a crosslinking agent when good resistance to chlorine was required (ref. 44). Cyclic six-membered ureas (propylene ureas), carbamates, and amides containing -N-CH$_2$OH or -N-CH$_2$OR groups usually react slowly with Cl$_2$ or OCl$^-$, and are thus recommended for use when chlorine resistance is important (ref. 44).

The mechanisms by which representative N-methylol and N-alkoxymethyl compounds react with cellulosics under acidic conditions and the relationship between the chemical constitution of these compounds and their reactivity towards cellulose have been reviewed by Petersen (refs. 46,47). Condensation of various -N-CH$_2$OH and -N-CH$_2$OR compounds with cellulose obeys first-order kinetics and follows the laws of general acid catalysis (Figure 4.17) with the rate-determining step being the cleavage of the C-O bond to form a carbocation.

Determination of the rate constants of a variety of urea-formaldehyde compounds and related structures afford a basis of classification of these substances with regard to their reactivity with cellulosics and their hydrolytic stability after reaction with the fiber (ref. 46). Reactivity of -N-CH$_2$OR compounds are governed by the nature of the leaving group and the chemical constitution of groups on the nitrogen atom.

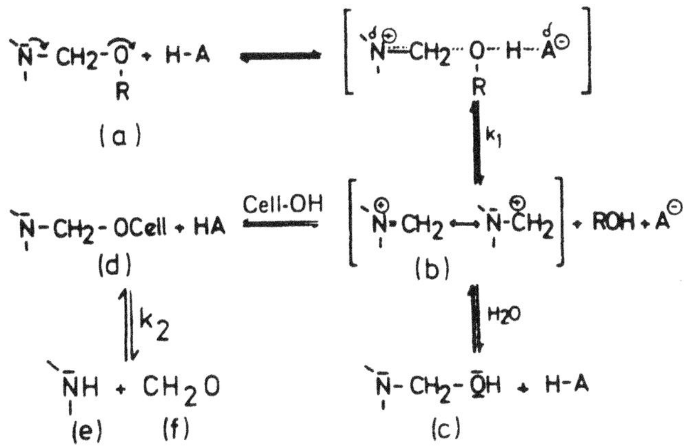

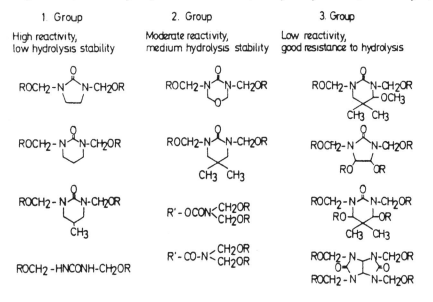

Figure 4.17. Reaction mechanisms for the general acid catalysis of the reaction of *N*-alkoxymethyl compounds (ref. 46). Courtesy of Marcel Dekker Publishers.

Substances that are highly reactive but only moderately stable after reaction with the fiber are shown in the first group in Figure 4.18. These include methylol and alkoxymethyl derivatives of urea and of ethylene and propylene ureas. Carbamates, amides, urones and some substituted propylene ureas are moderately reactive with cellulose with good stability to hydrolysis (second group in Figure 4.18). 4,5-Dihydroxy- and 4,5-dialkoxyethylene ureas and 4,6-dihydroxy- and 4,6-dialkoxy-5,5-

Figure 4.18. Reactivities of cross-linking agents (ref. 46). Courtesy of Marcel Dekker Publishers.

dimethylpropylene ureas containing -N-CH$_2$OH and -N-CH$_2$OR groups exhibit low reactivity (third group in Figure 4.18). However, they have excellent resistance to hydrolysis once they are covalently bound to cellulose.

The liberation of free and/or bound formaldehyde from wrinkle-resistant fabric has come under more stringent guidelines for health and safety aspects in occupational safety in textile processing and garment manufacture. Therefore, methods have been devised to reduce the level of formaldehyde in finished fabrics or use other types of reagents that do not contain formaldehyde, particularly in textile finishing plants. Petersen (ref. 47) considers the concern of formaldehyde exposure not to be particularly problematic, since it is ubiquitous and there have been enough human case studies to indicate that it does not have any long term adverse health effects at low levels of exposure. Nevertheless, there have been an increasing number of studies on determining the factors that influence the release of formaldehyde from treated fabrics. These factors include: free formaldehyde concentration in solution of the crosslinking agent or padding bath, concentration of the crosslinking agent and its resistance to hydrolysis, moisture content and temperature of the finished fabrics, degree and uniformity of fiber crosslinking and the nature and amount of catalyst employed (ref. 48). Approaches for reducing free formaldehyde in finished fabrics include an afterwash, addition of formaldehyde acceptors to the finishing bath or spraying them on finished fabrics (e.g., use of urea). Newer finishing agents (such as polyfunctional carbamates or 4-hydroxy- and 4-alkoxypropylene ureas) are also useful for reducing free formaldehyde in fabrics (ref. 48). Sello has recently reviewed (ref. 49) the varieties of crosslinking agents that are not formaldehyde-based yet still impart resiliency and wrinkle-recovery to cotton fabrics. These include: dimethylol-dihydroxyethylene urea (DMDHEU) etherified with polyhydric alcohols such as diethylene glycol and 2,3-propanediol and amide-glyoxal adducts such as the reaction product of dimethyl urea or of ethylene urea with glyoxal. Although these types of adducts generally mitigate the problem of the level of free formaldehyde, the resultant fabrics usually have lower resiliency and poorer smooth drying properties.

An alternative approach that has been gaining increasing popularity is the use of a non-formaldehyde reactant to crosslink cellulose-containing fabrics. Welch has critically reviewed recent investigations in this area, including his unique reaction of 1,2,3,4-butanetetracarboyxlic acid (BTCA) with cotton in the presence of acid catalysts to produce durable press properties that are retained even after 100 launderings (ref. 50). Although the BTCA is extremely effective as a new non-formaldehyde crosslinking agent, its cost is currently much higher than conventional crosslinking agents that are generators of low amounts of free formaldehyde. Other representative structures and agents described in this review are acetals, glyoxal, glyoxal-glycol adducts, glyoxal-amide adducts, other dialdehydes and phosphorylation (ref. 50). Although good durable press properties may also be attained with these systems, most suffer from disadvantages such as pronounced strength loss and yellowing after treatment. Commercialization of alternative agents that do not contain formaldehyde will ultimately depend on the cost-benefit analysis of potential adverse health and safety effects relative to expense and availability of replacement agents for the inexpensive and readily available formaldehyde-based resins.

Cotton fabrics undergo predictable and sizable losses in tensile strength and other mechanical properties irrespective of whether a crosslinking agent is non-formaldehyde or a conventional urea-formaldehyde structure. These losses range from 30 to 60% of the original strength of the fabric. When the cellulose is crosslinked, it inherently becomes less able to distribute stresses and strains imposed on the fiber during mechanical deformation. Thus, newer methods of obtaining durable press, smooth drying and resiliency of fabrics now use synthetic polymers that react in the presence of cellulosic fibers under mild conditions. Two examples of this approach are (a) the reaction of polyethylene glycols with DMDHEU in the presence of cotton to produce fabrics with excellent durable press properties, only moderate strength loss and outstanding increase in abrasion resistance (ref. 51) and (b) reaction of heat-reactive, self-crosslinkable copolymers (based on polysiloxanes and polyalkylenes) in the presence of cotton to produce fabrics with durable press fabrics that also have increases in tensile strength and abrasion resistance (ref. 52). Both polymer systems have the advantage of having low free formaldehyde (for the crosslinked polyols) or no formaldehyde (self-crosslinkable copolymers). Further advances in producing smooth drying and resilient fabrics will more likely result from

the discovery and application of specific polymers to cellulosic and other fibrous
substrates rather than traditional chemical approaches that only embody the chemical
modification of functional groups on fibers.

4.5.3 Flame-retardant textiles

The concept of protecting textiles from burning dates back to ancient times or at
least to the Middle Ages. However, it is generally agreed that the first systematic
attempt to make textiles flame-resistant was made by the eminent chemist Gay-
Lussac in 1821. He treated hemp and linen fabrics with various ammonium salts or
ammonium salts and borax (ref. 53). Since that time, there have been thousands of
investigations conducted to impart durable and semi-durable flame-retardance to
textiles and to produce inherently flame-retardant fibers. The flammability of textiles
is influenced by the inherent characteristics of different types of fibers and their

TABLE 4.6

Flammability characteristics of fibers (ref. 53). Courtesy of Technomic Publishing Co.

Fiber	Ignition temp. (°C)	Maximum flame (°C)	Flammability
Cotton	400	860	Burns readily with char formation and afterglow.
Rayon	420	850	Burns readily with char formation and afterglow.
Acetate	475	960	Burns and melts ahead of flame.
Triacetate	540	885	Burns readily and melts ahead of flame.
Nylon 6	530	875	Melts, supports combustion with difficulty.
Nylon 66	532	---	Melts, does not readily support combustion.
Polyester	450	697	Burns readily with melting and soot.
Acrylic	560	855	Burns readily with melting and sputtering.
Modacrylic	---	---	Melts, burns very slowly.
Polypropylene	570	839	Burns slowly.
Polyvinyl chloride and polyvinylidene chloride	---	---	Does not support combustion.
Wool	600	941	Supports combustion with difficulty.

blends. There are several mechanisms by which flame-retardant agents are believed to be effective. Flame-retardant finishes for textiles include a variety of chemical compounds and polymers that are appropriate for each fiber type, fiber blend, fabric weight and construction. Production of inherent flame-retardant fibers by polymerization of specific monomers represents the most recent technique in this area.

The flammability characteristics of commercially available fiber vary widely (Table 4.6). Cellulosics (cotton and rayon) burn readily with afterglow and the formation of char, while wool supports combustion only with difficulty. Synthetic fibers (excluding inherently flame-retardant fibers) may either melt slowly without burning (Nylon 6 and Nylon 66), burn and melt readily (polyester and acrylics), or not support combustion at all (polyvinyl- and polyvinylidene chlorides).

The thermal processes and combustion products of organic polymers, whether in fiber, film or in some other form, occur in a progressive and definable sequence (Figure 4.19) in which thermal decomposition precedes ignition and combustion of the material. After combustion, the polymer may burst into flame, melt, shrink, char or thermally degrade without flame. The thermal decomposition products determine the flammability of polymers (e.g., fibers or modified fibers). The role of flame retardants is to inhibit the formation of combustible products and/or to alter the normal distribution of decomposition products emanating from the original material.

The two predominant mechanisms by which flame retardants decrease the flammability of textiles and other polymeric substrates are considered to occur by solid- or condensed-phase and by gas-phase retardation of combustible products. Concomitant increases in the formation of char, carbon dioxide and water occur when a solid-phase mechanism is operative. Reduction of volatile, flammable products by the condensed-phase mode is affected by dehydration and/or crosslinking. Although these modes of action are well-established for cellulosic fibers, more detailed studies are needed to elucidate the precise modes of condensed-phase behavior in other fibrous polymers (ref. 55).

In gas-phase retardation, oxidation in the gas phase is inhibited by free radicals. The overall effect is to decrease the amount of fuel consumed in the flame and

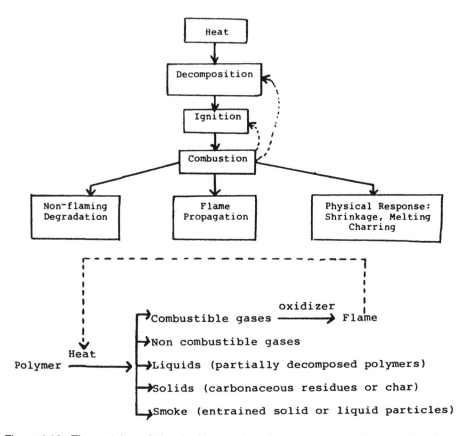

Figure 4.19. Thermal degradation, ignition, and combustion processes for organic polymers represented by general sequences. Thermally induced decomposition clearly precedes ignition (ref. 54). Courtesy of Marcel Dekker Publishers.

subsequent amount of heat generated. Halogen compounds exert a gas-phase inhibiting effect by reaction of HX with H · or OH · to produce hydrogen and water respectively plus X ·. When the hydrogen and hydroxyl radicals react with HX, they divert these radicals from reaction with oxygen or hydrogen to form combustible materials. Such combustion is thought to occur by the following reactions:

$$\cdot H \ + \ O_2 \ \underset{\longleftarrow}{\longrightarrow} \ \cdot OH \ + \ \cdot O \qquad (4.2)$$

$$\cdot O \ + \ H_2 \ \underset{\longleftarrow}{\longrightarrow} \ \cdot OH \ + \ \cdot H \qquad (4.3)$$

$$\cdot OH \ + \ CO \ \underset{\longleftarrow}{\longrightarrow} \ CO_2 \ + \ \cdot H \qquad (4.4)$$

Other modes of flame retardancy are also possible, such as the generation of noncombustible gases, endothermic reactions of thermal decomposition products and/or flame retardants and the formation of nonvolatile char and glassy coatings. The latter prevent diffusion of oxygen and heat transfer to the fibrous substrate (ref. 54).

Various finishing techniques have been used to impart flame resistance to textiles. These include insolubilization, coating, graft- or homopolymerization, crosslinking, chelation and covalent bond formation. Several classes of compounds and polymers have also been used to produce flame-retardant textiles. These include inorganic acids, acid salts and hydrates, organophosphorus and organobromine compounds, antimony salt/halogen systems and miscellaneous substances. A more comprehensive classification is given in Figure 4.20 that differentiates between primary and synergistic flame-retardants (such as phosphorus and nitrogen) and adjunctive or physical flame retardants (ref. 56).

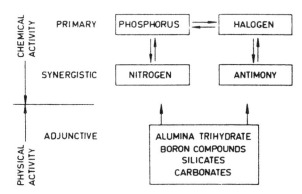

Figure 4.20. Classification of flame retardants (ref. 56). Courtesy of John Wiley and Sons Publishers.

Techniques and chemical approaches vary with the durability requirements of the flame-retardant textile and the fiber type. Most of the non-durable flame retardants for textiles are inorganic salts or hydrates such as borax, ammonium chloride or aluminum oxide trihydrate. These are used only in applications where durability to leaching or to laundering is not important. Durable flame retardants for natural and synthetic fibers and their blends usually are those that exhibit P/N and/or Sb/X synergism. Phosphorus-containing flame-retardants usually have functional groups that are phosphates, phosphonates, amido-phosphates, phosphazenes or phosphonium salts. Halogenated flame-retardants are usually polychlorinated or polybrominated aromatic or alicyclic structures such as polyhalodiaryl ethers and polyhalogenated Diels-Alder products (ref. 56). Representative and commercially important durable flame-retardants are shown in Table 4.7.

Cellulosic textiles are treated with retardants that function in the solid phase, i.e., those that increase char or non-combustible products. The FWWMR finish, a coating containing fungicides, chlorinated paraffins and antimony oxide to weather- and flameproof cotton tents, is considered to be the first commercial flame retardant treatment for textiles (ref. 57). Since that time, a variety of phosphorus compounds used in conjunction with urea and melamine resins, have gained commercial acceptance as flame-retardant treatments for cotton fabrics. THPC, a tetrahydroxy-methylphosphonium salt (ref. 53) and a phosophonamide--trade name of Pyrovatex CP (ref. 58) are both applied to cotton fabrics by a pad-dry-cure process. Both formulations use nitrogenous resins to enhance the flame resistance and durability of the finish to laundering. It is generally agreed that the presence of nitrogen promotes synergism with organophosphorus flame retardants, and that the latter are most effective in this regard when they are in the highest oxidation states(P^{+5}). Flame retardancy of modified cotton fabrics (or any other fiber type) is not as effective after prolonged laundering because ion-exchange processes occur at phosphorous groups that have acidic hydrogens or that may be slowly hydrolyzed to groups containing acidic hydrogens. More specifically, flammability of fibers is increased when the acidic hydrogens are exchanged with ions such as sodium, calcium and magnesium.

Other cellulosics such as rayon can be made flame retardant by incorporating appropriate additives into fiber spinning baths. Various phosphazene derivatives appear to be the most effective for producing rayon fabrics with durable flame resistance (ref. 59). Although unmodified wool possesses adequate flame resistance for most applications, it has been made much more flame resistant by application of specific finishing formulations. One such process that has gained commercial acceptance is the Zipro™ finish. This finish employs various metal complexes and transition metal salts such as zirconium and titanium (ref. 60).

TABLE 4.7

Examples of durable flame retardants applied to textiles

Compound	Fiber type	Finishing technique	Durability	Refs.
Chlorinated paraffins/Sb_2O_3	Cotton	Coating	Weathering	57
$(HOCH_2)_4$ P^+ Cl^-/urea/tri-methylolmelamine	Cotton	Homopolymer-ization/crosslinking	Laundering	53
N-methylol dimethyl-phosphonopropionamide/melamine resins	Cotton	Covalent bonds/crosslinking	Laundering	58
Alkoxyphosphazenes	Rayon	Insolublilization	Laundering	59
K_2ZrF_6 or K_2TiF_6	Wool	Chelation	Laundering	60
$(Br-CH_2-\underset{Br}{CH}-CH_2)_3$ $P=O$ [a]	Acetate, polyester, triacetate	Insolubilization	Laundering	61
Decabromophenyl oxide/Sb_2O_3	Cotton/polyester	Coating	Laundering	62

[a] No longer used because of its potential mutagenic and carcinogenic properties (commonly known as "tris").

Synthetic fibers are usually made flame retardant by copolymerization of flame-inhibiting monomers in the polymer chain, incorporation of an additive during fiber manufacture or topical application of a finish using strategies employed for natural fibers. Of the synthetic fibers, cellulose acetate, cellulose triacetate and polyester are the most flammable, and thus have received the most attention concerning application of flame-retardant finishes. This is typified by the greater burning rate of polyester compared to that of Nylon 66. Organobromine compounds that function as gas-phase inhibitors were commercially developed to impart such resistance to these fibers, but one of the more popular finishes was withdrawn from use because it contained tris (2,3-dibromopropyl) phosphate. This compound was suspected of having mutagenic and carcinogenic properties (ref. 61). The use of a mixed cyclic phosphonate (Antiblaze 19™) as a topical flame retardant for polyester afforded an alternative to the tris(2,3-dibromopropyl phosphate) because the former has low toxicity and good durability to laundering. Later efforts included the grafting of flame-retardant chemicals to polyester as well as the preparation of flame-retardant copolymers of polyesters. These copolymers normally contain organobromine or organophosphorus functional groups to attain the flame-resistance required.

Although aliphatic polyamides are less flammable than other synthetic fibers, there has been little commercial success in effective flameproofing of these types of fibers. The limited success has been generally due to high levels of add-on required, poor durability and stiffness of the resultant fabrics. Flame retardants for nylons that are effective lower the melting temperature of the polyamide and enhance dripping of the molten fabric during ignition. Some of the more effective flame-retardant additives or finishes for polyamide fibers are thiourea condensates, a graft copolymer of poly (2-methyl-5-vinyl pyridine) containing halogen and antimony agents and molybdenum disulfide (refs. 55 and 63).

Acrylic fibers burn readily and received adverse publicity in the early 1950's when several persons were killed or injured by burns when they wore apparel made of this type of fiber. Although acrylic fibers can be flameproofed by strategies used for other synthetic fibers, modacrylics are much less flammable and have replaced acrylic fibers in many applications that require flame resistance. Modacrylic fibers are usually copolymers of acrylonitrile with vinyl chloride and/or vinylidene chloride or other halogen-containing monomers.

Polypropylene fibers are flammable and drip during thermal decomposition. The most effective flame retardants are polyhalogenated compounds such as decabromo-diphenyl oxide in combination with antimony oxide. Addition of flame-retardants in the melt during the manufacture of the polypropylene fiber is the preferred technique. Higher loadings and amounts of flame retardants are normally required for polypropylene than for other synthetic fibers or natural fibers to impart good flame retardancy (ref. 55).

Fiber blends are more difficult to flameproof than fabrics composed of only a single fiber type. Changes in flame retardancy may be antagonistic, synergistic or neutral for a particular flame retardant system applied to a fiber blend relative to either fiber type alone. Cellulosic/polyester blends are by far the most important fiber blend and have received the most attention with regard to development of effective flame retardant systems. Combinations of known flame retardants effective for each type of fiber have been developed for application to cellulosic/polyester blends. Moreover, novel retardants for blends (such as polybrominated aromatics applied with acrylic latex and antimony oxide) have been developed to address this problem. Similar strategies have been developed to flameproof natural fiber/synthetic fiber and synthetic fiber/synthetic fiber blends (refs. 54, 62, 63).

Other factors that must be considered in the development and utilization of effective flame retardants for fibrous materials are smoke density of the modified fiber or fabric and the toxicity of the volatile products formed when the material burns. Since most fatalities in fires are caused by asphyxiation due to lack of oxygen and formation of carbon monoxide, suppression of smoke may be as important a consideration as protection against injury and death due to severe burns. It has been observed that aromatic structures and organohalogen compounds normally produce sizable quantities of smoke. However, it has also been observed that the pyrolytic path has a considerable influence on smoke formation and can vary markedly from one fiber

TABLE 4.8

Smoke emission from fabrics made from different fibers or blends (ref. 55). Courtesy of Marcel Dekker Publishers.

Sample	Decrease in visibility (%)	Optical density
Acrylic	97	1.5
Cotton	4	0.02
Cotton, flame-resist treated	98	1.7
Rayon	4	0.02
Wool	18	0.09
Nylon	6	0.03
Polyester	28	0.14
65% Polyester-35% cotton	99	2.00
55% Polyester-45% wool	98	1.70
Polyvinyl chloride	34	0.18

type to another or fiber blend even in the absence of flame-retardant finishes or additives (ref. 55). As Table 4.8 demonstrates fabrics composed of 100% wool, poly-polyester or cotton have relative low smoke density values.   In contrast, polyester/cotton and polyester/wool blends produce appreciable quantities of smoke and concomitant decreases in visibility.

Although the production of carbon monoxide gas is the major cause of fire fatalities, there are other dangerous gases that can be produced by a variety of fibers and materials containing flame retardants. Some of the more important gaseous products that have been detected are listed in Table 4.9. These include HCN, nitrogen and

TABLE 4.9

Gaseous products of the combustion of organic polymers (ref. 55).   Courtesy of Marcel Dekker Publishers.

Gas	Source
CO, CO_2	All organic polymers
HCN, NO, NO_2, NH_3	Wool, silk, nitrogen-containing polymers (PAN, ABS, polyurethanes, nylons, amino resins, etc.)
SO_2, H_2S, COS, CS_2	Vulcanized rubbers, sulfur-containing polymers, wool
HCl, HF, HBr	PVC, PTFE, polymers containing halogenated flame retardants, etc.
Alkanes, alkenes	Polyolefins and many other organic polymers
Benzene	Polystyrene, PVC, polyesters, etc.
Phenol, aldehydes	Phenolic resins
Acrolein	Wood, paper
Formaldehyde	Polyacetals
Formic and acetic acids	Cellulosic fibers

sulfur oxides, ammonia, hydrogen sulfide and carbon disulfide, and various organic compounds. More research is needed to determine the overall effectiveness of flame-retardant textiles from the composite effect that they produce with regard to smoke density and toxicity of gaseous products as well as protection from severe burns.

Inherently flame-retardant fibers and fabrics have also been developed as alternatives to additives and topical finishes. These inherently flame-retardant fibers are made by synthesizing monomomers or comonomers that polymerize to produce fibers that do not support combustion and usually only melt at extremely high temperatures. These fibers are also characterized by a high limiting oxygen index (LOI). The LOI is a measure of the minimum amount of oxygen in the environment that sustains burning under specified test conditions that are discussed in more detail in **Chapter 6.**

TABLE 4.10.

Inherently flame retardant fibers

Functional group or structure	Trade and/or common name
	Nomex
	Kevlar
	Kynol
Polybenzimidazole	PBI
Poly (*p*-phenylene sulfide)	Ryton
Crosslinked copolymer of acrylic acid/acrylamide	Inidex
Fully or partially carbonized fibers with aromatized structures	Curlon

Representative structures and functional groups that produce inherently flame-retardant fibers are given in Table 4.10. The aromatic polyamides or aramids are characterized by Nomex™ (a *meta* aramid) and Kevlar™ (a *para* aramid). Most other flame-resistant fibers are based on aromatic and/or aromatic structures. PBI or poly

[2,2'-(*m*-phenylene)-5,5'-bis(benzimidazole)] has recently been commercialized for use in various types of protective clothing. Other examples are Kynol (a phenol-formaldehyde polymer that thermally degrades to carbon fiber), Inidex (an acrylic acid/acrylamide crosslinked copolymer), Ryton (a melt-spinnable aromatic polymer of poly(*p*-phenylene sulfide) and Curlon, a carbon fiber (derived from cellulosic or acrylic precursors by stepwise thermolysis) that has highly aromatized structural features. Although these inherently flame-retardant fibers are now very expensive compared to conventional fibers that contain additives or are topically finished, they are becoming more common in protective clothing and other materials where resistance to heat and flame are required.

4.5.4 Absorbency and superabsorbency

The ability of fibers and fabrics to absorb liquid has progressed from an understanding of the mechanisms by which sorption and desorption occur for various natural and synthetic fibers to the modification of fibers by various techniques to render them highly absorbent or superabsorbent. A comprehensive and excellent monograph on absorbency and superabsorbency has been edited by Chatterjee (ref. 64). A more current publication critically reviews various approaches for imparting absorbency to a variety of fibers (ref. 65).

Absorption of a liquid by a fibrous material and transport of the liquid is due primarily to the inherent capacity of the fiber and the application of external forces such as capillary pressure and gravity. Initial adsorption occurs in the vapor phase and is characterized by various types of sorption isotherms (similar to those previously discussed in the sorption of dyes onto fibers) shown in Figure 4.21. The Langmuir type (Type I) relates to complete coverage of a monolayer, while Type II is most common for natural fibers and relates to multilayer formation at point B (where monolayer coverage is complete), while Types IV and V are indicative of saturation at fine capillaries via condensation. Although Type III is rarely observed for fibers, it is observed for hydrophilic polymers grafted or affixed to fibers to make the latter highly absorbent. Adsorption on porous surfaces such as fibers is also governed by hysteresis since desorption isotherms are also observed.

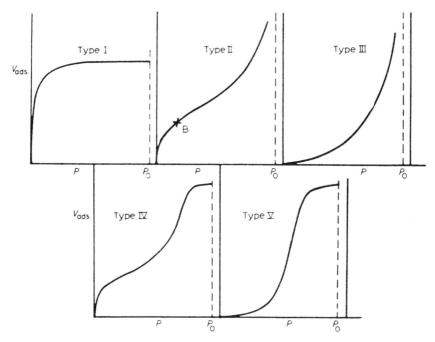

Figure 4.21. Five types of adsorption isotherms (ref. 64).

After initial adsorption, the mechanism of absorption of liquid through the fibrous solid may be characterized by flow through a system of capillary tubes having different porosities and pore size distributions. Several mathematical equations and models have been devised to characterize the nature of the flow. These include the Washburn equation (Eq. 4.5) and Darcy's law (Eq. 4.6):

$$L = (R_c \gamma \cos \theta / 2\eta)^{1/2} \cdot t^{1/2} = k_o t^{1/2} \tag{4.5}$$

$$q = -K \frac{\Delta P}{L_o} \tag{4.6}$$

The Washburn equation is valid for short time intervals (t) where L (capillary rise) is very small compared to equilibrium capillary rise (L_{eq}). R_c is the tube radius, η is the wetted length of the capillary tube, γ is the surface tension of the advancing liquid, θ is the contact angle at the interface and k_o is a constant. This mathematical expression is only valid for simple capillary flow. Darcy's law has been previously used in soil physics to describe flow of liquids, but is also valid to describe a linear and steady state flow through porous materials such as fibers. q is the volume flux in the flow direction, P is the net pressure head that causes flow, L_o the sample length in the flow direction and K_o a proportionality constant equal to the permeability of the medium (k) divided by the liquid viscosity (η). More complex derivations are required or valid when liquids exhibit turbulent and/or unsteady states of flow through porous fibers.

When the fibers contain highly absorbent or superabsorbent materials (usually hydrophilic polymers), the sorption phenomenon becomes even more complex because it is also governed by the swelling and partial dissolution of the absorbing material. There are five general models (Fig. 4.22) that depict swollen gel structures caused by high absorbency of liquids. The first model (a) is a prototype crosslinked

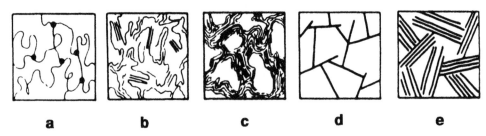

| a | b | c | d | e |

Figure 4.22. Five models of swollen gel structures (ref. 64). Courtesy of Dr. Bob Marchessault, McGill University.

network polymer in which the limit of swelling and dissolution is determined by a balance between thermodynamic factors from polymer-solvent interaction and entropy effects of coiled polymer chains. Model (b) is one with a modified crosslink morphology in the insoluble crystalline region. Most superabsorbents affixed to fibers can be described by (a) or (b). In model (c) the structure is that of a polymer-diluent matrix, while (d) consists of rigid anisotropically directed particles that could convert to a more ordered paracrystalline structure (e).

The absorbency and wettability of unmodified fibers vary with the type of fiber, its geometry, porosity, degree of accessibility and crystallinity and the type of liquid with which it comes into contact. Most absorbent evaluations are concerned with absorbency of water or biological aqueous solutions such as saline solutions, urine and blood. The moisture regains and sorptions of most natural and regenerated fibers are very high relative to that of most synthetic fibers. Wool is the most hydrophilic with an absorption regain of up to 18% at ambient temperature, followed by rayons (12-14%), mercerized cotton and jute (12%), then cotton (7-8%).   Poly-amide had sorption regains of about 4%, followed by polyester (0-4%) and acrylic (1-2%).   Unmodified poly(vinyl chloride), polyethylene and polypropylene fibers have essentially no wettability to water under normal end use conditions.

Although hydrophobic fibers are more frequently modified than hydrophilic fibers to improve their liquid absorbency and wettability, all types of fibers have been modified to increase their tendency to sorb and desorb liquids.   Approaches for increasing hydrophilicity or overall wettability and liquid retention of fibers may be classified into three categories (Table 4.11): (a) modification of the functional groups of the polymer that comprise the fiber; (b) physical and/or chemical attachment of hydrophilic polymers to the fibrous surface and (c) use of surfactants. Method (a) is the most permanent and usually most expensive, while method (b) confers limited durability of this effect and method (c) is inexpensive but requires reapplication to be useful.

TABLE 4.11

Methods for increasing absorbency of fibrous materials

General Methods	Specific Techniques	Examples
Introduction of hydrophilic groups into polymer units of the fiber	End group modification	Exposure to plasma or excimer lasers
	Formation of copolymers	Copolymers of polyamides and polyethylene oxides
Attachment of hydrophilic groups or polymers to fibers	Coating and/ or lamination	Poly(acrylic acid)
	Graft poly-merization	Poly(maleic anhydride) Poly(acrylamide)
Surfactants	-----------------	Detergents, wetting agents Softening agents Antistats, emulsifiers

There are several methods by which hydrophilic groups may be introduced into the repeating polymer units of a fiber. For natural fibers such as cotton and wood pulp that are already hydrophilic, this is usually accomplished by conversion into various cellulose esters and ethers (ref. 64). The most frequently used cellulose esters are cellulose phosphate and sulfate, while carboxymethylcellulose and hydroxyethyl-cellulose are the cellulose ethers employed for this purpose. To prevent the highly absorbent cellulose esters or ethers from complete dissolution in liquids, they are either reacted with crosslinking agents, treated with heat or converted into a partially ionic state (such as partial conversion of the sodium salt of carboxymethylcellulose into the free acid form). The technique that has been used for hydrophobic fibers is conversion of their end groups or incorporation of hydrophilic groups as an integral part of the polymer repeating unit to increase hydrophilicity and liquid absorbency. For example, there are numerous patents and publications that describe the hydrophilization of polyester fibers by their pretreatment in aqueous alkali at various times and temperatures. Copolymers of polyamides or of polyesters containing hydrophilic polyethylene oxide units also enhance hydrophilicity and absorbency of these fibers. More recently, exposure of fibers or fibrous surfaces to high energy sources such as glow discharge, plasma or excimer lasers alter the wettability of surfaces and generally make them more accessible and hydrophilic to various liquids.

Another technique for improving absorbency is the physical and/or chemical attachment of hydrophilic polymers to fabrics or fibrous surfaces. For durable end uses, grafting and/or homopolymerization onto the fibers is usually employed. For disposable products such as diapers and similar items, the polymer is incorporated as a fine powder dispersed throughout the fibrous mass. Poly(acrylic acid) is the most frequently used synthetic polymer, while various modified starches and natural gums as well as carboxymethylcellulose and hydroxyalkylcelluloses are natural polymers used for improving fiber absorbency.

Wetting agents, detergents and surfactants are primarily used to wet fibrous surfaces temporarily to improve their launderability or contact with active agents in solution during textile finishing processes. These agents include quaternary ammonium salts, phosphate and non-phosphate detergents, fatty and sulfonic acids, poly(oxyethylenes and ethylenes) and silicon polymers. These agents may be noionic, cationic, anionic or amphoteric.

4.5.5 Release and repellent finishes

For many applications, the release of solids and/or liquids from fibers and the ability of fibrous surfaces to repel, water, oil, soils, stains and chemical agents may be required. Such performance requirements may be discussed in terms of release-type behavior (usually of oily and other type of soils) and repellent finishes and surfaces (usually to water and/or oils).

4.5.5a Soil release finishes. Various compounds and polymers have been applied to textiles to facilitate removal of unwanted liquids and/or solid particles that are deposited from various sources onto fiber and fabric surfaces. These sources include soils, oily liquids and solids, food products, dyes and cosmetics. It is generally agreed that the removal or release of oily soils from fiber surfaces is primarily dependent on the hydrophobicity of the fibers. One manner in which this occurs is by a roll-up mechanism [Figure 4.23 (a)] in which the contact angle θ between the oil droplet and the fiber surface is increased to its theoretical maximum of 180° when roll-up is complete. This roll-up continues as long as the resultant force R is positive, being expressed by the equation:

$$R = \gamma_{FO} - \gamma_{FW} + \gamma_{OW} \cos \theta_d \qquad (4.7)$$

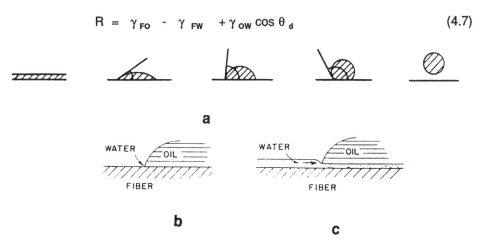

a

b   c

Figure 4.23. (a) Roll-up of oily soil, (b) diffusion of water into the oily-soil hydrophilic fiber interface without and (c) with a soil-release finish (adapted from ref. 66). Courtesy of Marcel Dekker Publishers for 4.23 (a) and of Textile Research Institute for 4.23 (b) and (c).

where γ_{FO}, γ_{FW}, and γ_{OW} respectively are, the surface tension values between oil/fiber, fiber/water, and oil/water. Other soil removal mechanisms that may also be operative include: penetration of soil and solubilization or emulsification by water and detergent and various forms of mechanical removal (hydrodynamic flow, fiber flexing, abrasion and swelling of fiber or finish) (ref. 66).

Hydrophobic fiber surfaces such as cotton treated with durable press agents or untreated polyesters are more prone to oily soiling than are hydrophilic fiber surfaces such as all cotton fabrics. Thus, hydrophobic fiber surfaces normally require soil-release finishes. These types of finishes function by allowing water or detergent solutions to diffuse between the oil/fiber interface, and thus facilitate removal of oily substances [Fig. 4.23 (b) and (c)].

In contrast to oily or liquid soils, deposition of solid soils or other particulate matter on fiber surfaces does not depend primarily on the hydrophobicity of the fiber. It depends on the adhesion of solid particles to the fiber that is caused by van der Waals' forces and on contact area between the fiber surface and the particle (ref. 66). The latter is influenced by surface texture, irregularities, fuzziness and weave of fibers, yarns and fabrics in the textile; it is also influenced by the relationship of their size to the size of the solid deposited on the fiber surface. Removal of solids from fiber surfaces requires breaking the adhesive bond between fiber and solid, followed by separation of the two surfaces by wetting.

Since soil-release finishes cause the fiber surface to become more hydrophilic, they are not effective in preventing deposition of solids onto the fiber. This limitation has been resolved by the development of finishes that contain both hydrophobic and hydrophilic segments or components. For example, fluorocarbons (hydrophobic component) with alternating oxyethylene groups in a polymer chain (hydrophilic component) are effective finishing agents against both liquid and solid soils and stains (ref. 66).

Soil-release finishes are usually applied to cotton/polyester blends treated with durable press agents, to hydrophobic fibers (such as polyester) and to a lesser extent, cotton treated with durable press agents to reduce static charge and the propensity (particularly of the synthetic fibers) to occlude various types of soils.

Various approaches have been used to impart soil-release to fabrics: (a) modification of fiber surfaces by grafting, hydrolysis, oxidation; (b) durable physical adsorption of surfactants at fiber surfaces; and (c) coating of fibers with polymers that have functional groups that promote removal of soil (ref. 66). Most commercial soil-release finishes are based on the third approach. These soil-release polymers have both hydrophilic (such as carboxyl, hydroxyl, oxyethylene groups) and lipophilic or hydrophobic components (such as alkyl or aryl groups). These functional polymers can be further classified as (a) anionic (usually vinyl polymers containing carboxyl groups); (b) nonionic with oxyethylene hydrophilic groups [based on polymers or copolymers of poly(ethylene oxide)]; (c) nonionic with hydroxyl hydrophilic groups (usually crosslinked starches, hydroxylalkyl celluloses or carboxymethyl celluloses) and (d) fluorocarbon block copolymers that increase surface hydrophilicity (due to oxyethylene or other hydrophilic segments) and also reduce fiber surface energy (due to hydrophobic fluorine structures). Selection of soil-release agents and method of application depend on the fiber type and prior finishing treatments. Different strategies are required to impart suitable soil-release to untreated cotton/polyester blends, durable-press treated cotton and cotton/polyester blends and all polyester fabrics.

Most soil-release finishes are generally applied in a separate bath after durable press treatments. A typical aftertreatment consists of application of a copolymer of acrylic or methacrylic acid. Although this two step process is more effective than a single step process containing both durable press and soil-release agents, it is not very economical. Thus, two or three component systems have been devised to impart durable press and soil-release properties to cotton/polyester fabrics in a single bath. These contain for example, a crosslinking agent with an acid catalyst and a fluorocarbon polymer and/or an acrylic acid copolymer.

A variety of approaches have proven successful for imparting soil-release to hydrophobic fibers such as polyester. One technique is to increase fiber hydrophilicity by selective alkaline hydrolysis to increase the number of carboxyl and hydroxyl groups. However, binding of nonionic polymers to the fiber surface network polymerization or crosslinking is more frequently used for most of the commercial processes. Many of these polymers are poly(ethylene oxides) or copolymers of poly(ethylene oxides).

4.5.5b Repellent finishes. Repellent finishes function by coating the fiber surface to lower its surface energy below that of liquids that would wet the fiber. If the liquid was an oil or had components that would stain the fiber, this limited wettability would prevent the unwanted liquid from residing on the fiber surface (ref. 67). Since stain-repellent finishes lower the surface energy of the fiber, they are also effective in preventing or minimizing adhesion of particulate matter to fibers. Modifications of fabric surfaces to repel water, oil and/or stains are the major focus of this discussion. Certain compounds and polymers are suitable for one kind of repellency while others provide repellency to various liquids and solids.

Stain and water repellent finishes are discussed together in many instances because the principles of lowering the surface tension of the fiber to repel water are the same as they are for oil and other liquids (ref. 67, 68). However, water-repellent finishes deserve to be discussed separately because many of the chemical approaches and strategies for imparting water repellency to fibers are different from those used to impart stain resistance. As with oil and other liquids, the advancing contact angle of water (θ_a) should be high (at least >90°) to produce a fiber with a low surface energy and wicking behavior, thus preventing wetting of the fiber surface. As noted in Table 4.12, several chemical approaches have been used to produce water repellent fabrics. Durability to laundering and dry cleaning and concentration of active ingredients in the finishing bath vary widely, as does the cost of applying various types of finishes. Although the fluorocarbon-based finishes impart both oil and water repellent properties to fabrics and have excellent durability to washing and dry cleaning, they are usually much more expensive than the other chemical classes of repellents listed.

Water repellent textile finishes based on wax and metal salts were first applied to cellulosics and to wool, and used various aluminum salts in wax emulsions to achieve such repellency. Later formulations employed zirconyl acetate or zirconyl oxychloride rather than aluminum salts to improve the washfastness of such finishes. Urea-formaldehyde and melamine-formaldehyde resins that have been modified with long

TABLE 4.12

Durabilities of hydrophobic finishes (ref. 68). Courtesy of John Wiley and Sons Publishers.

| Class | Durability | |
	Washing	Dry Cleaning
Pyridinium compounds	excellent	good
Organometallic complexes	fair	fair
Waxes and wax-metal emulsions	fair	poor
Resin-based finishes	good	good
Silicones	good	good
Fluorochemicals	excellent	excellent

chain alkyl substituents also produce hydrophobic polymers that are useful as water repellents for textiles.

Silicones that contain active hydrogens and thus condense with functional groups on the surface of fibers are more durable water repellents for textiles than those listed above. The silicones are believed to be effective because of the orientation of their alkyl groups along the fiber surface rather than being perpendicular to the fiber surface as they are with the fluorocarbons. Finishing baths usually contain mixtures of polymethylsiloxanes and polymethylhydrogensiloxanes ($-\overset{R}{\underset{CH_3}{Si}}-O-$, where R = CH_3 and H , respectively), but may also contain vinyl or other reactive groups to produce network or crosslinked structures that improve the washfastness and dry cleaning resistance of the finish on the fabric (refs. 67, 68).

Water-soluble pyridinium salts containing hydrophobic substituents, particularly stearamidomethypyridinium hydrochloride, are bound to fibers by a pad-dry-cure process to produce durable water repellent fabrics because of hemiacetal formation between cellulosic hydroxyl groups and the stearamidomethyl radical or deposition of insoluble N,N'-methylenebis(stearamide). Such compounds are also used as extenders with more expensive repellents such as fluorocarbons to produce fabrics that are both oil and water repellent.

A variety of fluorocarbon compounds and polymers may be used to produce water as well as oil repellent effects on fabrics. If only water repellency is required, fluorocarbon chain lengths as low as two are adequate.

Finally, it should be noted that the term "water-repellent finish" denotes one in which a hydrophobic substance is insolubilized onto fiber and yarn surfaces in such a manner that a discontinuous film is formed and the fabric is still permeable to water vapor. The term "waterproof finish" denotes treatment of a substrate such a textile to render it impermeable both to liquid and to water vapor by application of a continuous film on the fabric surface (ref. 67). Waterproof finishes for fabrics are usually applied by coating them with materials such as natural and synthetic rubber, polyvinyl chloride or polyurethanes by one of the various coating methods previously discussed. The differentiation between waterproof, impermeable finishes or coatings and water-repellent finishes have been somewhat obscured by the introduction of several fabrics with microporous coatings that are permeable to water vapor but impermeable to liquid water. The first fabric produced with this type of feature was Gortex[R] (ref. 69); it consists of a microporous poly(tetrafluoroethylene) film laminated to many different types of fibrous substrates. Since that time, a variety of microporous, breathable/waterproof fabrics have been produced; most of these are based on either polyurethane coatings or membranes or fluorocarbon membranes (ref. 70).

Stain repellent finishes for textiles (also initially called soil retardant finishes) are primarily based on the use of perfluoro compounds, polymers and/or copolymers with chain lengths of seven to eight carbon atoms (C_7F_{15} to C_8F_{17}). Fluorocarbons of this chain length have the lowest surface free energies and have the proper conformation to produce close packing to form a monolayer over rough textile surfaces. Fluoroacrylates and fluorocarbon-pyridinium salts (the latter exhibiting synergistic stain repellent effects) are two of the most effective types of fluorocarbons that impart stain repellency to textiles. The relationship between the structure of various perfluoro compounds and polymers to the repellent effects that they impart to fibers has been critically reviewed and discussed by Kissa (ref. 67). Antagonistic effects, i.e., loss of stain or oil repellency in fabrics, may occur when hydrophilic substances (such as softeners) or water repellents (such as siloxanes) are present even in trace amounts in the fiber.

Within the last decade, a new class of stain resistant agents have been commercialized for application and use on nylon carpets. These are called "stain-blockers" and have been critically reviewed with regard to their structure, mechanism of operation and mode of application (ref. 71, 72). From a structural perspective, there are four major classes of stain blockers: (a) sulfonated condensation products of phenolics and formaldehyde; (b) metallic derivatives of sulfonated condensation products of thiophenolics; (c) metallic derivatives of dihydroxydiphenylsulfones; and (d) nonaromatics such as polymeric alcohol sulfonic acids containing S-xanthogenate ester substituents and saturated and unsaturated aliphatic and cycloaliphatic sulfonic acids (ref. 72). The first three classes of stain blockers are also called syntans. Figure 4.24 shows the general structure of such a condensation product from phenols (also called a sulfonated aromatic condensation product or SAC).

General SAC Structure

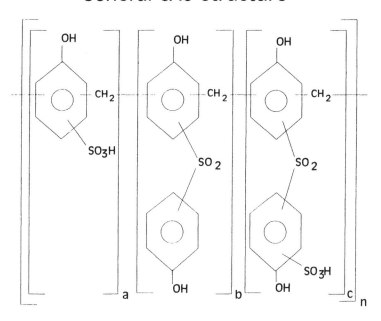

Figure 4.24. General SAC structure where a, b and c = representative mole fractions and n = number average molecular weight (ref. 71). Courtesy of American Association of Textile Chemists and Colorists.

The mechanism for fixation of the stain blockers to the polyamide fibers appears to be due to hydrogen bonding between amide groups of the fiber and hydroxyl groups of the stain resist structure and to electrostatic attraction between sulfonic acid groups in the stain resist structure and amine end groups of the fiber (ref. 71). Stain blockers may be applied to the polyamide fibers as they are spun or to the carpet after it is manufactured and dyed. The optimum stain resistance is usually obtained when the stain blockers are applied at a pH of less than 2.5. It is generally more advantageous to apply these materials to the carpet since they effective impart some dye resistance to fibers; it has also been confirmed that fixation of stain blockers occurs primarily near the fiber surface by a ring dyeing mechanism. Because of the complexity and variation of both the stain blocker structures and the morphology and geometry of polyamide fibers, commercial application of these materials usually consists of a composition of different molecular weight stain blockers appropriate for specific polyamide fibers used in carpets. Stain blockers are normally used in conjunction with a fluorocarbon additive to retard stain penetration into the fibers.

4.5.6 Antimicrobial and antiinsect finishes

Requirements for providing resistance to textiles against microorganisms and insects vary with the end use of the fibrous product and with the susceptibility of the fiber type to biodeterioration and/or infestation by microbes and insects. Textiles are treated with antimicrobial and antiinsect agents to: (a) protect the wearer and/or user of the textile product for aesthetic, hygienic, or medical purposes against bacteria, dermatophytic fungi, yeasts, viruses and other deleterious microorganisms; (b) protect the fiber and textile structure from biodeterioration caused by mold-, mildew-, and rot-producing fungi; and (c) protect the textile from insects and other pests for fiber preservation and/or protection of the wearer or user of the textile product (ref. 73, 74). Finishes that are employed for (a) may be classified as either antibacterial and/or antimycotic (antifungal). Those in the second end use group (b) are commonly called antimicrobial or rot- or mildewproofing finishes, while those in the third end use group (c) are denoted as antiinsect or more specifically mothproofing finishes when used to treat or protect wool fibers.

Different types of fibers have different susceptibilities to biological attack. Cellulosic fibers are usually the most prone to attack by rot- and mildew-producing fungi, while wool is the most susceptible to insect attack by moths and carpet beetles. Both natural and synthetic fibers of all types are susceptible to contamination and growth of pathogenic and parasitic bacteria, fungi, viruses and microorganisms. Susceptibilities of fibers to biological attack and the persistence of microorganisms and insects on fibers are discussed in more detail in **Chapter 6**.

Representative microorganisms and insects that may adversely effect textiles and/or their wearers and users have been critically reviewed and discussed (refs. 74-76). For antibacterial and antimycotic finishes, these include gram-positive bacteria (such as *Staphylococcus aureus*), gram-negative bacteria (such as *Escherichia coli*), dermatophytic fungi (such as Trichophyton interdigitale), yeasts (such as *Candida albicans*) and protozoa (such as *Trichomonoas vaginalis*). Protective clothing against contamination by pathogenic viruses such as hepatitis and HIV is also important. Representative deleterious fungi that cause fiber biodeterioration are *Aspergillus niger* and *Chaetomium globosum*. Protection against clothes moths (such as *Tineola bisselliella*) and mosquitoes (such as *Aedes aegypti*) are required to have useful insectproof textiles.

Mechanisms by which treated textiles are rendered resistant to biological attack include that of controlled release, the regeneration principle, and barrier or blocking action. Most antimicrobial and antiinsect finishes operate by a controlled release mechanism (ref. 77). For disease-causing bacteria, fungi, viruses and other microorganisms, a sufficient amount of moisture is required to release the antibacterial or antimycotic agent at a rate adequate for killing (-cidal) or inhibiting (-static) microbial growth on the fabric. However, to achieve suitable rot- and mildew resistance, fabrics treated with fungicides must leach off at a rate slow enough to be durable to long periods of outdoor exposure.

In the regeneration model of Gagliardi (ref. 77), a chemical finish on a fabric that produces an active antimicrobial agent continually regenerated during laundering should produce an unlimited supply of renewable antimicrobial agent. This regeneration may occur by hydrolysis or oxidation of functional groups present in the fiber or fiber finish or by exposure of the fabric to sunlight to generate active species. Although this principle has yet to be practically demonstrated, microencapsulation of antibacterial agents or pesticides (ref. 41) provides a reservoir of active agent that is effective for long periods of time, and thus approaches the concept of the regeneration model. A more elegant and recent technique is the preparation of acrylic fibers with varying pore size containing various antibacterial agents that can diffuse at different rates relevant to specific end uses (ref. 78).

Textile finishes that function by a barrier or blocking mechanism against microorganisms and insects do so by either forming an impervious film or coating on fabrics or by modifying the fiber surface so that it is biologically active when contacted by microbes or insects. Most finishes that are effective by this mechanism were applied many years ago by the first method, e.g., the development and commercialization of the FWWMR finish, a multi-purpose coating containing flame-retardants and fungicides (ref. 57). Since that time, the emergence of the insidious hepatitis and HIV viruses have led to the development and use of fabrics that do not allow body fluids and blood to have "strikethrough" of liquid to protect health care personnel (ref. 75). These fabrics have been based on primarily water repellent and/or waterproof coatings. However, microporous, breathable coatings such as Gortex[R] are also being considered as protective clothing to serve as a biobarrier to potentially pathogenic body fluids.

All representative physicochemical and chemical techniques for applying functional finishes to textiles (such as those described in Table 4.4) have been employed to produce fabrics with antimicrobial and antiinsect resistance. Application techniques and chemical compositions of such finishes vary with their durability requirements to laundering, dry cleaning or leaching. These also vary with regard to the specific microorganisms or insects that are problematic for a given textile product. Thus, multipurpose biological finishes (Table 4.13) have been devised and developed. Almost every major class of inorganic and organic compound, at one time or another, has been evaluated or applied as a specific or multipurpose biological finish for textiles. A brief list of these chemical classes include: inorganic salts, organometallics, iodophors (substances that slowly release iodine), phenols and thio-

TABLE 4.13

Finishes for the multi-biological protection of textiles (adapted from ref. 73).

Finishing techniques	Chemical agents	Organisms protected against	Fiber type
Insolubilization	$(n\text{-}Bu_3SnO)_2.TiX$[a]	Gram-positive bacteria, dermatophytic and mildew fungi	Polyamide, acrylic, polypropylene
Homo- and copolymerization	Acrylamido-8-hydroxy-quinolines and metal salts	Bacteria, mildew fungi	Cellulosics
Resin treatment	Hexachlorophene and DMEU[b]	Bacteria, mildew fungi	Cellulosics
Covalent bond formation	5-Nitrofurylacrolein	Bacteria, mildew and yeast fungi	Poly(vinyl alcohol)
Coatings	Hydrolysis product of $(CH_3O)_3Si\text{-}R$[c]	Bacteria;dermatophytic, yeast and mildew fungi; algae	All types
Microencap-sulation	Any bactericide, pyrethrins, pheremones	Bacteria and insects	Many types

[a] X = di(acetylacetonate), diisopropoxide, or dipropyl acetate.

[b] DMEU = N,N-dimethylolethylene urea.

[c] $R = \text{-}(CH_2)_3\text{-}\overset{+}{N}\text{-}CH_{18}H_{37}\ Cl^-.$
$\qquad\quad (CH_3)_2$

phenols, onium salts, antibiotics, heterocyclics with anionic groups, nitro compounds, ureas and related compounds, formaldehyde derivatives and amines.

Because antibacterial and antimycotic textile finishes usually require prolonged durability of biological activity to laundering and/or dry cleaning for their end uses (aesthetic, hygienic and medical), relative few finishes have been developed that meet these requirements. However, there are some promising approaches and commercial finishes that claim durable biological activity for at least 50 launderings (ref. 73). One such finish is Letilan-2, developed in the Soviet Union. It is a broad spectrum antimicrobial fiber produced by the reaction of poly(vinyl alcohol) and 5-

nitrofurylacrolein to form an acetal that slowly releases the aldehyde as the active antimicrobial agent. Another is the coating of fibers with a silicon polymer containing pendant quaternary ammonium groups to produce a surface that does not support microbial growth. The formation of a zinc peroxyacetate homopolymer on cellulosic and cellulosic blend fibers is achieved by the reaction of zinc acetate with hydrogen peroxide under mild curing conditions. The resultant fabrics are resistant to gram-positive and gram-negative bacteria for up to 50 launderings and are effective by a controlled release mechanism. A flame-retardant finish based on the application of the flame-retardant phosphonium salt THPC and trimethylolmelamine shows broad spectrum activity against various microorganisms after prolonged laundering. The treatment of synthetic fibers with resins and mercurated allyltriazines produces a washfast finish effective against dermatophytic fungi.

Textile finishes that provide resistance against rot and mildew must also usually be durable to the complex process of weathering (includes protection against sunlight as well as mildew), since they are more frequently used outdoors than indoors. Most of the important commercial rot- and mildew-resistant finishes for textiles were developed between 1940 and 1965 by the U. S. Army Quartermaster Corps and the U. S. Department of Agriculture, Agricultural Research Service, for cotton fabrics. This emphasis was due to the major use of cotton then as a textile fiber as well as the tendency of cellulosic fibers to undergo rapid biodeterioration.

Most earlier finishes consisted of insolubilization of copper or of other metal salts on the fiber, e. g., copper-8-hydroxyquinolinate, from wax or from aqueous emulsions. Organometallics, such as tributyltin oxide, have been applied to a variety of fiber types from petroleum solvents to impart fungicidal activity. As with antibacterial finishes halogenated phenols, such as dichlorophene, have been applied to textiles to render them mildew-resistant. However, unwanted side effects may occur, such as the accelerated phototendering of this fabric on outdoor exposure. Durability of fungicidal activity is generally improved by polymers or resins to bind the active agent to the fiber. The use of acid colloids of trimethylolmelamine for this purpose is an example of such improved durability. Although chemical modifications of cellulosic hydroxyl groups through acetylation and cyanoethylation are effective in preventing fungal attack by a blocking mechanism, commercial acceptance of such finishes has been negligible due to the cost of such modifications (ref. 73).

Most insect-resistant finishes for textiles have been developed to protect wool from damage by various species of clothes moths and carpet beetles. However, there have also been efforts to make textiles resistant to pests such as mosquitoes, lice and chiggers for the benefit of the wearer or user and to prevent infestation or attack of food and feed bags. Chlorinated hydrocarbons such as Dieldrin are extremely effective in protecting wool against insects. However, their use was discontinued in the 1970's because of their persistence in the environment and their toxicity to aquatic life. Thus, a variety of chemical agents to replace the chlorinated hydrocarbons have been evaluated and commercialized. Commercially available mothproofing agents, most of which may be advantageously applied from dyebaths, are shown in Figure 4.25.

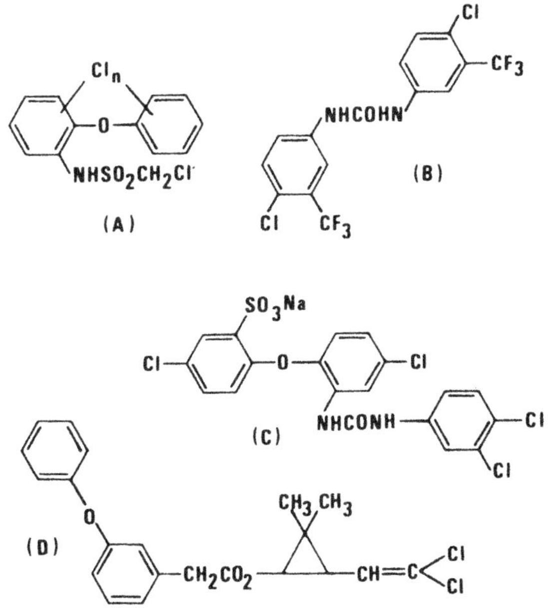

Figure 4.25. Chemical formulae of active constituents of some currently available industrial mothproofing agents (ref. 75). Courtesy of The Textile Institute.

The first three structures in Figure 4.25 (A-C) are aromatic amides or sulfonamides containing chloro- substituents, while structure (D) is Permethrin, representative of a synthetic pyrethroid. Other approaches that offer commercial promise have also been reviewed (ref. 75,76). These include: fiber-reactive agents that react with thiol groups in the wool fiber and release an active insecticide on digestion by the insect (usually phosphorothionolate and phosphorothiolate esters), various quaternary ammonium compounds and organotin compound such as triphenyltin chloride and acetate. As in the commercial finishes, many of these agents may be applied to the fiber from organic solvents or by exhaustion from dyebaths.

4.5.7 Environmentally protective finishes

Textiles may degrade, yellow and/or fade when they are subjected to prolonged exposure outdoors or storage indoors. Such degradation is primarily influenced by the environment to which they are exposed. Primary agents of degradation are physical, chemical and biological. These factors are discussed in more detail in **Chapter 6**. Physical agents include heat and ultraviolet light. Chemical degradation may be due to air pollutants that oxidize and/or hydrolyze dyed or undyed textiles. Biological degradation is primarily due to rot and mildew fungi and certain species of algae. The complex process of weathering requires that textiles be protected against heat, ultraviolet light, microorganisms, rain and air pollution. Effective environmentally protective finishes usually contain photostabilizers, antioxidants and antimicrobial agents. For purposes of discussion, these finishes can be subdivided into: (a) ultraviolet and heat resistant; (b) resistant to air pollutants and (c) weather-resistant.

4.5.7a Ultraviolet and heat resistant finishes. Degradation of textiles by heat and by ultraviolet light varies considerably, and is influenced by such factors as temperature, humidity, chemical structure of the fiber and the dyes applied to it. Detailed mechanisms by which various fibers are thermally and photochemically degraded are appropriately discussed in **Chapter 6** (physical and chemical agents of degradation).

For conventional textile fibers and normal textile end uses, degradation by heat is not usually a major problem, and thus development of heat-resistant finishes has not received much attention. In contrast, there has been much activity to develop heat-resistant fibers and fabrics for special end uses. These are primarily for use in the space program and for thermally protective clothing with requirements that the fibers be resistant to temperatures above 300°C.

These inherently heat-resistant fibers are also usually inherently flame-resistant and can be further subdivided into several categories. One category is the aromatic polyamides or aramids that are typified by Nomex™ and Kevlar™. The repeating units in these polymers are polyphenylene isophthalimide and phthalimide (see structures in Table 4.10 in section on **flame-retardant textiles**). The second group are heterocyclic polymers such as poly(benzimidazole) that were developed by the U. S. Air Force in the mid-1950's. The third group are those of the "ladder-polymers", typified by poly(quinoxalines). The fourth group are the various carbon and graphite fibers that are derived from precursors such as pitch, rayon or acrylic fibers by the controlled, stepwise heating of the precursor. The final group is composed of various inorganic and ceramic fibers. These heat-resistant textiles are not based on fiber finishes, but are an inherent part of the polymer backbone, and are used in flame-retardant as well as heat-resistant applications. These classes of compounds and polymers have been critically reviewed and evaluated (ref. 79), and more progress in this fast-growing area is likely.

Photodegradation of fibers and other polymeric substrates has been observed and verified by numerous investigators. However, it is only recently that reliable information has been published on general mechanistic pathways by which photodegradation occurs in polymers and how photostabilizers applied as finishes or additives function to prevent such degradation. Absorption of light and its emission are believed to lead to polymer degradation by two general pathways.

The first pathway is initiated by impurities in or at the end of polymer chains that provide chromophoric sites to absorb and emit light (Type A process). The second pathway is caused by chromophoric repeating units in the backbone of the polymer chain that absorb and emit light (Type B process). Both types of processes are shown in Figure 4.26. In the Type A process, impurities are believed to include -C=O, -C=C-, -OOH groups, dyes, cations and catalysts, and oxygen-polymer charge transfer complexes.

Subsequent breakdown of the polymer chain may be random with chain scission, and occur in some cases with crosslinking. Degradation may also involve "unzipping" or chain depolymerization. In all instances, photodegradation is considered to be

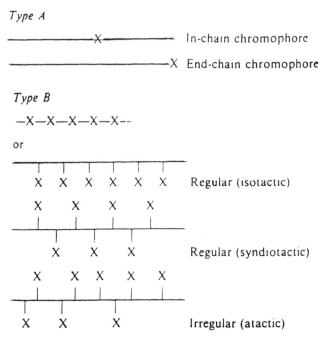

Figure 4.26. Pathways by which polymers absorb and emit light (adapted from ref. 80).

initiated by the formation of free radicals (R⋅) produced by breaking bonds homolytically in the polymer chain or produced by other chemical reactions that form free radicals. Reaction of the free radical to form hydroperoxides is the propagation step, with the termination step occurring by recombination of free radicals to form covalently bonded structures.

Photostabilizers prevent or minimize this degradative process for fibers and for other polymeric substrates, and are usually classified into four groups. The first group are the ultraviolet screeners. These are usually inorganic pigments such as the rutile form of TiO_2, ZnO, or chromium salts. They function by reflecting light away from the fiber and preventing or minimizing its penetration. The second group are the ultraviolet absorbers. Compounds such as o-hydroxybenzophenones or o-hydroxybenzotriazoles absorb ultraviolet light in the region of the spectrum where the polymer absorbs it, and then harmlessly dissipate the short wavelength radiation by various non-radiative processes. The third group is composed of excited-state quenchers. Compounds such as Ni^{II} oxime chelates function by deactivating the excited singlet or triplet states of the chromophore in the polymer formed after absorption of light.

The last group consists of antioxidants such as hindered amines or metal dithiocarbamates. These stabilizers function as free radical scavengers and/or hydroperoxide decomposers (ref. 80, 81). As might be expected, these photo-stabilizers have been observed to function by more than one of the four mechanisms.

Since natural cellulosic fibers were the first type used for outdoor applications (e.g., cotton, jute, hemp and linen), these fabrics were the first that were protected from ultraviolet degradation. Normally various inorganic pigments and mineral dyes were insolubilized or deposited on the fabrics. It was noted that the concentration and type of pigment present on the fiber could either dramatically impede or accelerate photodegradation. The classic example is titanium dioxide in which the *rutile* form protects the fiber and the *anatase* form accelerates fiber photodegradation. Various chromium salts have been co-applied with antimicrobial agents onto cellulosic fabrics to provide resistance to both sunlight and mildew (see section on **weather-resistant finishes**). Other approaches for protecting cellulosic fibers from photodegradation include graft polymerization, chemical substitution of hydroxyl groups and the application of ultraviolet screeners and excited state quenchers to the fabrics (ref. 82).

Wool and silk are more difficult to protect from photosensitized degradation and yellowing, since these may be caused by both ultraviolet light and fluorescent brightening agents. Also, photostabilization is very dependent on pH of the proteinaceous fibers. However, a variety of approaches have been used to provide such protection. These include application of: thiourea-formaldehyde resins, reducing agents such as thioglycolic acid, grafting of acrylonitrile and polymeric benzotriazole ultraviolet screeners on the fiber surface and synergistic combinations of ultraviolet absorbers and antioxidants (ref. 82, 83). The latter synergism is most effective when the ultraviolet absorbers are selected from the *o*-hydroxybenzotriazole class and the antioxidant from a class of hindered phosphonate derivatives of bisphenols (ref. 83).

Many of the finishing techniques and chemical approaches used for protecting natural fibers from photodegradation have also been used for this purpose to protect synthetic fibers. The representative synthetic fibers used in textiles vary widely in their susceptibility to photodegradation. Polyamides, particularly aliphatic types such as Nylon 66 are the most susceptible to ultraviolet light, while acrylic fibers are the least susceptible to photodegradation. Other synthetic fibers such as regenerated cellulosics (rayon and viscose), polyesters, cellulose acetate, cellulose acetate and polyolefins vary in their degree of resistance to ultraviolet light.

Incorporation of photostabilizers into synthetic fibers is usually achieved by adding them to fiber spinning baths. Polyamide fibers were the first type for which this technique was adopted. Typical photostabilizers for polyamide fibers include transition metal salts (e.g., manganese, zinc and nickel) and inorganic halides. These compounds probably function as free radical scavengers and/or hydroperoxide decomposers to protect the fiber from the adverse action of ultraviolet light. More recently, various hindered amines have been incorporated into synthetic fibers, particularly polypropylene, to act as photostabilizers by a free radical scavenger mechanism. Substituted tetramethyl piperidines are representative compounds in the hindered amine class. Figure 4.27 shows representative o-hydroxybenzophenones and two classes of effective polymeric hindered amine photostabilizers for polypropylene fibers based on piperidine structures--denoted respectively HALS-1 and HALS-2 (ref. 84). Phosphorylated poly(phenylene oxide) structures have been found to be equally effective photostabilizers and heat stabilizers for polypropylene fibers when compared to cyclic phosphite and phosphite esters used as additives in earlier investigations (ref. 85). The suitability of various types of photostabilizers for synthetic fibers and polymers has been reviewed and critically discussed by several authors (refs. 80, 81, 86).

4.5.7b Protection from air pollution. Until the late 1940's, the degradation of textiles outdoors by air pollutants was not recognized, and in many instances incorrectly attributed to degradation by sunlight. Since that time, it has been observed that the major air pollutants that degrade, discolor, and/or fade dyed and undyed fibers are sulfur dioxide, nitrogen oxides and ozone. The adverse effects produced by these gases are dependent on the chemical composition of the pollutant, the fiber type, and the chemical structures of dyes, additives and finishes on the fiber (ref. 87).

Dyed natural fibers (cellulosics and wool) are particularly sensitive to fading when exposed to nitrogen dioxide, but such fibers are only moderately affected by ozone

Abbrev.	Chemical Name	Trade Name	Structure
UVA-1	2-Hydroxy-4-n-octyloxy benzophenone	Cyasorb® UV-531	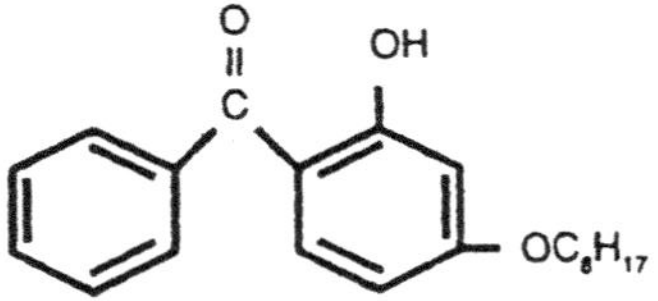
HALS-1	Dimethyl succinate polymer with 4-hydroxy-2,2,6,6,-tetramethyl-1-piperidine ethanol	Tinuvin® 622LD	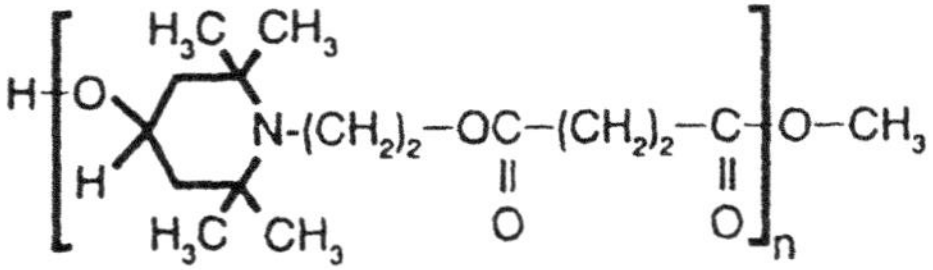
HALS-2	N-N'-bis(2,2,6,6,-tetramethyl-4-piperidinyl)-1,6-hexanediamine, polymer with 2,4,6-trichloro-1,3,5-triazine and 2,4,4-trimethyl-1,2-pentanamine	Chimassorb™ 944FL	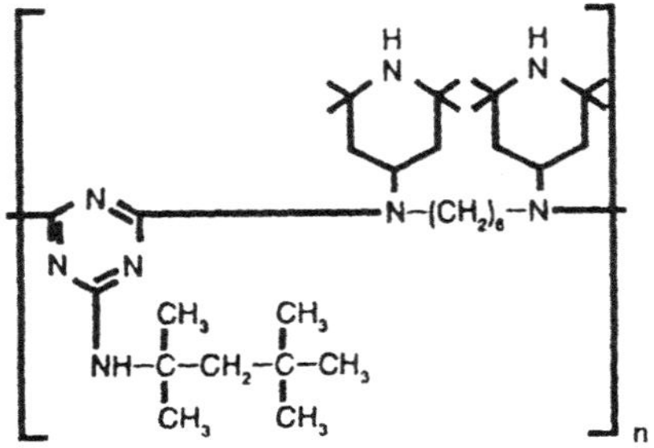

Figure 4.27. Light stabilizer structures for polypropylene fibers (ref. 84). Courtesy of American Chemical Society.

and by sulfur dioxide. In contrast, sulfur dioxide more readily degrades polyamides and undyed cellulosics than other types of dyed and undyed fibers. Although there have been few efforts to develop finishes for textiles that are resistant to air pollutants, the gas fume fading of cellulose acetate and of cellulose triacetate fibers was recognized as a serious air pollution problem several decades ago. This led to the development of dye structures and chemical additives (e.g., compounds containing amino groups) to prevent or minimize such fading. Since most air pollutants are oxidizing agents and photodegradation of fibers also occurs by oxidative processes, photostabilizers that are antioxidants or free radical scavengers should also be beneficial in protecting fibers from damage by air pollutants.

Atmospheric pollutants have also been observed to cause yellowing of textiles even when they are stored or packaged. In addition to yellowing by air pollutants (such as ozone, sulfur and nitrogen oxides), the incorporation of phenolic antioxidants may also cause yellowing of fibers due to the formation of quinoid-type structures (ref. 88).

More recent studies on the effect of air pollutants on fibers have focused on the beneficial effects that certain types of fibers have on reducing or minimizing indoor air pollution that is now considered to be significant due to the air tightness of residences. Although materials composed of proteinaceous fibers (such as wool carpets) appear to be quite useful in this regard, other studies indicate that cellulosics such as cotton and rayon would also sorb large quantities of indoor pollutants. These studies have been critically discussed and reviewed (ref. 89).

4.5.7c Weather-resistant finishes. Resistance of textiles to weathering outdoors varies considerably and is dependent on the climate to which the materials are exposed and the environmental variations due to sunlight, humidity, wind, rain, snow, particulate matter and air pollutants. Although there are few comparative studies on the ability of materials made from natural and synthetic fibers to withstanding weathering, it is generally agreed that acrylic fibers perform the best outdoors. Cotton, linen, polyester, polyolefins and other synthetic fibers vary in their performance. Aliphatic polyamides, jute, wool, and silk generally exhibit poor resistance to weathering (ref. 82).

Since the initial outdoor use for textiles involved only natural textile fibers, most weather-resistant finishes were first developed to protect cellulosics, particularly canvas or heavyweight cotton fabrics. The two most important ingredients or active agents for a weather-resistant finish are photostabilizers and antimicrobial agents, since damage due to ultraviolet light and to biodeterioration from mildew- and rot-producing fungi is the most extensive. Most successful weather-resistant finishes are thus based on strategies and finishing techniques previously discussed in the sections on **antimicrobial finishes** and **photostabilizers**, and are made durable for extended periods of use outdoors by co-applying these agents with resins or other binding agents. Some of the more important weather-resistant finishes developed for cellulosic fabrics are listed in Table 4.14.

TABLE 4.14

Weather-resistant finishes for cellulosic fibers (adapted from ref. 82).

Major ingredients		
UV resistance	Antimicrobials	Binders & other agents
Various pigments	copper salts	chlorinated paraffins
Various pigments	copper salts	urea-formaldehyde resins
Chromic oxide	copper salts	------------
Light-stable vat dyes	acetylation metal salts	------------ urea and melamine-formaldehyde resins
Chromic oxide	copper metaborate	zirconyl ammonium carbonate
Titanium oxide	----------	poly(vinyl chloride)

As previously noted, the application of some antimicrobial agents to fibers produces unwanted side effects, such as the accelerated photodegradation of the fiber. This is particularly true for certain copper salts and halogenated phenols. Conversely, more recent developments have led to the incorporation or application of a single chemical agent that is both antimicrobial and light-resistant. Examples include cadmium selenide bound to cotton fabrics with resin and a titanium-tin salt used as an additive in spinning baths for synthetic fibers (ref. 73).

4.5.8 Shrinkproofing

In addition to the dimensional stabilization of fabrics during processing by heat and by mechanical techniques, a variety of shrinkproof finishes for textiles have been developed to prevent or to minimize their shrinkage during use and particularly during laundering. Since the wool fiber is particularly susceptible to irreversible shrinkage by various processes (see section on **dimensional stabilization**), most shrinkproof finishes have been developed for producing washable wool fabrics. Shrinkage of wool due to felting, for which six mechanisms have been proposed (ref. 12), is the primary type of shrinkage for which such finishes are applied. These finishes function by reducing the frictional differences between wool fibers that cause felting shrinkage or by anchoring the scales so they cannot slide past each other.

There are essentially three general approaches that have been used to shrinkproof woolen textiles. These are pretreatment with oxidizing agents, reducing agents or solvents, treatment with oligomers or polymers and combinations of the first two methods. Approaches using oxidizing agents are the oldest method, and are effective because the cuticle of the wool fiber is softened or degraded by oxidation. This results in an increase in μ_w (coefficient of with-scale friction) and/or a decrease in the frictional difference $\mu = \mu_a - \mu_w$, where μ_a is the coefficient of against-scale friction (ref. 12). Oxidizing agents commonly used for rendering wool shrink-resistant include dry and wet chlorine, dichloroisocyanuric acid, potassium permanganate, and peroxymonosulfuric acid. Sodium sulfite is the most commonly used reducing agent. Polar organic solvents or organic solvent mixtures are also employed as pretreatments.

Application of various oligomers and/or polymers imparts shrink resistance to wool fibers by increasing fiber-to-fiber bonding to minimize fiber movement and friction, by encapsulation (sometimes called scale masking) of the fibers with a thin polymeric film, and/or by deposition of large particles or aggregates of polymers to hold the fibers apart or by penetrating into the area beneath the scales and thereby gluing the scales so they cannot telescope shut or be further agitated during processing (sometimes called the stand-off mechanism) (ref. 12, 90).

Earlier approaches for imparting shrink resistance to woolen fabrics with polymers used various crosslinking agents or homopolymers such as melamine and urea-formaldehyde resins, polysiloxanes and polyacrylates. Interfacial polycondensation was also carried out in the presence of the wool fiber to form polyamides and polyureas. More recent commercial approaches have improved shrinkproofing processes by first pretreating the fiber (preferably by oxidative chlorination), followed by subsequent reduction to form reactive sites ($-NH_2$ and $-SH$ groups). This reduces the critical surface tension, and renders the fiber surface more conducive to uniform spreading of the polymer. It also increases the durability of the finish by chemical reaction between reactive sites on the wool fiber with the polymer (refs. 12, 90, 91).

Examples of commercially useful processes for shrinkproofing wool with polymers are diverse with regard to the type of polymer and the method of application. One process (Hercosett, a chlorohydrin adduct to a polyamide backbone) first subjects the wool fibers to mild oxidation, then reacts with the wool under its scales to form C-N and C-S covalent bonds. Another process (Synthappret LKF) involves reaction of the wool with isocyanate-terminated prepolymers that are modified with bisulfite to make them substantive to wool fibers in aqueous baths. A third process (Lankrolan SR3) uses the reaction of self-crosslinking Bunte salt terminated polyethers that also react with functional groups in the wool fiber under a variety of dry and wet continuous and batch processes (refs. 90, 91). Most of the reactive prepolymers are based on poly(propylene oxide triols) whose terminal functional groups crosslink to affix the polymer to the wool fiber. Representative prepolymers of these types are shown in Table 4.15.

Since the introduction of the above commercial processes, several pretreatments and combinations of polymers have been proposed to improve the shrinkproofing performance of wool fabrics. Pretreatment with the reducing agent sodium sulfite in isopropanol/water containing a cationic surfactant (quaternary ammonium salt) was effective in reducing shrinkage of knitted wool fabrics (ref. 92). The addition of small amounts of poly(ethyleneimine) to a reactive epoxy-functional polyacrylate emulsion was also effective as a shrinkproofing treatment (ref. 93). When poly(ethyleneimine) or an azeotropic mixture of benzene/ethanol/water was used as a wool fiber pre-treatment, this greatly improved the shrinkproofing efficiency of various polymers applied to the wool (ref. 94). It was also observed that polysiloxanes containing only

TABLE 4.15

Reactive prepolymers manufactured commercially from poly(propylene oxide triols). (adapted from ref. 91). Courtesy of Marcel Dekker Publishers.

Functional group	Crosslink in cured polymer	Trade name
-SH	--S--S--	Oligan 500 or Oligan SW
$-S_2O_3-$	--S--S--	Lankrolan SHR3
-N=C=O	-NH-C=O NH-	Synthappret LKF
$-CH_2S^+(C_2H_4OH)_2$	with urea- $-CH_2NHCONHCH_2-$	Synthappret WF
-NH-C=O SO_3^-	-NH-C=O NH-	Synthappret BAP

small amounts of Synthappret BAP [water-soluble poly(carbamoyl sulphonate) based on a poly(propylene oxide prepolymer)] were as effective in preventing wool shrinkage as commercial prepolymers and their reaction products (ref. 95).

The shrinkproofing of other types of fibers with polymers, such as cotton and cotton/polyester knitted fabrics, is usually accomplished by the application of nitrogenous resins that are normally applied to impart crease and wrinkle resistance to fibers. Thus, this dimensional stability is usually achieved when cotton and cotton blend fabrics are treated with durable press resins.

4.5.9 Resistance to abrasion and pilling

Resistance of fabrics and of yarns and fibers in fabrics to abrasion, unacceptable wear that occurs when one surface is rubbed against another, is influenced by many factors. Thus, it cannot be readily or easily predicted, although numerous laboratory-type abrasion testers have been designed and used in an attempt to simulate this wear. Such simulation exposes the textiles to mechanical deformation, changes in resiliency, and to failure and fatigue phenomena that are essentially time-dependent. In spite of this ambiguity, it is generally agreed that the composite or overall abrasion resistance of polyamides or nylons are the greatest for major types of fibers. Cotton, acrylics and modacrylics, polyester, high wet modulus rayon and wool fabrics have good to moderate abrasion resistance. Silk, viscose rayon, and cellulose acetate fabrics and fibers generally have poor resistance to abrasion.

Abrasion is but one facet of the complex phenomena of fiber wear and fatigue that leads to fiber breakage and transverse splitting, fibrillation and other mechanical modes of deformation and failure (see section in **Chapter 6** on mechanical properties of fibers). However, the relationship of abrasion resistance to fiber type is usually proportional to its energy-to-rupture value, i.e., the greater the energy-to-rupture, the greater the abrasion resistance.

The proper choice of yarn, fabric construction and blending of fiber types has been advantageously employed to improve the abrasion resistance of textiles. Examples include the blending of fibers with high abrasion resistance such as the polyamides with those of moderate abrasion resistance such as cotton, increasing the thread count and the number of plied yarns in a fabric, and choosing the proper fabric geometry and weave to withstand specific types of abrasion (e.g., flat, flex and/or edge abrasion) relative to the performance requirements of the textile product.

Application of additives and/or finishes to textiles may have little or no effect on their abrasion resistance or may dramatically decrease or increase their abrasion resistance. Cellulosic fabrics finished with nitrogenous resins and other crosslinking agents to impart wrinkle resistance undergo marked reduction in abrasion resistance, particularly with respect to loss of edge abrasion. It has been hypothesized that such a reduction is due to loss of tenacity and elongation of the crosslinked fibers, thus reducing their ability to absorb energy from friction and from wear. Alternatively, the nonuniform distribution of crosslinks in the fiber causes its load bearing elements to bear unevenly the stress, and thus increases fiber breakage and failure.

Finishes that lubricate or coat the fiber improve its ability to withstand some or all types of wear or abrasion. Lubrication, with substances such as polysiloxanes, is usually only beneficial for improving flex abrasion. Polymeric coatings applied to fabrics (e.g, Teflon™, poly(vinyl chloride) and polyurethanes) afford overall improvement in abrasion resistance. These coatings have an inherently high resistance to abrasion and thus absorb wear and friction rather than such mechanical stresses being absorbed by the fiber. More current approaches have utilized water-soluble polymers that are durable after curing or can become affixed to fabrics by insolubilization or crosslinking under mild processing conditions. For example, water-soluble polyamides that were dried onto cotton fabrics enhanced the abrasion resistance of the fabrics as much as 400% with weight gains of as little as 6% of the bound polyamide. The modified fabrics also appear to have good durability to washfastness (ref. 96). Investigations with crosslinked polyols that are discussed in the section on **multifunctional property improvements** provided similar wear enhancement when applied to different fabrics and fiber types.

Pilling is the entanglement of loose fiber ends to form balls or pills on the surface of fabrics. It is particularly problematic with high tenacity fibers such as polyester, since the fibers do not wear off due to abrasion, but instead remain on the surface and become entangled. Pilling may be reduced during processing and weaving by singeing (see discussion in **Chapter 1**) protruding fibers from the fabric and by using tightly woven fabric constructions when possible for specific fabric end uses. Most chemical approaches for minimizing or reducing pilling of fiber such as polyester parallel those used for preventing felting shrinkage of wool. These include hydrolysis, alcoholysis or ammonolysis to alter the surface of the polyester fibers. Alternate methods include the application of film-forming polymers to reduce the mobility of the fibers and thus minimize entanglement and subsequent pilling (ref. 97).

4.5.10 Antistatic fibers and finishes

All materials, including fibers, vary in their ability to acquire a static charge and transfer this charge from one material to another. The moisture content of fibers and relative humidity to which they are exposed significantly influences their ability to dissipate such charges (see section on electrical properties of textiles in **Chapter 6**). In most instances, it is not advantageous to rely totally on moisture-related fiber properties to dissipate or reduce static charge on textile fibers and fabrics. Thus, a variety of durable and nondurable finishes as well as inherently antistatic fibers have been developed.

Most antistatic agents applied to textiles function by increasing their rate of charge dissipation, although some agents also function by reducing the generation of static charge on the fiber. A variety of compounds and polymers have been used to impart static resistance to textiles. This is accomplished by either making the fibers more hydrophilic and thus more prone to dissipate static charge and/or by treating them with a good conductor or polyelectrolyte to achieve such an effect. Most antistatic finishes or fibers were developed for synthetic fibers with low to moderate moisture content such as polyamides, acrylics and polyesters.

General classes of antistatic agents used for nondurable and for durable finishes and for producing inherently static free fibers have been discussed and reviewed (refs. 98 and 99). These include; nitrogenous compounds with long aliphatic chains (such as amines, amides, and quaternary ammonium salts), sulfonic and phosphoric acid derivatives, various polyhydroxy compounds and their derivatives (such as cellulosics, starches, poly(vinyl alcohols) and poly(ethylene glycols)), esters of fatty acids and their derivatives, metals and carbon black. Most nondurable antistatic finishes for textiles involve application of surfactants such as quaternary ammonium salts. Durable antistatic finishes involve treatment of fabrics with polyelectrolytes such as ion-exchange resins or modified starches and cellulose derivatives bound to the fiber with urea-formaldehyde resins. Inherently static free fibers are usually prepared by either incorporating hydrophilic polymers into the fiber during spinning or by blending the static-prone fiber with a conductive element composed of aluminum, silver or carbon black.

There are a couple of newer developments in the theory and application of antistatic agents to textiles. An ion-transport mechanism for poly(ethylene oxides) has been experimentally confirmed by demonstrating that reduction of static charge is greater with aromatic polyethers than for aliphatic polyethers; it has also been demonstrated that the hydrophilicity of these polymers is not responsible for their excellent antistatic performance. Figure 4.28 shows the transport of electrostatic charges through polyether linkages. Good antistatic properties for polyamide, polyester and polypropylene fibers were observed when they were treated with phosphoric acid derivatives of poly(ethylene oxides); this approach offers an advantage over previous poly(ethylene oxide) systems and warrants further investigation and development (ref. 101).

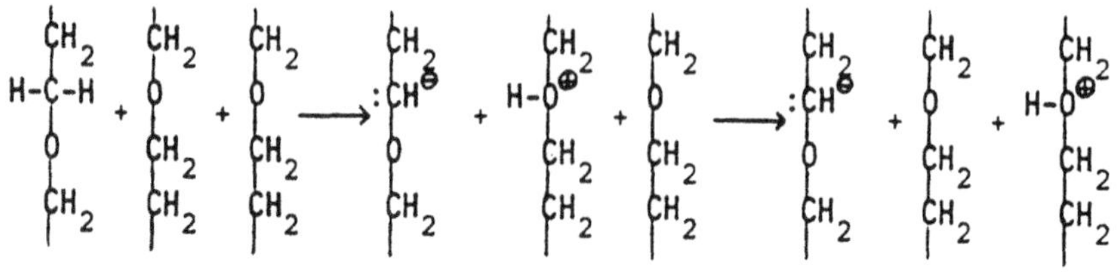

Figure 4.28. Ion-transport mechanism for the antistatic effect of polyethers (ref. 100). Courtesy of Melliand Textilberichte.

4.5.11 Multifunctional property improvements

The numerous functional and aesthetic property improvements of textiles have traditionally been achieved by focusing on a single property or at most two or three important property improvements. However, there has been a progressive development from imparting several important property improvements in fabrics by using several types of finishing agents in a treatment bath to using single polymer coatings and fixation to achieve these improvements. Weatherproof finishes (discussed earlier in this chapter) represent the earliest approach to impart multifunctional property improvements to textiles. The combination of flame retardants, water repellents and mildew-resistant agents applied to cotton tents and canvas in the early 1940's is a prototype example of imparting multifunctional properties to fabrics (ref. 57).

The second important stage for imparting multifunctional properties to textiles occurred in the 1970's. A combination of flame-retardants and other ingredients produced wool fabrics that were flame-retardant and resistant to oil, water and strong acids (ref. 91). The third important stage was the realization that fixation or attachment of specific polymers or copolymers to a variety of fibers produce a desired set of multifunctional property improvements. Such multifunctional property improvements are not readily attainable with several ingredients in a formulation nor economically feasible with multistep finishing processes. One example of this strategy is the application of aqueous emulsions of alkylpolysiloxane elastomers to fabrics comprised of various fiber types. The modified fabrics were noted to have improved stretch, dimensional stability, resilience or crease recovery, sewability and durability to washing and dry cleaning (ref. 102). A more diverse set of functional improvements was later disclosed by Vigo and his colleagues for insolubilization of polyethylene glycols by reaction with crosslinking resins under mild conditions. The modified fabrics can be produced from diverse fabric constructions and fiber types. This process results in the attachment of crosslinked polyethylene glycols to fibers to impart several improved functional properties. While the thermal adaptability of the modified fabrics is unique because of the latent heat provided by the bound polyols, many other properties are also improved. Other fabric properties that were enhanced include: resistance to wind, static charge, pilling, oily soils, bacteria and fungi and improved resiliency, flex life and absorbency (ref. 103). It is probable that such approaches will become more common for imparting improved functional properties by formulations of appropriate polymeric agents.

4.6 ENERGY CONSERVATION PROCESSES

4.6.1 Overview

The high cost of energy and materials used in textile processing has led to the development of techniques that consume less energy and materials. Since the major cost of energy in textile processing (preparatory steps, dyeing and finishing) is usually the removal of water or other liquids from the fiber, most effort has been directed to this area. However, some attention has also been given to using fewer textile auxiliaries, dyes, and/or finishing agents to achieve specific process objectives. Such special techniques may be broadly classified into three areas. The first group is that of low wet pickup methods, in which less water or other liquids are used in textile processing. Utilization of novel agents and/or recycling of dyes and other agents constitutes the second group. The third group is drying methods that expedite the removal of water or other liquids from the fiber during many or all the required drying steps in textile processing. These special techniques, their applicability to various textile processes, benefits derived from their use, and examples of specific types of techniques are listed in Table 4.16. (ref. 104).

4.6.2 Low wet pickup methods

Low wet pickup methods to conserve energy and to use less dye and/or chemical agents are diverse in scope and have been summarized in an excellent and

TABLE 4.16

Special textile techniques to conserve energy and use of materials (compiled and condensed from several chapters in ref. 104).

Type of technique	Textile applications	Benefits
Low wet pickup methods		
Special padding rolls and methods		
Special extractors and applicators	Scouring, bleaching, various finishing,	Less energy to remove liquids
Sprays and aerosols	dyeing, printing	
Foams		Less textile chemicals, dyes
Non-aqueous solvents		More uniform treatment
Radiation curing		
Novel reagents or their use		Less dye, water, energy used
Reactive catalysts	Dyeing and printing	
Dyebath reuse		Increase speed of processing
Drying methods	Any textile process (prep.,	Less energy to
Convective	dyeing/printing and/or	remove liquids
Conductive	finishing) where removal of	
Radiant	water or any other liquid	Increase speed
Mechanical removal	is required	of processing

informative critical review (ref. 105). These methods consist of low add-on expression and topical techniques and foam application techniques. Earlier efforts focused on use of emulsions composed of organic solvents with small amounts of water to increase the evaporation rate of liquid from the fiber after fixation or application of dyes or finishing agents. Except for solvent scouring of wool and of cotton/polyester blends (usually with solvents such as perchloroethylene), the use of organic solvents to reduce wet pickup on textiles has not gained commercial acceptance because of problems associated with toxicity, flammability, storage and disposal of these solvents.

For low wet pickup methods, a variety of special padding rolls, extractors and applicators have been developed to dispense or apply dyes and finishing agents. In expression techniques, excess liquor is removed from fabrics by squeezing or other methods. Removal is normally achieved by use of special rollers, vacuum extraction and air-jet ejectors. The Machnozzle (Figure 4.29) is a variation of an air-jet ejector in which steam passes at high speed through a narrow channel at the nozzle (B) as the fabric passes over guide rollers with mechanically driven tension rollers (A) keeping the fabric under tension against the nozzle. The amount of wet pickup can be as low as 5% for hydrophobic fiber types (polyester or polyamide) and as low as 25-30% for wool or cotton fabrics.

In topical techniques, only a limited amount of liquor is allowed to contact the fabric surface. Padding techniques include the use of engraved rollers, nip padding with a doctoring device, transfer padding systems such as lick or kiss rolls or loop-transfer, liquor transfer by wicking systems and spray systems with special nozzles.

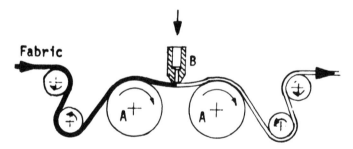

Figure 4.29. Machnozzle system for low wet pickup (ref. 105). Courtesy of The Textile Institute.

Although expression and topical techniques are used as low wet pickup methods, the application of dyes and functional finishes to textiles by contact with stable and semi-stable foams has evolved as the most commercially viable technique for achieving low wet pickup. There are several excellent reviews that discuss the properties of foams and their application to fabrics (refs. 106-108).

For a foam to form, there must be a reduction of surface tension due to oriented adsorption and air must be introduced into the liquor by mechanical action. All foams become unstable; however, certain foam structures are more stable than others.

Unstable foams are usually spherical [Fig. 4.30 (a)] but may become more stable by conversion into a honeycomb or polyhedral structure [Figs. 4.30 (b) and (c) respectively]. Thus, critical parameters in foam dyeing and finishing are the stability of the foam and its density. The latter is achieved by using different blow ratios, i.e., the ratio of liquid to air necessary to produce the foam. Blow ratios may vary in textile applications from approximately 5:1 to as high as 500:1.

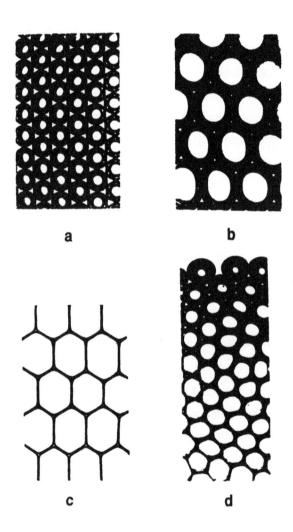

Figure 4.30. Structure of foams: (a) spherical; (b) honeycomb; (c) polyhedral; (d) change from a spherical foam via a honeycomb into a polyhedral configuration (ref. 106).

A variety of techniques are suitable for applying foam to one or both sides of a fabric; these techniques have been discussed earlier for foam dyeing and are illustrated in **Figure 3.19** in the previous chapter. Doctor blades, rollers, pressure or suction may be used for a single side application to fabrics. For two-sided applications, a horizontal padder or a pair of transfer or kiss rolls may be employed to apply the foam. The FFT process (schematic of the applicator shown in Figure 4.31) is the most widely used foam finishing and dyeing process for fabrics. The applicator is designed to distribute foam uniformly across the width of the fabric as it continuously passes through the machine. The foam collapses on contact with the fabric and the wet pick-up may be controlled by the fabric processing speed, fabric weight and liquor flow rate. This process minimizes insufficient or excess delivery of foam to the fabric.

There are numerous variables that must be controlled or monitored for successful application of foam to fabrics to obtain desired property improvements. Most processes consist of the following four step sequence: (a) formulation to achieve foamability; (b) generation of foam; (c) application of foam to the substrate; and (d) destruction of the foam. Several variables (shown in Table 4.17) must be considered for the treating liquor, foam, fabric and process. Perhaps the two most important are preparation of a suitable formulation to generate foam and the foam absorption rate on the fabric (ref. 108). Further refinements in foam application may make it possible to combine or have tandem processes for preparation, dyeing and finishing.

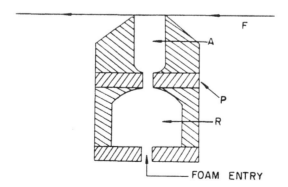

Figure 4.31. Schematic drawing of FFT process foam applicator. A = foam application chamber; P = distribution plate; R = foam relaxation chamber; F = fabric (ref. 108). Courtesy of Marcel Dekker Publishers.

TABLE 4.17

Variables to be considered in a foam finishing system (ref. 108). Courtesy of Marcel Dekker Publishers.

Treating Liquor	Foam	Fabric	Process
Composition	Drainage rate/ half-life	Composition	Dry add-on
Foamability	Wetting rate	Wettability	Foam-fabric contact time and pressure
Total solids	Density	Cleanliness/preparation	Foam residence time
Wetting rate	Viscosity	Construction	Wet pickup
Viscosity	Bubble size	Porosity	Fabric speed
	Shear stability	Weight	Foam generation variables
		Moisture content	Temperature

Another low wet pickup technique that has been explored is the low temperature curing of textiles with radiation sources such as ultraviolet light or electron beams (refs. 109-111). Ultraviolet radiation sources require less capital investment than electron accelerators. However, the latter is being used more frequently because the electron beams penetrate more deeply into the fibrous substrate than ultraviolet rays and result in more efficient and complete polymerization of monomers. Electron beam methods also do not generally require solvents nor heat and materials can be processed for extremely short times at high production rates. Nevertheless, in some cases working in an inert gas atmosphere or vacuum is required and the costs of suitable polymers or monomers to modify the fabrics are generally higher than conventional finishing, coating or dyeing processes. A variety of dyeing and finishing operations may be conducted by electron beam processing with diverse types of polymers. Polyester and polyether acrylates, epoxy resins, silicon acrylates and other monomers may be used to coat textiles, improve resiliency, absorbency, antistatic behavior, flame resistance and water repellency. An example of radiation curing with ultraviolet light is the pigment printing of cotton/polyester fabrics ultraviolet light with pastes that do not contain water or other liquids (ref. 109). It is probable that electron beam devices will be more frequently employed when they become more economically competitive with conventional finishing processes.

4.6.3 Novel reagents and their use

Developments of agents, dyes, and textile auxiliaries (applied to textiles in low concentration and at mild fixation or processing temperatures and the recycling of such materials for further use in textile processes) represent another strategy that has been explored for conserving energy and materials (ref. 104). A promising development for processing under mild conditions involves the use of highly reactive catalysts for pigment printing of cotton and cotton/polyester fabrics under mild curing conditions. When a highly active catalyst system (equimolar amount of $MgCl_2.6H_2O$/ citric acid) was incorporated into a pigment padding formulation instead of a conventional catalyst (e.g., NH_4Cl), the pigment was fixed in less than a minute at 100°C compared to fixation at 3-4 minutes at 150°C with the conventional catalyst.

The reuse of dyes in baths is another development of interest. Dye reuse is a technique suitable for the atmospheric disperse dyeing of nylon and polyester carpets, nylon hosiery and pressure disperse dyeing of polyester packaged yarns (ref. 104). Good lot-to-lot shade correlations were obtained with reuse of a variety of dye types and fiber blends. Dyebath reuse is monitored spectrophotometrically, and in addition to using less dye, has also been shown to save energy by using less liquor, reusing hot liquors and requiring less treatment of stream effluents.

4.6.4 Drying methods

The drying of textiles during processing and manufacture is usually achieved by some form of heat transfer (convection, conduction and/or radiation) and less frequently by mechanical removal of water or other liquids from the fiber (ref. 112, 113). It is generally agreed that the moisture content of the textile decreases exponentially with drying time [Figure 4.32 (a)]. It has also been established that the drying rates of porous materials such as textiles are characterized by a decrease of the moisture content in three distinct stages: an initial warm-up period, a constant drying rate period, and a falling drying rate period [Figure 4.32 (b)].

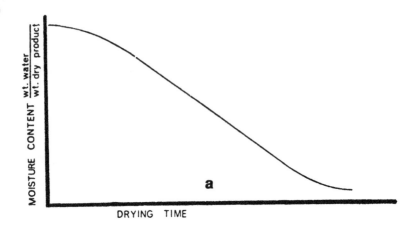

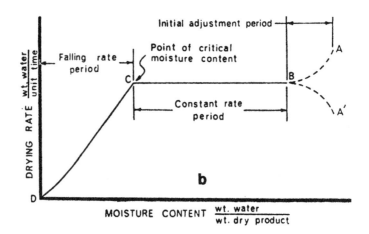

Figure 4.32. Conventional heating: (a) typical drying curve, (b) typical drying rate curve (ref. 112). Courtesy of Textile Research Institute.

Since convective drying is still the most commonly employed technique (e.g., the use of tenter frames), there have been several mathematical and conceptual models advanced that compute and describe the minimum energy needed to dry textiles by this mode of heat transfer. Beard (ref. 114) has developed three differential equations that can be simultaneously solved with a digital computer. These equations describe the drying process in a tenter frame in terms of the enthalpy balance of dry and moist layers in the fabric. Chakravatry and Slater (ref. 115) have empirically derived a unified drying equation that describes both the constant and falling rate portions of the drying rate curve and that is useful over the entire drying range and is valid for wide variations in fiber type and textile moisture regain values.

Drying by conduction is also commonly employed in textile processing by contact of the textile with a hot nonporous surface (such as steam cans). However, since the energy and moisture flow occur in the same direction in conduction drying (Figure 4.33), the surface of the fabric will tend to dry much more rapidly than the interior of the fabric. Thus, overdrying could cause damage to the fabric surface. In contrast, the use of radiofrequency and microwave drying in textile processing appears to occur by a different mechanism, i.e., uniform vaporization of the liquid through the textile or porous material (Figure 4.34). Use of low frequency radiation to dry textiles

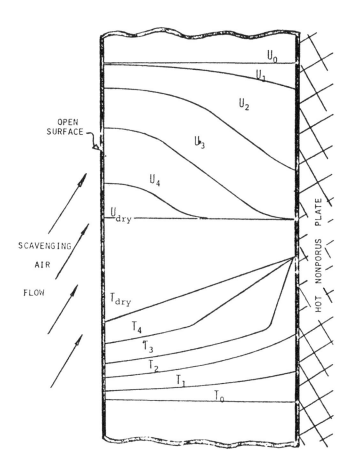

Figure 4.33. Typical internal temperature T and moisture U profiles for drying by conduction heating (ref. 112). Courtesy of Textile Research Institute.

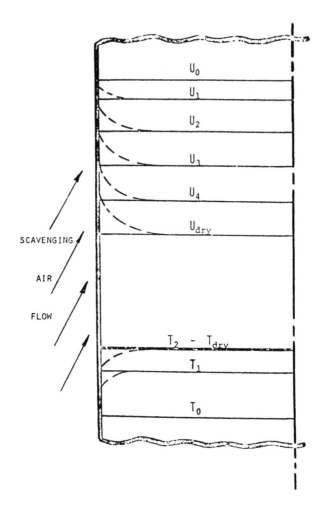

Fig. 4.34. Typical internal temperature T and moisture U profiles for drying by microwave heating; the dashed lines represent the profiles when the surface is cooled by the scavenging air (ref. 112). Courtesy of Textile Research Institute.

offers great promise because of its selectivity, but is still used to only a minor extent compared to convective and conductive drying methods. Other types of radiant drying, e.g., infrared and dielectric heating, have also received some attention as alternative methods of drying textiles in a reproducible and energy efficient manner.

Mechanical methods of removing water or other liquids from textiles have been employed to some extent. These include conventional methods such as mangling (passing the fabric through one or more pads or rollers) and newer methods such as the use of vacuum extractors or other devices discussed earlier such as the

Machnozzle; the latter creates a pressure differential across the fabrics to facilitate the removal of liquids. Although mechanical dewatering by deliquification of textiles during processing is relatively inexpensive and readily performed, it is limited to the removal of unbound water from textiles and is profoundly influenced by the surface tension between the liquid to be removed and the fiber surface (ref. 104).

GENERAL REFERENCES

Fachhochshule Darmstadt and N. Verse (Eds.), **Flock '82. 7th International Flock Symposium**. Darmstadt, Germany, Sept. 27-29, 1982. Fachhochschule Darmstadt, Fachbereich Kunstofftechnik, Darmstadt, Germany (FRG), 1982, 252 pp.

M. Lewin and S. B. Sello (Eds.), **Handbook of Fiber Science and Technology, Vol. II. Functional Finishes, Pt. A, Chemical Processing of Fibers and Fabrics**, Marcel Dekker, New York, 1983, 431 pp.

M. Lewin and S. B. Sello (Eds.), **Handbook of Fiber Science and Technology, Vol. II. Functional Finishes, Pt. B, Chemical Processing of Fibers and Fabrics**, Marcel Dekker, New York, 1984, 515 pp.

M. Lewin and S. B. Sello (Eds.), **Handbook of Fiber Science and Technology: Vol. III. High Technology Fibers. Pt. A.**, Marcel Dekker, New York, 1985, 397 pp.

TEXT REFERENCES

1 J. W. S. Hearle, **The nature of setting**, in: J. W. S. Hearle and L. W. C. Miles (Eds.), The Setting of Fibres and Fabrics, Merrow Publishing, Watford Herts, England, 1971, Chapter 1.

2 J. Lindberg, **Fabric setting**, in: J. W. S. Hearle and L. W. C. Miles (Eds.), The Setting of Fibres and Fabrics, Merrow Publishing, Watford Herts, England, 1971, Chapter 8.

3 J. S. Heaton, **Techniques of fabric setting**, in: J. W. S. Hearle and L. W. C. Miles (Eds.), The Setting of Fibres and Fabrics, Merrow Publishing, Watford Herts, England, 1971, Chapter 9.

4 R. M. Crow, T. J. Gillespie and K. Slater, **The drying of textile fabrics. Part I: The relationship between regain, dewpoint temperature, and air temperature**, J. Text. Inst. 65 (2) (1974) 75-81.

5 J. A. Medley, **The drying of textile fabrics**, J. Text. Inst. 65 (8) (1974) 450-451.

6 W. J. Morris, **The use of heat in the production of synthetic-fibre textiles**, Textiles 10 (1) (1981) 1-10.

284

7 V. B. Gupta, C. Ramesh and A. K. Gupta, **Structure-property relationship in heat-set poly(ethylene terephthalate) fibers. I. Structure and morphology**, J. Appl. Poly. Sci. 29 (1984) 3115-3129.

8 V. B. Gupta, C. Ramesh and A. K. Gupta, **Structure-property relationship in heat-set poly(ethylene terephthalate) fibers. II. Thermal behavior and morphology**, J. Appl. Poly. Sci. 29 (1984) 3727-3739.

9 V. B. Gupta, C. Ramesh and A. K. Gupta, **Structure-property relationship in heat-set poly(ethylene terephthalate) fibers. III. Stress relaxation behavior**, J. Appl. Poly. Sci. 29 (1984) 4203-4218.

10 V. B. Gupta, C. Ramesh and A. K. Gupta, **Structure-property relationship in heat-set poly(ethylene terephthalate) fibers. IV. Recovery behavior**, J. Appl. Poly. Sci. 29 (1984) 4219-4235.

11 P. N. Peszkin and J. M. Schultz, **Kinetics of fiber heat treatment. I. Experimental apparatus**, J. Appl. Poly. Sci. 30 (1985) 2689-2693.

12 K. R. Makinson, **Shrinkproofing of Wool**, Marcel Dekker Inc., New York, 1979, pp. 113-118.

13 T. Shaw and M. A. White, **The chemical technology of wool finishing**, in: S. B. Sello and M. Lewin (Eds.), Functional Finishes, Pt. B, Vol. II, Marcel Dekker, New York, 1984, pp. 329-346.

14 S. De Jong, **Linear viscoelasticity applied to wool setting treatments**, Text. Res. J. 55 (11) (1985) 647-653.

15 N. G. Trollip and F. Raabe, **The use of sulphur dioxide gas in the setting of wool**, SAWTRI Tech. Rept. No. 590 (Feb., 1987) 1-10.

16 W. Hoff, **Tendencies in the design of calenders - The universal calendar for mangling, smoothing, matt-finishing and schreiner effects**, Text. Praxis Intl. 38 (7) (1983) 684-686; 38 (12) (1983) XXI-XXIII.

17 S. W. Poser, **Calenders regain the finishing limelight**, Textile World 135 (8) (1985) 53-56.

18 B. Ashkenazi, **Latest developments in calenders and their effects**, Colourage 37 (12) (1990) 37-40.

19 H. Neuhausen, A. Wandke and H.-G. Wroblowski, **Mechanical process in dry finishing: raising vs. emerizing**, Intl. Text. Bull. Dyeing/Printing/Finishing Ed. 35 (4th Quarter) (1989) 41-52.

20 A. Wehlow, **The technique of flocking - resurgence of a textile sector**, Intl. Text. Bull. Dyeing/Printing/Finishing Ed. 33 (2nd Quarter) (1987) 60-79.

21   D. Brokmeier, **New findings in flocking textile substrates**, Chemiefasern/Textileind. 30/82 (1980) 956-961/E111-E113.

22   G. Brenner and W. Rompp, **Textile flocking**, Textilveredlung 20 (4) (1985) 129-135.

23   V. A. Semenov, S. P. Hersh and B. P. Gupta, **Maximum fiber-packing density in electrostatic flocking**, Text. Res. J. 51 (12) (1981) 768-773.

24   P. J. Kassenbeck, **Silent discharge process and its application to wool**, Bull. Inst. Textile France 18 (110) (1963) 7-33.

25   W. J. Thorsen, **Improvement of cotton spinnability, strength, and abrasion resistance by corona treatment**, Text. Res. J. 36 (5) (1963) 455-458.

26   A. E. Pavlath, **Plasma treatment of natural materials**, in: J. R. Hollahan and A. T. Bell (Eds.), Tech. Appl. Plasma Chem. 1974, Wiley, New York, 1984, pp. 149-175.

27   M. S. Aronoff, **Differential dyeing via laser irradiation**, Appl. Poly. Sci. 31 (1977) 397-405.

28   W. Rakowski, M. Okoniewski, K. Bartos and J. Zawadzki, **Plasma treatment of textiles- -potential applications and future prospects**, Mell. Textilber. 63(4) (1982) 307-313.

29   F. N. Teumac, **Coated fabrics**, in: M. Grayson (Exec. ed.), Kirk-Othmer Encyclopedia of Chemical Technology, Vol. 6, 3rd Ed., Wiley-Interscience, New York, 1979, pp. 377-396.

30   T. H. Ferrigno, **Fundamental principles of formulating coatings for fabrics**, J. Coated Fabrics 17 (1988) 265-271.

31   S. C. Zink, **Coating processes**, in: M. Grayson (Exec. ed.), Kirk-Othmer Encyclopedia of Chemical Technology, Vol. 6, 3rd Ed., Wiley-Interscience, New York, 1979, pp. 397-426.

32   B. M. Latta, **Coating industrial fabrics - a review**, J. Coated Fabrics 11 (1982) 111-121.

33   R. R. Grant, **Coating and laminating industrial fabrics**, J. Coated Fabrics 12 (1983) 196-212.

34   J. A. Pasquale III, **Coating machinery and techniques - a review**, J. Coated Fabrics 15 (1986) 263-274.

35   J. A. Brocks, **Modern coating technology - synthesis of processing technology and chemistry**, Intl. Text. Bull./Dyeing, Printing and Finishing Ed. 3 (1987) 64-78.

36   F. Kassack, **New developments in application technology and chemistry of polyurethanes for fabric coating**, Text. Chem. Color. 5 (8) (1973) 21-29.

37   J. T. Renshaw, **Vinyl plastisol coated fabrics**, J. Coated Fabrics 10 (1980) 151-166.

38 E. Reiss, **A survey of carpet back-coating materials. Pt. II**, Text. Inst. & Ind. 16 (5) (1978) 114-119.

39 G. Nelson, **Microencapsulates in textile coloration and finishing**, Rev. Prog. Color. Rel. Topics 21 (1) (1991) 72-85.

40 M. H. Gutcho, **Microcapsules and Microencapsulation Techniques**, Noyes Data Corp., Park Ridge, N. J., 1976, p. 339.

41 Anonymous, **Getting ready to tape the bugs,** Chem. Week 115 (8) (Aug. 21, 1974) 43-44.

42 S. Davies, **Japanese high-technology yarns and fibres for fashion**, Text. Horizons 8 (12) (1988) 45-47.

43 J. T. Marsh, **Self-smoothing Fabrics**, Chapman and Hall Ltd., London, England, 1962, pp. 3 and 353.

44 **BASF Textile Finishing Manual**, BASF A. G., Ludwigshafen, W. Germany, 1973, pp. 24 and 57.

45 J. J. Willard, G. C. Tesoro and E. I. Valko, **Effect of radial location of cross links in cellulosic fibers on fabric properties**, Text. Res. J. 39 (5) (1969) 413-421.

46 H. Petersen, **Cross-linking with formaldehyde-containing reactants**, in: M. Lewin and S. B. Sello (Eds.), Handbook of Fiber Science and Technology, Vol. II. Functional Finishes, Pt. A, Chemical Processing of Fibers and Fabrics, Marcel Dekker, New York, 1983, Chapter 2.

47 H. Petersen, **The chemistry of crease-resist crosslinking agents**, Rev. Prog. Color. Rel. Topics 17 (1987) 7-22.

48 H. Petersen and P. S. Pai, **Reagents for low-formaldehyde finishing of textiles**, Text. Res. J. 51 (4) (1981) 282-302.

49 S. B. Sello, **Easy-care finishing of cellulosic and cellulose-containing textiles with low- or zero-formaldehyde cross-linkers**, J. Soc. Dyer. Color. 101 (3) (1985) 99-105.

50 C. M. Welch, **Formaldehyde-free durable press finishes**, Rev. Prog. Color. Rel. Topics 22 (1992) 32-41.

51 J. S. Bruno and T. L. Vigo, **Temperature-adaptable textiles with multifunctional properties**, Proc. AATCC Natl. Tech. Conf. (1987) 258-264.

52 P. Maurer, **Alternatives to conventional resin finishing processes: the Pretavyl process**, Text. Praxis Intl. 46 (3) (1991) 235-239.

53 W. A. Reeves, G. L. Drake, Jr. and R. M. Perkins, **Fire Resistant Textiles Handbook**, Technomic Publishing Co., Westport, CT., 1974, pp. 45 and 86.

54 G. C. Tesoro, **Chemical modification of polymers with flame-retardant compounds**, J. Poly. Sci., Macro. Rev. 13 (1978) 283-353.

55 M. Lewin, **Flame retardance of fabrics**, in: M. Lewin and S. B. Sello (Eds.), Handbook of Fiber Science and Technology, Vol. II. Functional Finishes, Pt. B, Chemical Processing of Fibers and Fabrics, Marcel Dekker, New York, 1984, Chapter 1.

56 D. W. Van Krevelen, **Flame resistance of chemical fibers**, in: M. Lewin (Ed.), Fiber Science, Wiley-Interscience, New York, J. Appl. Polym. Sci. Appl. Polym. Symp. 31 (1977) 269-292.

57 E. C. Clayton and L. L. Heffner, **Impregnating cellulosic materials such as cotton duck, to render them resistant to fire, water and mildew**, U. S. Pat. 2,299,612; Oct. 20, 1942.

58 R. Aenishänslin, C. Guth, P. Hofmann, A. Maeder and H. Nachbur, **A new chemical approach to durable flame-retardant cotton fabrics**, Text. Res. J. 39 (4) (1969) 375-381.

59 L. E. A. Godfrey and J. W. Schappel, **Alkoxyphosphazenes as flame retardants for rayon**, Ind. Eng. Chem. Prod. Res. Dev. 9 (4) (1970) 426-436.

60 L. Benisek, **Improvement of the natural flame resistance of wool. Part I: metal-complex applications**, J. Text. Inst. 65 (2) (1974) 102-108.

61 H. Stepniczka, **Flame-retardant polyester textiles. Part I**, Textilveredlung 10 (3) (1975) 188-200.

62 G. L. Drake, Jr., **Flame retardants for textiles**, in: M. Grayson (Exec. ed.), Kirk- Othmer Encyclopedia of Chemical Technology, Vol. 10, 3d Ed., Wiley-Interscience, New York, 1980, pp. 420-444.

63 A. R. Horrocks, **Flame-retardant finishing of textiles**, Rev. Prog. Color. Rel. Topics 16 (1983) 62-101.

64 P. K. Chatterjee (Ed.), **Absorbency**, Elsevier Publishers, Amsterdam, 1985, 333p.

65 P. Hardt, **Chemistry and properties of hydrophilic finishes**, Text. Praxis Intl. 45 (4) (1990) 387-392.

66 E. Kissa, **Soil-release finishes**, in: M. Lewin and S. B. Sello (Eds.), Handbook of Fiber Science and Technology, Vol. II. Functional Finishes, Pt. B., Chemical Processing of Fibers and Fabrics, Marcel Dekker, New York, 1984, Chapter 3.

67 E. Kissa, **Repellent finishes**, in: M. Lewin and S. B. Sello (Eds.), Handbook of Fiber Science and Technology, Vol. II. Functional Finishes, Pt. B., Chemical Processing of Fibers and Fabrics, Marcel Dekker, New York, 1984, Chapter 2.

68 M. Hayek, **Waterproofing and water/oil repellency**, in: M. Grayson (Exec. ed.), Encyl. Chem. Tech., Vol. 24, Wiley-Interscience, New York, 1984, pp. 442-465.

69 D. J. Gohlke and J. C. Tanner, **Gore-Tex^R waterproof breathable fabrics**, J. Coated Fabrics 6 (July 1976) 28-38.

70 E. Reagan, **Comparing the high-tech fabrics**, Outside Business (April 1989) 43-49.

71 P. W. Harris and D. A. Hangey, **Stain resist chemistry for nylon 6 carpet**, Text. Chem. Color. 21 (11) (1989) 25-30.

72 T. F. Cooke and H.-D. Weigmann, **Stain blockers for nylon fibres**, Rev. Prog. Color. Rel. Topics 20 (1991) 10-18.

73 T. L. Vigo, **Protection of textiles from biological attack**, in: M. Lewin and S. B. Sello (Eds.), Handbook of Fiber Science and Technology, Vol. II. Functional Finishes, Pt. A., Chemical Processing of Fibers and Fabrics, Marcel Dekker, New York, 1983, Chapter 4.

74 T. L. Vigo, **Protective clothing effective against biohazards**, in: M. Raheel (Ed.), Protective Clothing Systems and Materials, Marcel Dekker, New York, 1994, Chapter 9.

75 R. J. Mayfield, **Mothproofing**, Textile Progress, 11 (4) (1982) 1-11.

76 D. M. Lewis and T. Shaw, **Insectproofing of wool**, Rev. Prog. Color. Rel. Topics 17 (1987) 86-94.

77 D. D. Gagliardi, **Antibacterial finishes**, Am. Dyest. Reptr. 51 (1962) P49-58.

78 C. D. Potter, **Porous acrylic fibres for controlled release applications**, J. Coated Fabrics 18 (2) (1989) 259-272.

79 M. Lewin and S. B. Sello (Eds.), **Handbook of Fiber Science and Technology: Vol. III. High Technology Fibers. Pt. A.**, Marcel Dekker, New York, 1985, 397p.

80 J. F. McKellar and N. S. Allen, **Photochemistry of Man-Made Polymers**, Applied Science Publishers, Essex, England, 1979.

81 B. Rånby and J. F. Rabek, **Photodegradation, Photo-oxidation and Photo-stabilization of Polymers**, Wiley-Interscience, New York, 1975.

82 T. L. Vigo, **Preservation of natural textile fibers-historical perspectives**, in: J. C. Williams (Ed.), Preservation of Paper and Textiles of Historic and Artistic Value, Adv. Chem. 164, American Chemical Society, Washington, D. C., 1977, pp. 189-207.

83 C. M. Carr and I. H. Leaver, **Photoprotective agents for wool. Synergism between uv absorbers and antioxidants**, J. Appl. Poly. Sci. 33 (1987) 2087-2095.

84 R. L. Gray, **Stabilization of polypropylene fiber and tape for geotextiles**, in: T. L. Vigo and A. F. Turbak (Eds.), High-Tech Fibrous Materials: Composites, Biomedical Materials, Protective Clothing and Geotextiles, Am. Chem. Soc. Symp. 457, American Chemical Society, Washington, D. C., 1991, Chapter 21.

85 K. Chandler, R. C. Anand and I. K. Varma, **Degradation of polypropylene. Effect of phosphorylated poly(2,6-dimethyl-1,4-phenylene oxide)**, Angew. Makromol. Chem. 127 (1984) 137-143.

86 S. H. Zeronian, **Conservation of textiles manufactured from man-made fibers**, in: J. C. Williams (Ed.), Preservation of Paper and Textiles of Historic and Artistic Value, Adv. Chem. 164, American Chemical Society, Washington, D. C., 1977, pp. 208-227.

87 J. B. Upham and V. S. Salvin, **Effect of Air Pollutants on Textile Fibers and Dyes**, EPA-650/3-74-008, Research Triangle Park, N. C., 1975.

88 H. R. Cooper, **Yellowing of textiles due to atmospheric pollutants**, in: P. W. Harrison (Ed.), Update on Yellowing, The Textile Institute, Manchester, England, Text. Prog. 15 (4) (1987) 1-6.

89 B. Walters, B. C. Goswami and T. L. Vigo, **Sorption of air pollutants onto textiles**, Text. Res. J. 53 (6) (1983) 354-360.

90 D. M. Lewis, **Continuous dyeing and polymer shrink-resist processes for wool**, J. Soc. Dyer. Color. 93 (4) (1977) 105-113.

91 T. Shaw and M. A. White, **The chemical technology of wool finishing**, in: M. Lewin and S. B. Sello (Eds.), Handbook of Fiber Science and Technology, Vol. II. Functional Finishes, Pt. B, Chemical Processing of Fibers and Fabrics, Marcel Dekker, New York, 1984, Chapter 5.

92 P. Erra, J. García Domínguez, R. Juliá and R. Infante, **Influence of an organic solvent and a cationic surfactant on sulphite treatments for preventing shrinkage of wool fabrics**, Text. Res. J. 53 (11) (1983) 665-669.

93 J. H. Brooks and M. S. Rahman, **Effect of shrinkproofing pretreatments with solvents and polyamines on the surface energy of wool**, Text. Res. J. (56) (1986) 473-475.

94 T. Jellinek, **Shrink-resist treatment of wool with epoxy-functional acrylic emulsions -- polyethylene polyamines as curing agents**, Text. Res. J. 53 (12) (1983) 763-771.

95 J. R. Cook, **Shrink-resisting wool with aqueous silicon polymers: the effect of added Synthappret BAP**, J. Text. Inst. 75 (3) (1984) 191-195.

96 W. B. Achwal, M. K. Praharaj and A. S. Thakur, **New techniques in finishing**, Colourage 32 (3) (1985) 15-22.

97 J. Hürten, **How can the pilling behaviour of PES fabrics be controlled by finishing?**, Text. Praxis Intl. 33 (7) (1978) 832-835.

98 S. B. Sello and C. V. Stevens, **Antistatic treatments**, in: M. Lewin and S. B. Sello (Eds.), Handbook of Fiber Science and Technology, Vol. II. Functional Finishes, Pt. B, Chemical Processing of Fibers and Fabrics, Marcel Dekker, New York, 1984, Chapter 4.

99 D. M. Brown and M. T. Pailthorpe, **Antistatic fibres and finishes**, Rev. Prog. Color. Rel. Topics 16 (1986) 8-15.

100 R. Garvanska, K. Draganova and K. Dimov, **Mechanism of the antistatic effect of polyether compounds**, Mell. Textilber. 64 (12) (1983) 921-924.

101 A. Polowinksa, L. Szosland and R. Jantas, **New phosphoric derivatives of poly (ethylene glycols) as antistatic agents for man-made fibres**, Acta Polymerica 38 (2) (1987) 125-131.

102 J. V. Isharani, **Ultratex-new breed of textile fibers**, Proc. AATCC Natl. Tech. Conf. 1982, pp. 144-153.

103 T. L. Vigo and J. S. Bruno, **Improvement of various properties of fiber surfaces containing crosslinked polyethylene glycols**, J. Appl. Poly. Sci. 37 (1989) 371-379.

104 T. L. Vigo and L. J. Nowacki (Eds.), **Energy Conservation in Textile and Polymer Processing**, ACS Symp. Series No. 107, American Chemical Society, Washington, D.C., 1979, Chapters 10-17.

105 G. H. J. van der Walt, et. al., **Low liquor dyeing and finishing**, Text. Progress 14 (2) (1986) 60pp.

106 D. Fiebig, **Chemical and physical properties of foams**, Text. Praxis Intl. 37 (4) (1982) 392-398.

107 T. F. Cooke, **Foam wet processing in the textile industry**, Text. Chem. Color. 15 (5) (1983) 74-85.

108 G. M. Bryant and A. T. Walter, **Finishing with foam**, in: M. Lewin and S. B. Sello (Eds.), Handbook of Fiber Science and Technology, Vol. II. Functional Finishes, Pt. A, Chemical Processing of Fibers and Fabrics, Marcel Dekker, New York, 1983, Chapter 3.

109 W. K. Walsh and W. Oraby, **Radiation processing**, in: M. Lewin and S. B. Sello (Eds.), Handbook of Fiber Science and Technology, Vol. II. Functional Finishes, Pt. B, Chemical Processing of Fibers and Fabrics, Marcel Dekker, New York, 1984, Chapter 6.

110 U. Einsele and H. Herlinger, **Electron-beam induced polymerization reactions and their application in textile finishing processes. Pt. I. State of the art and potential applications**, Text. Praxis. Intl. 42 (8) (1987) 848-852.

111 U. Einsele and H. Herlinger, **Electron-beam induced polymerization reactions and their application in textile finishing processes. Pt. II. Physico-chemical properties**, Text. Praxis. Intl. 42 (11) (1987) 1371-1373.

112 D. W. Lyons and C. T. Vollers, **The drying of fibrous materials**, Text. Res. J. 41 (8) (1971) 661-668.

113 L. W. C. Miles, **The drying of textile materials**, Rev. Prog. Color. Rel. Topics 15 (1985) 21-24.

114 J. N. Beard, Jr., **More efficient tenter frame operations through mathematical modeling**, Text. Chem. Color. 8 (3) (1976) 30-33.

115 R. K. Chakravarty and K. Slater, **A unified drying equation for automatic control in stentering**, J. Text. Inst. 68 (12) (1978) 370-378.

Chapter 5

TEXTILE PERFORMANCE DETERMINED BY CHEMICAL AND INSTRUMENTAL METHODS

5.1 INTRODUCTION

The end use performance of textiles is influenced by many factors, and attempts have been made to simulate and predict their performance as well as to assess damage caused by their exposure to physical, chemical and biological agents or influences or to combinations of these factors. There are numerous well-known standardized tests that have been specifically devised for evaluating and predicting textile performance that will be separately discussed in **Chapter 6**. However, such test methods cannot alone be used to assess or predict the actual performance of textiles concerning enhancement or deterioration of their functional and aesthetic characteristics. Application of various wet chemical methods and of instrumentation used in chemistry, physics, and polymer science are usually necessary to augment information obtained from standard test methods for proper evaluation of textile performance.

Such methods and techniques for evaluating, predicting, and/or simulating the performance of textiles with regard to their functional and aesthetic properties may be broadly classified into: (a) wet chemical analysis; (b) instrumental methods such as chromatography, spectroscopy, X-ray diffraction and other techniques commonly employed in the physical and biological sciences and in engineering; and (c) color theory and measurement of various types of colorfastness behavior of textiles. It is not the intent of this book to discuss the principles of chemical and instrumental analysis (with the exception of color theory and measurement), but to indicate the utility of these methods for characterizing fibrous materials and assessing their performance. Table 5.1 gives an overview of the various techniques in each of these three categories.

5.2 WET CHEMICAL METHODS

5.2.1 <u>Elemental analysis</u>

Wet chemical methods and analyses have been adopted to identify textile fibers, dyes and finishes and to evaluate and predict their performance. Elemental analysis is frequently used to determine additives and/or finishing agents that have been applied to textiles during their processing and manufacture to ascertain whether these substances are retained during exposure and use to various agents and environments. For example, many textiles are treated with flame retardants containing phosphorus and nitrogen; changes in their phosphorus/nitrogen (P/N) ratio and the accumulation of certain cations in the fiber during laundering (e.g., calcium and/or magnesium) may adversely affect the flammability of the fabric, and thus are amenable to analysis for these particular chemical elements.

TABLE 5.1

Representative techniques and examples for evaluating end-use performance of textiles and dyes by wet chemical methods

Wet chemical methods	Examples of applications
Elemental analysis	P/N ratio for flame-retardants
End group analysis	Determine average chain length
Changes in solubility and swelling	Increased swelling in wool indicates damage to that fiber
Viscosity	Determine molecular weight of sizing material such as poly(vinyl alcohol)
Changes in dyeing behavior	Differentiate cellulosic from polyester fibers by response to a mixture of direct and disperse dyes
Colorimetry	Violet complexes formed when polyester and/or cellulose react with hydroxamic acid in presence of $FeCl_3$
Column, paper and thin-layer chromatography	Determination of acid dyes Structure of lipids in wool

5.2.2 <u>End group analysis</u>

End group analysis is used to determine the average molecular weight in fibrous polymers and to determine the extent of depolymerization due to chain scission in the polymer caused by physical, chemical and/or biological degradation. End group analyses were initially developed many years ago when only natural fibers were available. This technique was and to some extent is still used to determine the number of carboxyl groups in cellulosic fibers associated with chain scission and oxidation of anhydroglucose units. Such an increase has been correlated with an increase in uptake of basic dyes (such as Methylene Blue) and with an increase in the uptake of copper salts (called the copper number). Similar end group analysis methods, used in conjunction with colorimetry, potentiometric and acid-base titrations and thin layer and paper chromatography, have been devised and developed for synthetic fibers such as polyesters, polyamides and acrylics. These methods include the determination of carboxyl end groups in polyester fibers by reaction with hydrazine, of basic or primary amino end groups in polyamides by reaction with 2,4-dinitrofluorobenzene and of hydroxyl end groups in polyester fibers by reaction with 4-nitro-1-naphthyl isocyanate (ref. 1).

5.2.3 Changes in solubility and swelling

The solubility, sorption and swelling of fibers in various organic solvents, in acidic and basic solutions and in other reagents, have been employed to identify fiber types and the composition of fiber blends, to determine the uniformity and degree of crosslinking in fibers produced by application of resins and other polyfunctional materials or caused by degradative processes, to ascertain the amount and type of oligomers present after manufacture of synthetic fibers and to detect changes in polymer chain length caused by scission and/or oxidation during processing, use or exposure to various environments.

The old chemical axiom "like dissolves like" is frequently used to advantage to rank or classify fibers based on their differences in solubility (e.g., use of *N,N*-dimethylformamide as a solvent for polyamides). Detailed and useful separation schemes for identification of most fiber types in blends by differential solubilities and of only one fiber type in specific solvents have been published by the Textile Institute in the book "Identification of Textile Materials" (ref. 2). Table 5.2 is an abbreviated version of the comprehensive solubility scheme that may be used to differentiate specific types of fibers and their blends from each other. For example, Group 2 consists of six different

TABLE 5.2

Identification method for fibers and fiber blends based on difference in solubilities[a] (adapted from ref. 2). Courtesy of The Textile Institute.

Grouping reagents Fibers soluble in	Fiber type	Solvents/Group 2					Solvents/Group 3	
		70% v/v acetone	100% HOAc	4.4N HCl	35% w/w H$_2$SO$_4$	5N HCl 65°C	Trypsin test 40°C	1% NaOH boil
1:Boiling 0.25% Na$_2$CO$_3$	Calcium alginate						sol	slowly sol
	Diacetate	sol 1 min	sol					
	Triacetate		sol 4 min					
2: Within 15 min in 1:10 CaCl$_2$/90% HCOOH	Nylon 6			sol 2 min	sol	sol		
	Nylon 66				sol 2 min	sol		
	Vinylal					sol 5 min		
	Bombyx silk							
3: Within 5 min in 1N NaOCl + 0.5N NaOH	Regen. protein						sol 15 min	
	Wool							
	Tussah silk							sol 18 min
4: Within 8 min in 1,4-butyro-lactone	Copolymer vinyl acetate/chloride							
	Chlorinated PVC							
	Modacrylic							

[a] Solvents at 25°C unless otherwise noted. Blank space = fiber insoluble in solvent after 30 min.

types of fibers (cellulose diacetate, cellulose triacetate, Nylon 6, Nylon 66, vinylal and Bombax silk) because each of these fibers is soluble within 15 min in a 1:10 solution of $CaCl_2$/90% formic acid. Further differentiation is possible because each of these fibers has a different response to the solvents listed for Group 2. Cellulose diacetate can be distinguished from cellulose triacetate because the former dissolves in 70% aq. acetone within a minute while the latter is insoluble in that solvent under the same conditions.

Because of the popularity and diversity of cellulosic/polyester blend fabrics, there have been several analytical methods devised to determine the amount of each fiber type in these blends by preferential solubility. Trichloroacetic acid in methylene chloride is claimed to be an effective and versatile room temperature solvent for determining the amount of polyester in all types of cellulosic/polyester blends. It preferentially dissolves only the polyester, and gives comparable analytical results to those obtained using standard sulfuric acid methods (ref. 3).

The extraction of oligomers or low molecular weight fractions from synthetic fibers by choice of proper solvent allows one to determine quantitatively the amount of oligomer present in the fibers. This is important and useful because the low molecular weight fraction may vary considerably due to differences in processing and manufacturing even for the same type of fiber. The amounts of oligomers in aliphatic polyamide, aramid and polyester fibers vary and can be removed by extraction with appropriate solvents. The low molecular weight fraction in polyamides varies from trace amounts in aramids (removed with various solvents) to 11-12% in Nylon 6 (removed by extraction with methanol). Similar variations exist with polyester fibers for which solvents such as dioxane or dichloromethane are used for extraction of low molecular weight fractions (ref. 4).

The types of solvent separation schemes used in qualitative organic analysis have been adapted to the separation of the major classes of dyes. Goetz (ref. 5) describes a useful scheme for identifying and separating dyes and pigments by their solubility in a series of ready available solvents and various reagents. As noted in Figure 5.1, about half of the dye classes (by application) are soluble in water and the remainder are insoluble. The insoluble dye classes can be further distinguished by their solubility in ethanol, acetone, toluene and various reagents. Water-soluble dyes, such as direct, acid and basic, can be differentiated (scheme not shown) by their ability to dye multifiber test cloth different colors.

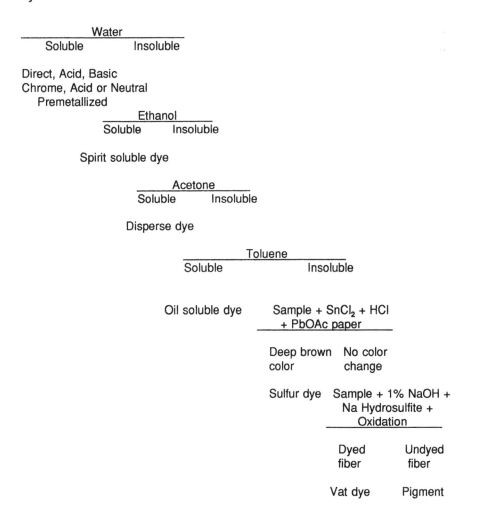

Figure 5.1. Scheme for separation of dyes into classes by solubility (ref. 5). Courtesy of American Association of Textile Chemists and Colorists.

Disperse dyes, which are widely used to dye fibers such as polyester and polyamides, can only be distinguished from each other structurally with difficulty by using static methods based on their limited solubility in water. However, the generator column technique, a dynamic method for measuring solubility of highly insoluble compounds, appears to be suitable for measuring solubility of disperse dyes since it provides the required sensitivity and precision to determine equilibrium processes of dyes and accurate ratios of monomer to dimer ratios of these dyes at ambient temperature (ref. 6).

In addition to solubility of fibers in specific solvents or solvent mixture, changes in the solubility of fibers also indicate changes in the structure or accessibility of the fiber caused by some external or internal influence. For example, an increase in the solubility of wool in either alkali or urea-bisulfite solution indicate it has undergone some form of physical, chemical and/or biological degradation.

Sorption of specific chemical reagents on fibers and swelling of fibers (the latter in essence limited solubility in a solvent) have also been advantageously used to characterize differences in the fine structure of fibers, to determine fiber damage and to determine the extent of fiber crosslinking. The nature and degree of sorption of iodine onto polyester fibers have been used to detect orientation and other structural changes that occur when it is drawn, textured, heat set or subjected to a variety of mechanical and thermal treatments during processing and manufacture (refs. 7, 8).

Many early studies used the swelling and/or dissolution of cellulosic fibers in cuene (cupraethylenediammonium hydroxide) or cadoxen (cadmiumethylenediammonium hydroxide) as an indication of the extent of crosslinking of these types of fibers. However, a more recent study has indicated that hydrolysis of cellulosics by the enzyme *cellulase*, in conjunction with scanning electron microscopy (ref. 9), is more analytically precise and useful than the commonly used cuene or related solubility tests.

5.2.4 Viscosity of fibrous polymers

Various solvents are also advantageously employed to dissolve fibers or extract polymers that have been applied to fabrics as sizing agents or finishes to measure their viscosity. Viscosity, or the resistance of a liquid to flow, is related to the molecular weight of the polymer or substance of interest. Generally, solvents or mixtures of solvents that are used in identification of fibers or finishes may also be used to dissolve the material of interest for viscosity measurements. This measurement is most frequently used to determine viscosity-averaged molecular weights (M_v) of the polymer in the fiber or polymers that are used in processing and finishing textiles. In a typical fiber or polymer sample, M_v approximates the weight-average (M_w) rather than the number-average, as depicted in Figure 5.2. (ref. 10). A variety of viscometers are available for determining the viscosity of macromolecular materials. Selection of the most appropriate viscometer and description of experimental techniques for measurement of viscosity may be found in most basic polymer texts and reference books such as Odian (ref. 10).

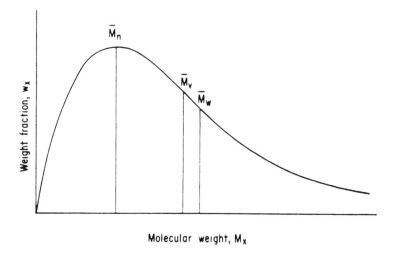

Figure 5.2. Distribution of molecular weights in a typical polymer sample (ref. 10). Courtesy of John Wiley and Sons Publishers.

5.2.5 Colorimetric methods

The property of color (visible spectrum region) has been used in at least two different ways to identify fiber types and finishes and/or to detect changes in fiber accessibility to various reagents. In the first technique, this is accomplished by observing changes in the dyeing behavior and relating this to changes in fiber type or functionality. A relevant example is the dyeing of cellulose acetate fibers with varying acetyl content with a mixture of direct (blue) and disperse (yellow) dyes. This procedure, known as differential dyeing, indicates the proportion of esterified and free hydroxyl groups in the cellulosic fibers. Fibers with the highest acetyl content dye yellow (having affinity for only the yellow disperse dye), while hydrolyzed or saponified acetate fibers with low acetyl content dye blue (having affinity for only the blue direct dye); fibers with intermediate acetyl content dye varying shades of green, since both both hydroxyl and acetyl groups are present.

A comprehensive review on how dyeability tests may be used on all types of textile substrates discusses many other applications besides their use to differentiate between fiber types. It also provides relevant information on how fiber damage and changes in fine structure and the uniformity and type of fabric finish may be detected (ref. 11). An example of the latter is determining the distribution of resins in fabrics by staining with a fluorescent brightening agent, then examining the treated fabric under a fluorescent microscope. Another example is the use of acid dyes to detect the extent of damage in wool fibers and of highly sulfonated direct dyes to detect the extent of surface damage in rayon fibers produced by the viscose process.

In the second technique, numerous colorometric spot tests are used (essentially, modified procedures used to detect functional groups or species in organic and inorganic compounds). One example is the colorimetric determination of titanium dioxide (a delusterant frequently incorporated into polyester and polyamide fibers and fabrics) by its conversion to a soluble yellow solution containing the peroxytitanium species. Another example is the direct colorometric determination of formaldehyde in textiles and other fibrous materials by its reaction with indole-3-acetic acid or tryptophan (ref. 12). Qualitative spot tests for cations, anions, surfactants and a variety of organic compounds and polymers commonly found on or used to process and finish textiles are basically adaptations of these tests used in analytical wet chemistry; these have been summarized (ref. 13) and are periodically updated as needed.

5.2.6 Column, paper and thin layer chromatography

With the advent of chromatography (separation of substances through stationary and mobile phases by adsorption and/or partition), wet methods such as column, paper- and thin layer chromatography have been adapted to identify various fiber additives, finishing agents and products of fiber degradation. Paper chromatography has been employed for separating amino acids present in wool and thus explaining differences in the physical and the chemical properties of these fibers derived from different sources and relating changes in their amino acid content to various types of fiber damage. Thin layer chromatography is frequently employed to identify dyes, fluorescent brighteners, mothproofing and durable press agents applied to textiles and byproducts of these substances after the fibers have been exposed to various environments or exposed to conditions that simulate long term performance requirements. One specific example is the use of reverse phase thin layer chromatography to differentiate between many commercial acid dyes denoted by numbers 2 through 32 in Figure 5.3. In this instance, the solvent system is a 3/1.5 ratio of methanol/water; other systems can be used with these groups of dyes to further differentiate them by their R_f values (calculated as distance of center of spot from origin/distance of solvent front from origin) (ref. 14). Other examples use thin layer techniques to differentiate reaction products of the hydrazinolysis of polyester fibers (ref. 1) and to separate and detect the internal lipids in wool by combining thin layer chromatography with flame ionization methods (ref. 15).

5.3 INSTRUMENTAL METHODS OF ANALYSIS

5.3.1 Spectroscopy

Numerous types of spectroscopic techniques have been used for identifying fibers and fiber blends, finishes, dyes and for assessing textile end use performance with regard to physical, chemical, biological agents and influences. These spectroscopic methods include: ultraviolet-visible, infrared, Raman, nuclear magnetic resonance (NMR), X-ray photoelectron spectroscopy (XPS; also known as ESCA or electron spectroscopy for chemical analysis), mass, electron spin resonance (ESR), atomic emission and absorption and Mössbauer spectroscopy.

5.3.2 UV and visible spectroscopy

Antioxidants, ultraviolet stabilizers, fluorescent brighteners and other fiber additives and finishing agents as well as fiber and fiber finish degradation products that absorb light in the ultraviolet and visible region have been identified and elucidated by ultraviolet spectroscopy. In Figure 5.4 (ref. 16) the differences between thermally aged nylon 6,6 yarns and unexposed yarns are ascertained by the differences in the absorption spectra of the polyamide fibers. Increase in absorption is caused by formation of a mixture of unidentified chromophores from oxidation of polyamide fibers.

302

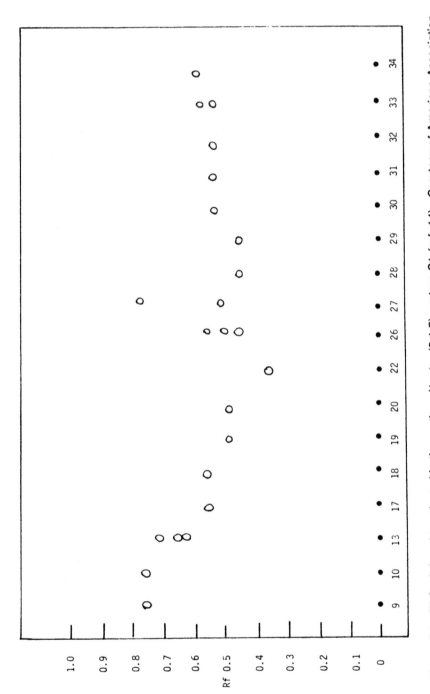

Figure 5.3. TLC of dyes investigated in the methanol/water (3:1.5) system S1 (ref. 14). Courtesy of American Association of Textile Chemists and Colorists.

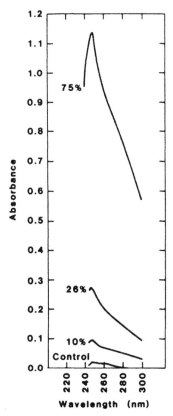

Figure 5.4. Absorbance spectra of aged yarn samples dissolved in 2,2,2-trichloroethanol. Tensile strength loss of each yarn is listed besides its spectrum. Concentrations are 1.42 mg/mL (ref. 16). Courtesy of John Wiley and Sons Publishers.

The fluorescence spectra of wool exposed to mid- and near ultraviolet light were used to characterize its ultraviolet and visible fluorescence. Visible fluorescence has maxima at 354 and 435 nm and can be correlated with weathering, oxidation or chemical modification of the wool fiber (ref. 17).

5.3.3 Infrared and Raman spectroscopy

Infrared spectroscopy has also proven useful for differentiating various fibers and fiber blends, determining the crystallinity of cellulosic fibers by measuring the ratio of free and hydrogen-bonded hydroxyl groups, and detecting finishes, additives or oxidation products. Attenuated total reflectance (ATR) has also been employed to detect changes occurring on the surfaces of fibers, films, and other polymeric materials. Table 5.3 highlights some of the characteristic absorption bands that can

TABLE 5.3

Some frequencies useful in the interpretation of infrared spectra of textile materials (adapted from ref. 18). Courtesy of American Association of Textile Chemists and Colorists.

Frequency (cm^{-1})	Functional group assignment	Fibrous materials with IR absorption bands
3700-3300	O-H stretch (free and bonded)	Cellulosics
3500-3200	N-H stretch	Polyamides, modified cellulosics
2967-2857	C-H stretch	Most fibrous polymers
2252-2062	C≡N	Acrylics, modacrylics
1750-1735	C=O stretch (ester)	Acetates, polyesters, acrylics
1725-1700	C=O stretch (saturated aliphatic acids)	Oxidation products of many fibers
1725-1705	C=O stretch (saturated aliphatic ketones)	Degraded or oxidized fibers
1680-1630	C=O stretch (amide)	Polyamide, wool, resin-treated cotton
1600,1500	C=C (aromatic)	Polyester, aramids
1570-1515	N-H (secondary amide)	Aliphatic and aromatic polyamides
1250-1150	C-O (ester)	Polyester
ca. 1250	P=O	Flame-retardant fibers containing phosphorus compounds
720-730	C-H rock	Polyethylene, polypropylene

be used to identify and differentiate fiber types, modified fibers or materials that have been degraded or oxidized by weathering or other types of end use.

A few pertinent examples illustrate the scope and utility of infrared spectroscopy for analysis of fibrous materials. Fourier-transform infrared spectrometers (FTIR) spectrometers were used to provide differential spectra to characterize polymer blends and copolymers such as Nylon 6 and Nylon 66 (ref. 19). A photoacoustic detector was used in combination with a FTIR spectrometer to characterize such diverse fibers as cotton yarns sized with polyurethanes and chemically modified polyester (ref. 20). Figure 5.5 shows how a depth-profiling technique using different optical velocities allows one to detect the presence of polyurethane sizing on the

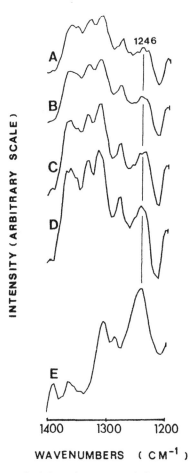

Figure 5.5. The photoacoustic infrared spectra of the cotton yarn sized with the poly-urethane collected at (A) velocity 0; (B) velocity 3; (C)) velocity 6; (D) velocity 9; (E) The photoacoustic infrared spectrum of the polyurethane sizing agent (ref. 20). Courtesy of Society for Applied Spectroscopy.

cotton at an absorption of 1246 cm^{-1}. Photoacoustic techniques have also been used for determining the structures of reactive or acid dyes applied to wool fabrics (ref. 21).

FTIR has also been used to determine the surface characteristics of glass fabrics before and after treatment with a typical silane coupling agent; differences in the infrared spectra of the two fabrics below 1500 cm^{-1} and the appearance of a weak peak in the modified glass fabric (due to the coupling agent) at 1610 cm^{-1} were observed (ref. 22). Near infrared spectroscopy (NIR) was suitable for determining the amount of a conventional durable press resin (DMDHEU) present on cotton fabrics. Good correlations were obtained between infrared data and Kjeldahl nitrogen analyses on the modified fabrics (ref. 23).

Raman spectroscopy has been employed only sparingly for characterization of fibers compared to infrared spectroscopy. This lack of attention is probably due to the greater expense of the Raman spectrometers, the limited amount of instrumentation available, and some sample damage by visible lasers required in dispersive Raman methods. Nevertheless, Raman spectroscopy, based on the scattering phenomenon that changes with the frequency of incident light falling on a sample, has some distinct advantages over infrared spectroscopy for characterizing subtle structural features of compounds, polymers and fibers. The simplicity of the spectra obtained (no overtones or combination bands usually observed in infrared spectra) and detection of low-frequency molecular vibrations are two major advantages. Raman spectroscopy in the low frequency region (less than 200 cm^{-1}) has been employed to determine changes in crystallinity of polyester with different thermal histories (ref. 24). This type of spectroscopy has also been used to determine the amounts of load-bearing, extended-chain disordered structures present in different high-tenacity, gel-spun polyethylene fibers (ref. 25). More recent instrumentation (Fourier-transform Raman spectrometers) is now available that use near-infrared red laser sources that cause little damage to fiber samples. Thus, it is probable that Raman spectroscopy will become more commonly employed as its potential and advantages for structural characterization of fibers are recognized.

5.3.4 Nuclear magnetic resonance spectroscopy

Since nuclear magnetic resonance (NMR) spectrometers and techniques are extremely valuable for determining numerous types of structural features, it is pertinent to describe briefly principles and current instrumentation available. When nuclei have a spin quantum number of $I = 1/2$, signals can be generated by exposure to an applied magnetic field that are specific to the environment of the nuclei. Nuclear magnetic resonance (NMR) spectra generated by such exposure have been used routinely over the past thirty-five years to identify numerous organic compounds and other structures. The first group of NMR spectrometers available determined the proton (^{1}H) environment of various structures. Such characterization required that the structure of interest be dissolved in a non-proton solvent such as CCl_4 or deuterated chloroform ($CDCl_3$). It is only in the last fifteen years that instrumentation and techniques were developed to obtain high resolution spectra of solids (such as polymers and fibers). These improvements included the use of techniques such as cross-polarization (CP), magic angle spinning (MAS) and dipolar decoupling (DD) and the availability of NMR spectrometers that determined the ^{13}C and ^{19}F spectra of structures of interest.

NMR spectroscopy may be routinely used to determine the structure of soluble components of finishes, dyes, oligomers and degradation products of fibers. However, more significant structural data can be obtained by using solid state NMR to characterize fibers. Solid state NMR may be used to correlate mechanical strength and physicochemical properties of fibers with molecular motion in polymer chains. This technique may also be used to determine the degree of crystalline order and orientation and chemical changes occurring in the fiber. Crystallinity and orientation values obtained from NMR spectra may also be compared to values obtained from other analytical techniques discussed later in this chapter (e.g., X-ray diffraction and thermal analysis). A few selected examples follow to illustrate the utility of NMR for determining characteristics of fibrous materials.

Quantitative analysis of the ^{13}C-NMR spectra has been performed to characterize molecular chain conformation of the crystalline and noncrystalline components of various cellulosic fibers. Fibers used in the study included cotton, ramie and rayon (ref. 26). A later study by the same investigators (ref. 27) used solid state ^{13}C-NMR to determine differences in amorphous and crystalline regions of never-dried cotton fibers (Fig. 5.6) relative to those that were dried and those that were rewet.

Solid state ^{13}C NMR has also been used to determine increase in -C=O and -C=C- content that occurs when wool is heated at temperatures above 150°C (ref. 28). The amount and types of fluorochemicals on carpets and home furnishings have been determined by a ^{19}F-NMR spectrometric method (ref. 29).

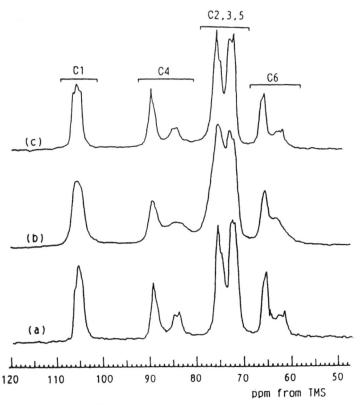

Figure 5.6. 50 MHz CP/MAS ^{13}C NMR spectra of (a) never-dried cotton; (b) the same dried cotton; and (c) wetted cotton. Subscripts for carbon atoms correspond to positions in an anhydroglucose unit of the cellulosic polymer (ref. 27). Courtesy of John Wiley and Sons Publishers.

5.3.5 X-ray photoelectron spectroscopy (XPS)

X-ray photoelectron spectroscopy (XPS) is an analytical technique that provides surface analysis of polymers and fibers. It is also frequently called ESCA (electron spectroscopy for chemical analysis) in the scientific literature. The scope and applications of XPS for polymers and fibers have been critically reviewed (refs. 30, 31). XPS is an analytical technique that can be used to provide information on chemical structure and bonding at sampling depths of less than 10 nm.

Table 5.4 gives principal features of ESCA or XPS spectra of polymers. These features are also applicable to fibers and polymeric coatings or finishes on fibers. Besides determining the type of atoms on the surface and their oxidation states, ESCA is very useful in determining π to π^{*} transitions that occur when polymers and fibers are exposed to ultraviolet light and other sources of destructive radiation.

TABLE 5.4

Principal features in the ESCA spectra of polymers (ref. 30). Courtesy of John Wiley and Sons Publishers.

Spectral feature	Information
Main peak position	Atom identification
Chemical shift	Oxidation state
Peak-area ratios	Stoichiometry
Shake-up satellites	π to π^* transitions

XPS has been used for a variety of applications in evaluation of modified and unmodified textiles. However, the expense to obtain and maintain this type of instrumentation has limited its use. This technique has been used primarily to determine differences in oxidative states of atoms contained on fiber surfaces. Since phosphorus in its highest oxidation state (P^{5+}) imparts the best flame retardancy to modified textiles, early XPS studies were used to determine different oxidation states of phosphorus in flame-retardant textiles. A later study used XPS to determine the oxidation state of sulfur in wool fabrics before and after polymer modification with shrinkproofing agents (ref. 32). As noted below (Fig. 5.7), the unmodified wool has a disulfide crosslink that is oxidized after chlorination or chlorination followed by treatment with a shrinkproofing polymer. XPS was used in conjunction with Fourier transform infrared photoacoustic spectroscopy to determine the location of a copolymeric finish in polyester fibers (ref. 33).

5.3.6 Mass spectroscopy

Mass spectroscopy or spectrometry is a method that produces ions from a sample that is bombarded with electrons. The mass spectrum that is recorded is a result of separation of the ions according to their mass-to-charge (m/e) ratios. This technique has been used to characterize dyes, additives and particularly changes in fiber structure resulting from various types of physical and chemical degradation.

Exposure of nonwoven polypropylene fabrics to a xenon arc light source (simulation of ultraviolet degradation) produces different mass spectra in unexposed and

310

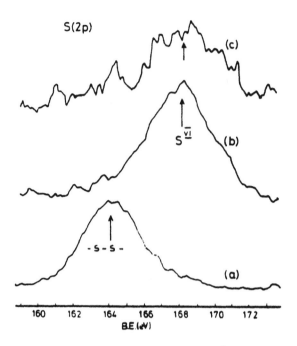

Figure 5.7. SD (2p) spectra at 100 ev (electron volts; 10³ c/s) for (a) untreated fibres; (b) Kroy-chlorinated fibers, 1.8% available chlorine, ; and (c) Kroy-chlorinated + 2% Hercosett 125 polymer. All analyses are based on final weight of modified or coated fibers (ref. 32). Courtesy of The Textile Institute.

fabrics exposed to the light source. As shown in Figure 5.8, the appearances of new m/e peaks at 57, 71 and 85 are consistent with carbonyl groups formed by scission of the polymer chain (ref. 34). Mass spectrometry has also been used to determine the structure of indigoid vat dyes used in historic textiles; this technique could be important for characterization of modern indigoid vats that are extensively used to dye cotton denim (ref. 35).

5.3.7 Electron spin resonance spectroscopy

When unpaired or free electrons are placed in a magnetic field, signals are generated that are characteristic of certain types of structures containing free radicals. Since electron spin resonance (ESR) detects and differentiates between various types of free radicals, it is particularly suitable for the detection of peroxy and of other radical species formed in the photooxidation and the photodegradation of fibers and in ascertaining the subsequent reaction of these radicals to produce crosslinking and

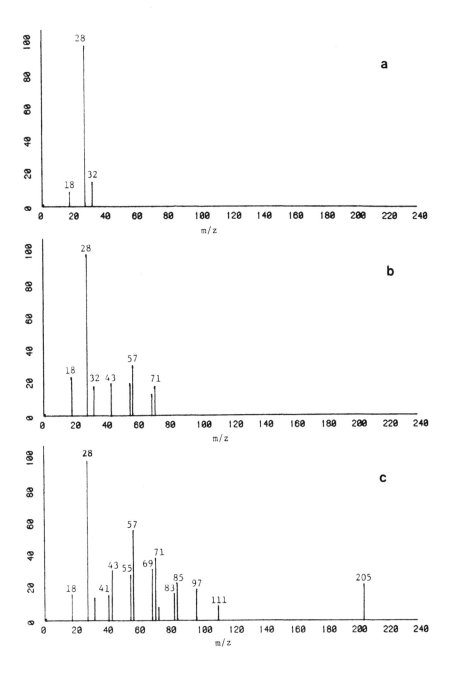

Figure 5.8. Mass spectra (70 eV) of nonwoven polypropylene fabrics at a probe temperature of 50°C, previously exposed to a xenon arc light source (Weatherometer) for (a) 0; (b) 24; and (c) 48 hrs (ref. 34).

/or depolymerization (ref. 36). Figure 5.9 shows typical ESR spectra for isotatic polypropylene at two different temperatures after exposure to an ultraviolet light source. Generation of free radicals caused by the mechanical fracture of aramid fibers (such as Kevlar 49) has also been characterized by ESR spectroscopy (ref. 37).

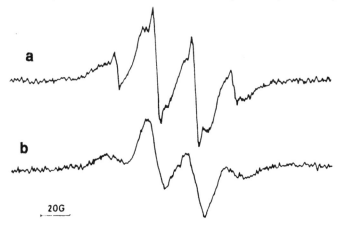

Figure 5.9. ESR spectra of isotatic polypropylene after irradiation with 254 nm UV light at 77°K in a vacuum, recorded: (a) at -160°C; (b) after warming up to -140°C (ref. 36). Courtesy of Academic Press.

5.3.8 Chromatographic instrumentation

Various types of instrumental chromatographic techniques have been adapted for characterizing fiber additives, impurities, dyes, and finishing agents and for assessing overall textile performance. Gas (GC) and high performance liquid (HPLC) chromatography have been primarily employed to identify low molecular weight fiber additives, dyes, finishing agents and fiber degradation products. Gas chromatography is generally useful for substances that volatilize below 400°C, while liquid chromatography (particularly HPLC) is employed for characterizing less volatile substances or substances that decompose on heating. Gas chromatography, HPLC and thin layer chromatography (latter discussed earlier under wet methods in **5.2.6**) have been critically reviewed for their applications in characterization of textiles (ref. 38). Gel permeation (GPC) or size exclusion chromatography, in which substances are separated on the basis of the size of molecules and thus their ability to pass through, is extremely useful for determining the molecular weight average (number and weight averages) and molecular weight distribution of polymeric and oligomeric substances that are present in fibers, either from unreacted monomers or additives and/or from various degradation processes. The most critical aspects of gel permeation chromatography are the proper choice of solvents or solvent mixtures to dissolve fibers and other polymeric substances and the proper calibration of such instruments with polymers and/or oligomers of known molecular weight. Selected examples that follow illustrate the diversity of chromatographic techniques and applications suitable for textile characterization.

Liquid chromatography has been successfully used to determine the concentrations of unreacted resin DMDHEU in its crosslinking reaction with cotton cellulose. The chromatographic data allows one to calculate rate constants and energy of activation for the reaction of the resin with the cellulose (ref. 39). Inverse gas chromatography was used to determine adsorption isotherms and heats of sorption of n-alkanes on fibers. These determinations were then used to compare surface areas of representative synthetic and natural fibers; surface areas determined by this approach were in good agreement with those determined by other analytical methods (ref. 40). Mothproofing agents on wool fabrics have been analyzed by a variety of GC and HPLC methods (ref. 41). As noted in Table 5.5, specific solvents or solvent mixtures are required for extraction of mothproofing agents before their chromatographic analysis.

HPLC methods have also been used to identify various disperse dyes extracted from polyester fibers (ref. 42) and to separate a variety of basic dyes (ref. 43) used in the dyeing of acrylic fibers. Good separation of dyes are achieved in the chromatogram shown in Figure 5.10.

Gel permeation chromatography was used to determine the molecular weight distribution of a variety of cationic polymers especially a polyamide-epichlorhydrin resin frequently used in shrinkproofing processes for wool (ref. 44). More recent studies have used fibrous polymers as the stationary phase rather than the mobile phase in size exclusion chromatography to determine pore volume and pore size distribution in fibers. A recent method describes the use of fabric as the stationary

TABLE 5.5

Chromatographic methods for the analysis of insect proofing agents on wool (ref. 41). Courtesy of Marcel Dekker Publishers.

Mothproofing agent	Method	
	Extraction solvent	Chromatography[a]
Eulan WA New/Eulan U33	2-methoxyethanol in soxhlet	GC/ECD
	methanol/ammonia in sealed ampoules	HPLC
Dieldrin	2-methoxyethanol in soxhlet	GC/ECD
Mitin LP	methanol/ammonia in sealed ampoules	HPLC
Mitin FF, high conc.	Hydrolysis with KOH, simultaneous extraction into isooctane	HPLC
	Acetonitrile-methanol in soxhlet	GC/FID
Permethrin	2-methoxyethanol in soxhlet	GC/ECD
	Isooctane-HCl in sealed ampoules	HPLC

[a] GC, gas chromatography; ECD, electron capture detector; FID, flame ionization detector; HPLC, high performance liquid chromatography.

phase to relate the analytical data more closely to end use performance than earlier studies that used chopped or ground fibers as the stationary phase (ref. 45).

5.3.9 Measurement of molecular weights by other methods

Molecular weight and molecular weight distribution of fibrous materials influence their overall end use performance. Thus, a variety of other techniques (in addition to viscosity measurements and gel permeation chromatography) have been occasionally employed to determine number-average (M_n) and weight-average (M_w) of fibrous polymers. These techniques are described at some length in individual chapters in Happey's three volume treatises on applied fiber science (see **General References** at end of this chapter). Number-average methods are based on methods that measure colligative properties and include techniques such as vapor- phase osmometry, cryoscopy and ebulliometry. Light scattering and ultracentrifugation are examples of representative methods for determining M_w.

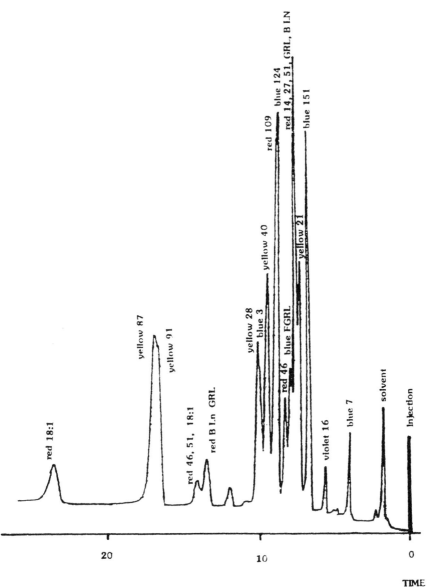

Figure 5.10. Chromatogram illustrating the separation of twenty-one basic dyes (ref. 43).

5.3.10 Thermal analysis

Thermal analysis of fibers or any other substance consists primarily of evaluation by differential thermal analysis (DTA), differential scanning calorimetry (DSC), thermogravimetric analysis (TGA) and thermomechanical analysis (TMA).   DTA measures temperature-dependent differences between a reference material and a sample while DSC measures the amount of energy absorbed or released from samples with variance of time or temperature. TGA measures the change in the weight or mass of a sample as a function of time or temperature. TMA measures changes in the dimensions or mechanical properties of a sample as a function of temperature (ref. 46).

DTA and TGA were the most frequent techniques used for fiber characterization when thermal instruments became readily available in the 1970's (ref. 47). However, DSC is currently the most popular thermal analysis method because one can readily obtain a variety of quantitative data not available from DTA instruments. In earlier studies, DTA was used to correlate flame retardancy of textiles with lowering of the thermal decomposition temperature of fibers and to qualitatively study annealing and crystallization processes occurring in synthetic fibers. DSC affords quantitative data on second order transitions in fibers and polymers such as determination of the glass transition temperature T_g (temperature above which amorphous, flexible regions become hard and relatively brittle) and on first order transitions such as melting (T_m) and crystallization (T_c) processes. Figure 5.11 is a prototype DSC scan of polyester that illustrates both first and second order thermal transitions. TGA has been primarily been used for determining the thermal stability of fibers and their additives

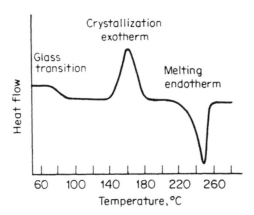

Figure 5.11. Three transitions of polyethylene terephthalate obtained on a DSC trace. Sample size: 5.1 mg. Program rate: 20°C.min⁻¹. Atmosphere: nitrogen (ref. 46). Courtesy of McGraw-Hill Publishers.

or finishes before and after exposure to various environmental conditions. TMA has proven useful for relating mechanical changes in textiles to changes in temperature and humidity associated with their end use. Special instrumentation has also been devised to measure thermal events in materials under various conditions. A few examples illustrate the diversity of analytical data that can be obtained with fibers by thermal analysis techniques.

Thermal stress analysis (TSA), a variation of TMA, was used to determine the heat history and effective heat set temperatures of nylon 66 yarns in carpets; the thermal stress analyzer measures the shrinkage force that develops in fibers as a function of temperature (ref. 48). DSC has been used to determined the amounts of bound water in wool (ref. 49) and in rayon (ref. 50). Thermal scans in both studies were effective in determining types and amounts of water in the fibers as a function of moisture content and regain.

The heat of wetting and differential heat of moisture sorption of viscose rayon and nylon 6 fibers were determined at ambient temperatures with a heat transfer type calorimeter. Use of this instrument allowed calculation of thermodynamic parameters of the textiles at various moisture contents and regain levels (ref. 51). The structure and thermal profiles of polyester fibers melt spun at different speeds were evaluated by DSC over a broad temperature range (0-200°C). A three-phase model of the molecular structure of the fibers under different processing conditions was proposed to explain thermal analysis and other analytical data (ref. 52). DSC has also been recently used for determining the degree of removal of oily soil from fabrics (ref. 53).

5.3.11 X-ray diffraction and related methods

X-ray (wide and small angle) diffraction and electron diffraction methods have been used to determine various structural features of fibers and polymers. Wide angle X-ray diffraction methods are the most frequently employed while small angle X-ray and electron diffraction are used less frequently because the former instrumentation is relatively expensive and the latter is limited to analysis of single crystals or ultrathin surfaces. When X-rays or electrons strike a polymer or fiber, a pattern of concentric arcs is generated characteristic of several structural aspects of the polymer chains.

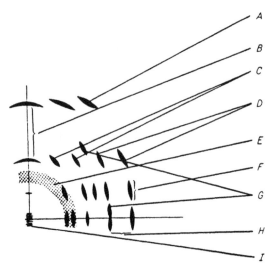

Figure 5.12. One quadrant of a cylindrical film fiber photograph and the information obtainable from it: (A) diffraction breadth gives measure of crystallite size or perfection; (B) layer line spacing gives chain repeat distance and hence information on chain configuration and mode of polymerization; (C) fade-off of diffraction effects with increase in diffraction angle gives information on thermal motion of atoms; (D) arc intensities may enable complete crystal structure to be deduced; (E) amorphous scattered intensity provides information on amount, arrangement and orientation of amorphous material; (F) arc lengths and distribution of intensity within the arcs allow distribution of orientation of crystallites to be determined; (G) tilted crystal effects help to confirm choice of unit cell; (H) lateral spacings give lateral unit cell dimensions; (I) continuous and discrete low angle scattering effects provide information on the long range order and size, shape and orientation of the scattering particles (ref. 54). Courtesy of John Wiley and Sons Publishers.

Structural features that are most usually determined are polymorphism (different crystalline forms of the same polymer or fiber), the degree of crystalline order and perfection in the polymer and orientation of crystallites in the polymer. Figure 5.12 shows the type of information (caption self-explanatory) that can be obtained from a representative film fiber photograph. Figure 5.13 is a prototype 2θ scan obtained from wide angle X-ray diffraction of polyester fibers. The peaks at 18, 23 and 26° correspond to reflections from the 010, 110 and 100 planes of a unit cell containing the polymer chains.

The pioneering work of Professor Herman Mark and his colleagues in the late 1920's established the dimensions of the unit cell for native cellulose (e.g., cotton) by wide angle X-ray diffraction methods. Numerous diffraction studies of fibers and polymers have been conducted since that time. The four polymorphic forms of cellulose are described in a concise review on the application of X-ray techniques to

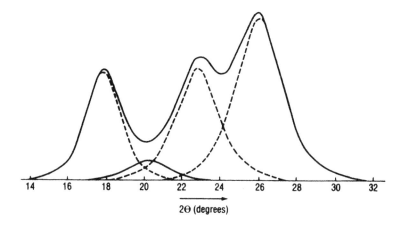

Figure 5.13. Equatorial scan from PET [poly(ethylene terephthalate)] fibers resolved into crystalline and amorphous peaks (ref. 55). Courtesy of Intl. Science Communications Inc.

fibers (ref. 54). An excellent tutorial and review by Murthy (ref. 55) demonstrates how x-ray diffraction data from polymers may be analyzed by using a two-phase model (crystalline and amorphous components) to resolve peaks in diffraction scans. The application of electron diffraction methods to ultrathin sections of polyester and polyamide fibers is described in a comprehensive treatise on fiber diffraction methods. Preliminary results indicate that subtle structural features of crystalline order and orientation are detectable with this technique (ref. 56).

Wide-angle X-ray diffraction was used to determine the orientation and crystallinity of nylon 6 fiber blends containing various amounts of polyacrylic acid, polyolefins and polyacrylamide (ref. 57). Previous X-ray diffraction studies have been summarized and new computer-generated calculations were made to standardize the unit cell dimensions of native cellulose (Cellulose I lattice). Ramie was used as the source of native cellulose because of its high degree of crystallinity and orientation (ref. 58). The pore size distribution and porosity of polyester fibers have been determined by small angle X-ray scattering and related to their dyeing behavior with disperse dyes (ref. 59). A new model of the superstructure of nylon 6 fibers was proposed based on data obtained from small angle x-ray scattering; the model incorporates paracrystalline structural features and lamellar stacks and a distribution function to describe cluster orientation of crystallites about the fiber axis (ref. 60).

5.3.12 Electroanalytical methods

Electroanalytical techniques have been only sparingly used to characterize fibrous materials relative to other types of analytical techniques. These types of techniques involve detection of electrically active species or redox reactions. The current, resistance and/or potential are varied or controlled under either steady-state (equilibrium) or transient conditions. Steady-state methods such as potentiometric, coulometric and amperometric titrations are rarely used for fiber characterization. In contrast, dynamic methods such as various forms of voltammetry (measurement of current as a function of time after applying a potential to the polarizable or indicator electrode) have been used for determining concentrations of various species in fibrous materials. Polarography (type of voltammetry that normally uses a dropping mercury electrode) has been used to evaluate the light stability of cotton fabrics by analysis of an alkaline extract of the dyed and exposed fabrics (ref. 61). Concentrations of several heavy metals (Fe, Cu, Co, Mn and Zn) in textiles that cause detrimental effects on bleaching with hydrogen peroxide have been measured by anodic stripping voltammetry (ref. 62). Very low levels of formaldehyde in resin-treated cotton and cotton/polyester blend fabrics were determined by square wave voltammetry through derivatization of the formaldehyde to a hydrazone (ref. 63).

5.3.13 Methods using low frequency radiation

The properties of textiles have also been characterized by their exposure to low frequency or long wavelength energy sources. Low frequency radiation sources include sound, microwaves and radio frequency. A comparison of molecular orientation of heat-drawn polypropylene fibers was made by sonic velocity measurements and birefringence (optical method discussed below). There was good correlation between these methods for crystalline regions of the fiber but higher values were always obtained by sonic techniques for the amorphous regions (ref. 64). The acoustical properties of various carpets and their backing materials were determined by measurement of sound absorption and improvement in insulation on impact. The average overall absorption coefficient correlated well with overall carpet thickness in the range of 250-2500 Hz (ref. 65). A nondestructive method (polarized microwaves) was used to determine fiber orientation of nonwoven fabrics. Rapid determination (30 sec) of this parameter was possible due to the angular dependence of transmitted microwave intensity (ref. 66). Radio frequency sensors were successfully used to control automatically a pilot-plant scale infrared drying range for fabrics. This type of sensor can detect changes in the water content of the textile due to changes in the dielectric content of the wet fabric (ref. 67).

5.3.14 Optical methods

A variety of optical methods have been used to characterize fibers and their changes after dyeing, finishing and exposure to various types of physical, chemical and biological influences. These optical methods can be subdivided into: (a) refractive methods; (b) various forms of microscopic evaluation; (c) image analysis and (d) miscellaneous optical techniques and instruments such as speckle photography and laser scan devices.

5.3.14a Refractive methods. The velocity of light transmitted or refracted through a solid such as a fiber varies with the medium through which it passes. The ratio of the phase velocity of light in a vacuum to the velocity of light of the fiber in a specified medium is called the refractive index (n). Since fibers are usually anisotropic, light will pass at different velocities through the x, y and z directions. Two principal indices of refraction (parallel and perpendicular to the fiber axis) are used to determine inherent refractive fiber properties. The difference between the refractive index parallel to the fiber axis ($n_{//}$) and perpendicular to the fiber axis ($n_\perp$) is called birefringence and is given by the expression:

$$\Delta n = n_{//} - n_\perp \tag{5.1}$$

Because orientation of crystalline regions in the fiber strongly influences the refractive properties of a fiber, birefringence measurements are usually made to determine orientation in fibers. Although earlier methods used simple refractometers to calculate fiber birefringence, more modern techniques use polarizing microscopes and other advanced optical equipment to obtain more precise measurements of this property.

A simple, inexpensive method to determine birefringence of a variety of synthetic fibers has been developed. It consists of using a basic microscope, two polarization foils and a green light filter ($\lambda = 547$ nm) of fibers prepared by a V-cut method (ref. 68). The birefringence of highly ordered fibers such as aramids cannot be accurately measured by the same techniques used to measure conventional fibers. Thus, a spectrophotometric method was developed that gave suitable birefringence values for various aramid fibers and conventional fibers (ref. 69). Improved techniques and instrumentation also gave accurate birefringence values of wool fibers (at various relative humidities) that were not readily measured in previous studies (ref. 70).

5.3.14b <u>Microscopy</u>. Several types of microscopy have been and are routinely used to characterize fibers at different stages of processing and end use. The many standard tests, characterization techniques and applications of microscopy are more appropriately discussed in detail in **Chapter 6**. The current discussion focuses primarily on the types of microscopic techniques and instrumentation available to evaluate fibrous materials.

Optical or light microscopy (ref. 71) and scanning electron microscopy (ref. 72) for fiber characterization and other fiber applications have been critically reviewed. A more concise and current review of the use of microscopy for textiles is given by Hesse (ref. 73). The major types of microscopy that are normally employed for fiber evaluation are summarized in Table 5.6. All types were used to identify fibers and determine changes from processing or exposure to physical, chemical or biological agents during use and wear. However, polarizing microscopy is usually used to assist in the determination of fiber birefringence and fiber properties related to that measurement. Hot-stage microscopy has been used primarily to differentiate fiber types and blends by their characteristic melting point. However, the rapid and more accurate determination of melting point by DSC has become the method of choice for this mode of fiber characterization. Fluorescence microscopy is becoming increasing popular because it can determine trace amounts of fluorescent brightening agents in fiber as well as natural fluorescence in fibers such as wool. The clarity of photomicrographs and high magnifications attainable with scanning electron microscopy (SEM) have led to its widespread use for textile applications since the introduction of the first commercial microscope of that type in 1965. Finally, the development of an environmental scanning electron microscope, i.e., one that allows evaluation of fibers in the wet state (ref. 74), should provide scientists with information not previously available by conventional electron microscopy techniques.

TABLE 5.6

Microscopy techniques and instruments for textile evaluation

Technique or instrument	Salient features	Representative fiber applications
Normal light microscopy	Best resolution at magnifications 10-700x	Identification and end-use performance
Hot-stage microscopy	Heating attachment to conventional microscope	Fiber identification by melting point
Polarizing light	Fibers viewed in polarizing light	Birefringence, orientation, related properties
Phase-contrast	Allows better identification of transparent features of solids	Fiber identification, surface analysis properties
Interference	Same as phase-contrast but incident and diffracted waves not separated	Birefringence, fiber morphology, other applications
Fluorescence	Admits only light that subsequently fluoresces	General use
Scanning electron microscopy (SEM)	Electron beam sweeps over sample; excellent resolution at high magnifications	All applications
Transmission electron microscopy (TEM)	Similar to SEM, but thinner samples required for light transmission	All applications
Environmental SEM	Special pressure-limiting aperture allows wet fiber evaluation	All fibers in the wet state

5.3.14c Image analysis. Although microscopic observations are extremely useful for determining numerous structural features and influences on fibers, careful sample preparation usually required is time-consuming and tedious. Thus, image analysis was introduced as an alternative optical method to determine structural changes in fibers (refs. 75, 76). The sample is first optically scanned with either a densitometer, microscope, television camera or other suitable device. Transmitted light viewed on a screen shows the salient structural features or changes in the fiber. This technique was used to detect damage in the cortex and in the cuticle of wool fibers processed under various conditions (ref. 75). In a more extensive series of studies Spivak and his colleagues (refs. 76-78) used an integrated image analysis system to quantify

324

textural changes in carpets. Goniophotometry (measurement of light reflected from a surface at different angles) and densitometry (measurement of reflected light from a photographic negative) techniques were also used to determine textural differences. Figure 5.14 is a representative image analysis system used in their studies. Computer software allowed the construction of histograms to examine more subtle variations in visual texture and relate them to carpet wear, appearance and pile density.

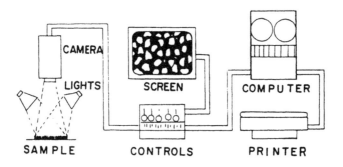

Figure 5.14. Schematic illustration of the image analysis system (ref. 76). Courtesy of Textile Research Institute.

The appearance evaluation of durable press fabrics has also determined by digital image analysis. This technique is offered as an alternative to subjective fabric appearance ratings of trained observers. Good correlation of appearance ratings of the fabrics was observed in gray level histograms from image analysis relative to typical subjective evaluations (ref. 79).

5.3.14d Miscellaneous optical methods. The advent of various practical laser systems makes it feasible to use them as sensors and characterization devices for textiles. Speckle photography (derived from exposure of a sample to a coherent light source such as a laser) is recommended as a method for characterizing surfaces of films and fibers (ref. 80). A modern laser scan system has been recently used to detect and locate various types of fabric faults during processing. The light source is an infrared laser diode that allows dyed as well as undyed fabrics to be inspected (ref. 81).

5.4 COLOR AND ITS MEASUREMENT

The structural characteristics of dyes that produce color were discussed in the chapters on textile dyeing and printing. However, the effect of the light source and the response of the detector (viewer or optical measuring device) in determining the colors of an object were not considered. In this context, for textiles or for any other materials to which colorants are applied, it is important to describe color in some orderly arrangement. It is also important to measure color and color differences to obtain reproducible shade matching, to measure visual fabric characteristics such as luster, hue, yellowness and brightness, and to have representative tests that measure and predict colorfastness of dyed materials on their exposure to various physical and chemical agents.

5.4.1 Color and color-difference measurement

Classification schemes or orderly arrangement of colors and their measurement have evolved from many different disciplines and are discussed in detail in many books, monographs and journal articles. The historical development of three-dimensional color space systems, color difference equations, differences between visual and instrumental observations and other related topics are comprehensively discussed in two books (refs. 82, 83). The authors of these books, who are experts in this area, are quick to point out that there has not been or may ever be a perfect system devised to measure color or color differences because of the complex interaction of factors that determine visual and/or instrumental perception and measurement of color.

There are several three-dimensional "color space" or "color order" systems that have been devised to classify colors. However, the Munsell, Ostwald and CIELAB systems (Table 5.7) are the three most widely used color order systems. The Ostwald system is employed most frequently for production of colorants, dyes and paints and emphasizes depth and brightness. The Munsell system was devised in 1905 by the artist A. H. Munsell. In the Munsell color system (Figure 5.15), the perceived color of objects is described in three-dimensional terms with regard to their hue, value and chroma. Hue is defined as the quality of color described such as green or red. Value is defined as the quality of color described by light or dark and

TABLE 5.7

Systems of three-dimensional color space (ref. 83). Courtesy of John Wiley and Sons Publishers.

System Use	Arrangement of Attributes in Color Space	Terms for Attributes
Designers and colorists in fields such as interior decorating, product packaging. (Munsell)		Hue Chroma (saturation) Value (lightness)
Color formulators and color chemists in the creation of colored products such as paints, dyes. (Ostwald) Swedish "Natural Color System"		Hue Depth Brightness (vividness)
Paper machine foremen or textile bleach plant foremen. (Opponent-colors) Hunter L,a,b CIE 1976 L*a*b* (CIELAB)		Yellowness- blueness Redness- greenness Lightness- darkness

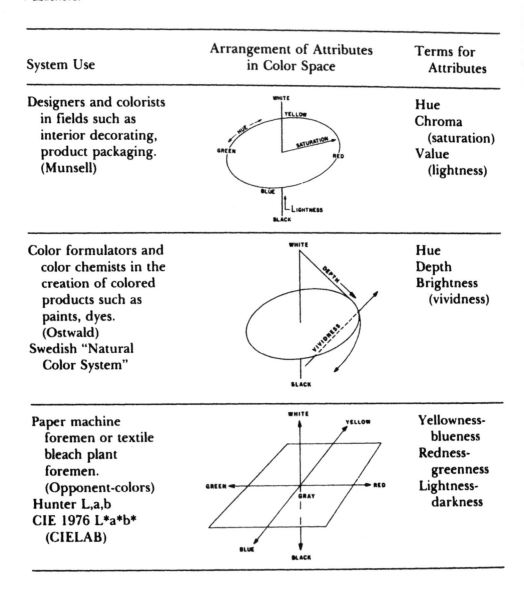

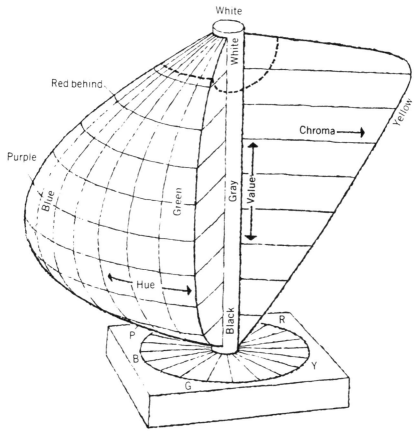

Figure 5.15. Munsell System is based on the perceived colors of objects; white and black have unique positions at the top and bottom, respectively, of the value axis. All other colors lie between them, with the Munsell color solid bulging out to accommodate those that are farther away from the neutral value axis (adapted from ref. 82). Courtesy of John Wiley and Sons Publishers.

related to a gray scale while chroma is the degree of difference between a color and a gray color with the same value. The value axis in the Munsell color system is the y axis, with the top being white and the bottom being black. Chroma is the color difference of an object on the z axis, while the other colors are represented on the x axis by different hues.

The CIE (Commission International l'E'clairage) tristimulus value system was formulated in 1931 and is not based on the perceived color of the object but on the principle that the stimulus of color is dependent on the interaction and proper

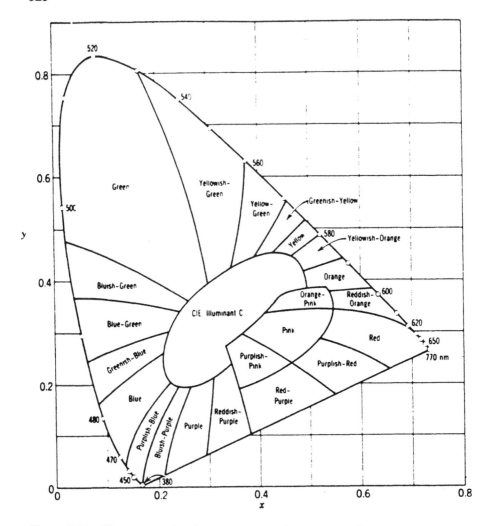

Figure 5.16. The names of various colors are shown on this CIE chromaticity diagram. They are **approximately** correct for colors viewed in daylight by an observer whose eyes are adapted to that daylight (ref. 82). Courtesy of Deane B. Judd.

combination of a light source, an object, and a standard observer. By using the CIE chromaticity diagram shown (Figure 5.16), one can tell whether two colors match. This system does not readily tell how two colors differ if they do not match. However, one can reasonably assign color names to specimens by knowing their exact location on the CIE chromaticity diagram. The CIE tristimulus values are the mathematical values X, Y and Z that represent the spectrum of the primary colors red, green and

329

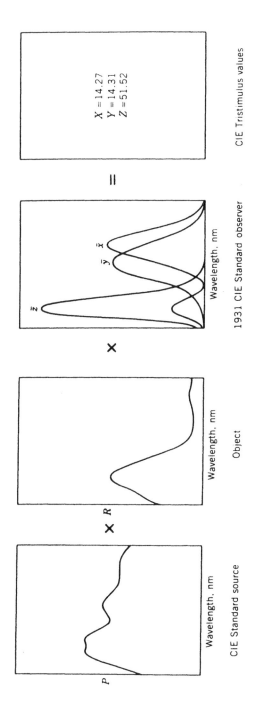

Figure 5.17. The CIE tristimulus values X, Y, and Z of a color are obtained by multiplying together the relative power P of a CIE standard illuminant, the reflectance R (or transmittance) of the object, and the standard observer functions x, y, and z. The products are summed up for all the wavelengths in the visible spectrum to give the tristimulus values, as indicated in the diagrams above and by "chromaticity coordinates" x, y and z (adapted from ref. 82). Courtesy of John Wiley and Sons Publishers.

blue, respectively. They are derived (as shown in Figure 5.17) from multiplication of a standard illuminant (either A = incandescent light; B = simulated noon sunlight; and C = simulated overcast sky daylight) times reflectance R or transmittance of the object times standard observer evaluation of standardized red, green and blue light sources (called x, y and z, respectively). The CIE system (or its improvements or modifications) is the one most frequently used to describe color in dyed textiles. In 1964 and 1976, some changes were made in the system to more correctly define the viewing angle and other conditions for a standard observer and include the L, a, b values (L = lightness, a= redness-greeness and b = yellowness-blueness) shown in Figure 5.18 to produce a more uniform color scale (called CIELAB).

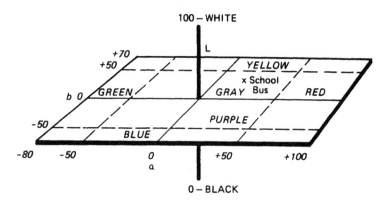

Figure 5.18. The Hunter L,a,b diagram. The location of the yellow school bus is shown. This system is considerably more uniform than is the CIE system (ref. 83). Courtesy of John Wiley and Sons Publishers.

To determine the color difference in materials that is mathematically represented by ΔE and usually computed spectrophotometrically, a variety of color difference equations have evolved over the past sixty years that are based on color space systems and chromatic difference equations are widely employed to measure color difference. The CIE $L^*a^*b^*$ and the analogous CIE $L^*u^*v^*$ color difference equations **5.2** and **5.3** are the most commonly used for evaluating textiles. The CIE $L^*u^*v^*$

$$\Delta E^*_{ab} = [(\Delta L^*)^2 + (\Delta a^*)^2 + (\Delta b^*)^2]^{1/2} \tag{5.2}$$

$$\Delta E^*_{ab} = [(\Delta L^*)^2 + (\Delta u^*)^2 + (\Delta v^*)^2]^{1/2} \tag{5.3}$$

uniform color space system and corresponding color difference equation were adopted at the same time as the CIE L*a*b* system and equations. However, it has been noted that they are no more reliable than a visual observer in the number of wrong decisions about shade matching. Alternative color difference equations have been proposed that presumably give fewer wrong decisions on shade matching than those that are made by visual observations. These will be discussed in more detail in **Section 5.4.3**.

5.4.2 <u>Corrections for reflectance and metamerism</u>

In addition to color scales and color difference equations, there are geometric attributes of objects (such as fabric surfaces) that affect our perception. Luster and opacity are the two most important attributes for textiles and cannot be readily quantified in three-dimensional space like color. These geometric attributes could also affect our perception of the color of an object if it is viewed under different light sources. This phenomenon is called metamerism. It is defined by the CIE as two specimens having identical tristimulus values for a given reference curve that differ within the visible spectrum.

The development of the Kubelka-Monk equation in 1931 where K = absorption

$$K/S = (1-R)^2/2R \tag{5.4}$$

coefficient, S = scattering coefficient and R = reflectance, allows one to control and measure the opacity and reflectance of dyes and dyed textiles. The utility and application of this important equation has been critically reviewed by Nobbs (ref. 84). It can be used in conjunction with CIELab systems or other related color systems to match a given color standard.

Metamerism has been reviewed with emphasis on the usefulness of various indices for its quantification. These mathematical indices are derived from a combination of different light sources in conjunction with representative color difference equations (ref. 85). Metamerism was further characterized in terms of discrepancies due not only to different light sources but also due to differences in the observer, field size (change from small to large or reverse) and geometry (change in viewing angle of specimen) (ref. 86).

5.4.3 Newer color difference equations and measurements

Revised color difference equations have been formulated in an attempt to improve further the reliability of the CIELAB and CIELUV equations. The first set of color difference equations is based on the JPC79 formula, a modification of the older ANLAB derivation. However, anomalous results were obtained with very dark and near-neutral color specimens. Thus, The Society of Dyers and Colourists Colour Measurement Committee (CMC) suggested further modification of this formula to allow for different lightness or chroma weightings for different circumstances, i.e., for perceptibility rather than acceptability evaluation of different substrates. These formulas were called CMC (l:c) for lightness:chroma. One formula was used for perceptibility data (CMC (1:1) and the other for acceptability data (CMC 2:1) (refs. 83, 87). The CMC color difference formula appears to give substantially better correlation with visual judgments in both perceptibility and acceptability decisions with small to medium color differences than the CIELAB formula (ref. 88). However, with large color differences it was poorer than the CIELAB system. There has however been some adoption of the CMC color system as a practical method for consistent reproduction of color, accurate assessment of color and proper response to pass/fail judgments on textile color differences (ref. 89).

A further refinement of the CMC color difference equations was made by Luo and Rigg (refs. 90, 91). This refinement is called the BFD (l:c) after the University of Bradford where the research was performed. These investigators claim those perceptibility results and small color differences are far superior with these color difference equations compared to CMC and other equations such as CIELAB. They recommend BFD (1:1) for perceptibility data and BFD (1.5:1) for acceptability data. However, this system is again not particularly reliable for large color differences in samples. It is probable that the CMC and/or the BFD color difference equations may be adopted to either supplement or replace the CIELAB or CIELUV systems. It is equally probable that further refinements will be offered in an attempt to discover the "ultimate equation" suitable for the determination of color differences for all types of samples under different conditions.

5.4.4 Classification by shade sorting methods

Shade sorting techniques of dyed fabrics have been practiced for many years to insure good quality control in large scale fabric production. However, it is only recently that several mathematical and geometric parameters have been used to assist in improved shade sorting. The 555 graphical sort system proposed by Simon (ref. 92) consists of the division of color spaced into blocks of fixed shape with the population grouped by assigning each sample with the block into which it falls. The 555 block is usually centered on the standard and other blocks oriented in Munsell color space ($L^*C^*H^*$ where L = lightness, C = chroma and H = hue). The Munsell color space is preferred to the CIELAB color space system because the former is spaced according to regular visual color differences. Nevertheless, the relative position of the shade sorting blocks is determined by the shape of the block; a rhombic dodecahedron (RD) or truncated octahedron (TO) are the preferred shapes (ref. 93).

Alternative shade sorting methods have been proposed that can be related to a variety of color space systems (CIELAB and CMC as well as the Munsell system). These methods are based on a truncated octahedron (TO) shape and on a sphere and are respectively called the TO microspace method (ref. 94) and Clemson Color Clustering (CCC) system (ref. 95, 96). The microspace concept defines the size of the block only by maximum permissible color differences between any two specimens in the same block. In conjunction with the TO shape, this method is claimed to be very useful and give excellent shade sorting results (ref. 94). In the CCC method, a minimal number of spheres of a specified size required to close all sample points are directly determined. This results in fewer groups or clusters. Each cluster has samples that are within a certain range of maximum color differences. As shown in Figure 5.19, this produces half the number of groups by other shade sorting methods. Another advantage of the CCC method is elimination of effects associated with edges and corners in polyhedron shapes. The only disadvantage noted was the lack of correspondence between color groups obtained from samples of one lot compared to clusters obtained from a second lot (ref. 95).

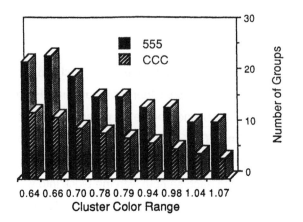

Figure 5.19. Comparison of number of groups produced by 555 and CCC shade sorting methods applied to 301 samples of a blue navy twill. Color differences measured using the CMC (2:1) formula. The CCC method produces an average 51% fewer groups for this data (ref. 95). Courtesy of American Association of Textile Chemists and Colorists.

5.4.5 Status of instrumental methods of measurement

Instrumentation has been increasingly used to measure color and color differences in textile products. However, the human eye is still more accurate than instruments in determining differences between two samples. Instruments do have the advantage of better precision, i.e. more reliable reproduction of a color measurement than a group of human observers.

The CIE and other groups historically have done an excellent job of standardizing various light sources used and correcting these and all types of instrumentation for metamerism and whiteness discrepancies. A critical review of the accurate measurement of whiteness of textiles and other opaque objects discusses the importance of correcting for the presence of fluorescent brightening agents. Flexibility for a particular whiteness preference by a group of observers or a culture or country is advocated. A variety of mathematical formulas for whiteness such as the Ganz Linear Whiteness Equations are also discussed (ref. 97).

Color measuring instruments can be traditionally subdivided into spectrophotometers (instruments that measure spectral reflectance or transmittance) and colorimeters. Most commercial colorimeters (such as the Hunter Color Difference Meter) are based on the CIELab color space and color difference equations. These colorimeters approximate the X, Y and Z tristimulus values that are generated by a light source, object and standard observer.

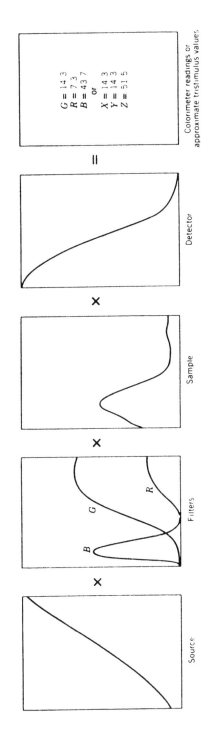

Figure 5.20. In **colorimetry** the tristimulus values of numbers closely related to them are obtained from measurements with the instrument's response (light source times filter times detector) adjusted to that required by the CIE system and sketched in Figure 5.17. It is important to note that it is difficult to do this with absolute accuracy (adapted from ref. 82). Courtesy of John Wiley and Sons Publishers.

The interfacing of color measuring instruments to computers offers additional sophistication and reliability in areas such as computer color matching, color quality monitoring, colorant or dye solution evaluation and color communication. A simple and useful system includes a monitor with a graphic adapter, processor, magnetic disks, a matrix printer, a suitable operating system and appropriate application software (ref. 98). Improvements in color communication have been recently attained by development of color visualization software that incorporates a color vision model. This enables colors to be accurately displayed on a variety of media including CRT (cathode ray tube) displays and printers. This minimizes or resolves two basic problems in color communication: (a) transmitting specification of color of an actual sample to another location and (b) communication of the idea of a color that is only a concept of a color designer or selector (refs. 99, 100). A recent attempt to replace totally visual color matching with instrumental color matching was not successful. However, the investigators did determine that instrumentation can dramatically reduce the number of visual matches required and still believe that future systems for color matching will ultimately be completely instrumental (ref. 101).

5.4.6 Tests and standards for colorfastness of textiles

Colorfastness of textiles towards a variety of agents and parameters is determined, and numerical ratings (1 to 5 or 1 to 9) are assigned to describe the colorfastness of the fabric. There are several standard and laboratory tests to determine the colorfastness of textiles to various natural and artificial light sources. There are also standard and laboratory tests to determine the colorfastness of textiles to abrasion, crocking, pleating, acids and alkalis, chlorine and peroxide bleaches, washing, dry cleaning, gas fumes and ozone, and water from various sources (e.g., sea water, swimming pools, etc.). Until recently, a blue wool fabric was the standard against which the lightfastness of other fabrics was compared. However, it has been recently been replaced by a violet polyester fabric purported to have more linear fading behavior. There are some investigators that still consider the blue wool fabric to be a more reliable lightfastness standard. Other important considerations are the use of reproducible and representative lighting conditions and measuring color or color differences of textiles by appropriate visual and instrumental methods.

The relationship between dye structure and colorfastness is comprehensively reviewed by Giles and co-authors (ref. 101). Fastness of dyes on textiles is subdivided into wet and dry conditions. Colorfastness to washing of dyes depends to some extent on the type of dye, its molecular weight and the rate of diffusion of the dye into the fiber. Fastness of dyes under dry conditions is even more complex and is influenced by factors such as different responses to atmospheric pollutants, rubbing, dry heat and ultraviolet light. Response of dyed textiles to air pollutants and photochemical events are discussed in detail in **Chapter 6** under chemical influences and agents.

The numerous standard tests performed in many countries for colorfastness of dyed textiles have been summarized and compared in two reviews (ref. 102, 103). The increasing involvement and importance of the ISO (International Standards Organization) in determining colorfastness standards and the close relationship of these ISO methods to BSI (British Standards Institution) are noted (ref. 102). Numerous categories for testing colorfastness of textiles are listed. These include: colorfastness to light, washing or laundering, dry cleaning, various aqueous solutions (such as seawater, perspiration and chlorinated water), air pollutants, bleaching agents, heat, miscellaneous influences (such as rubbing and organic solvents) and special standards for floorcoverings and carpets (ref. 102). Increased standardization of colorfastness tests and standards by the ISO with tests devised by groups such as the American AATCC and British BSI are likely to occur.

GENERAL REFERENCES FOR TEXTILE PERFORMANCE DETERMINED BY CHEMICAL AND INSTRUMENTAL METHODS

F. W. Billmeyer, Jr. and M. Saltzman, **Principles of Color Technology**, 2d Ed., Wiley, New York, 1981, 240 pp.

C. A. Farnfield and R. D. Perry, **Identification of Textile Materials**, The Textile Institute, Manchester, England, 1975, 262 pp.

P. H. Greaves, **Fiber Identification and the Quantitative Analysis of Fibre Blends**, Rev. Prog. Color. Rel. Topics 20 (1990) 32-39.

F. Happey (Ed.), **Applied Fibre Science. Vol. 1**, Academic Press, New York, 1978, 562 pp.

F. Happey (Ed.), **Applied Fibre Science. Vol. 2**, Academic Press, New York, 1979, 511 pp.

F. Happey (Ed.), **Applied Fibre Science. Vol. 3**, Academic Press, New York, 1978, 553 pp.

R. S. Hunter and R. W. Harold, **The Measurement of Appearance**, 2d Ed., Wiley Interscience, New York, 1987, 411 pp.

J. W. Weaver (Ed.), **Analytical Methods for a Textile Laboratory**, Am. Assoc. Text. Chem. Color., Research Triangle Park, N. C., 1984, 411 pp.

TEXT REFERENCES

1 D. Nissen, V. Rossbach and H. Zahn, **Chemical analysis of synthetic fibers**, Angew. Chem. Intl. Ed. 12 (8) (1973) 602-616.

2 C. A. Farnfield and R. D. Perry, **Identification of Textile Materials**, The Textile Institute, Manchester, England, 1975, pp. 182-183.

3 S. Sivakumaran and S. Rajendran, **A quick, room temperature method for polyester-cellulosic blend analysis**, Text. Res. J. 56 (3) (1986) 213-214.

4 V. Rossbach, **Analysis of low molecular weight homologues of fiber-forming polycondensates**, Angew. Chem. Intl. Ed. 20 (10) (1981) 831-840.

5 A. Goetz, **Procedures for identifying dyes and pigments**, Text. Chem. & Color. 17 (9) (1985) 171-176.

6 T. A. Perenich and G. Baumann, **Measuring the solubility of disperse dyes**, Text. Chem. & Color. 21 (2) (1989) 33-37.

7 J. Gacén, J. Maillio and A. Naik, **Iodine sorption and tensile parameters of LOY, MOY, POY and FOY polyester yarns**, Mell. Textilber. 69 (12) (1989) 867-870.

8 J. Gacén, J. Maillio and F. Bernal, **Iodine sorption by polyester-theory, variables and sensitivity**, Mell. Textilber. 70 (10) (1990) 737-742.

9 S. Aravindanath, K.M. Paralikar and S. M. Betrabet, *Cellulase* **dissolution: electron microscope technique for the study of chemically modified cotton**, Ind. J. Text. Res. 4 (3) (1979) 17-24.

10 G. Odian, **Principles of Polymerization**, Wiley-Interscience, New York, 2d Ed., 1981, p. 24.

11 J. Park and J. Shore, **Dyeability tests for textile substrates**, Rev. Prog. Color. Rel. Topics 12 (1982) 43-47.

12 J. Chrastil and R. M. Reinhardt, **Direct colorimetric determination of formaldehyde in textile fabrics and other materials**, Anal. Chem. 58 (13) (1986) 2848-2850.

13 J. H. Hoffman, **Qualitative spot tests**, in: J. W. Weaver (Ed.), Analytical Methods for a Textile Laboratory, Am. Assoc. Text. Chem. Color., Research Triangle Park, N. C., 1984, Ch. 8, pp. 155-206.

14 T. A. Perenich, **Reversed phase TLC of acid dyes**, Text. Chem. & Color. 14 (3) (1982) 60-64.

15 I. H. Leaver, D. M. Lewis and D. J. Westmoreland, **Analysis of wool lipids using thin-layer chromatography with flame ionization detection**, Text. Res. J. 58 (10) (1988) 593-600.

16 C. L. Renschler and F. B. Burns, **Monitoring of degradation in thermally aged nylon 6,6. I. UV-Visible absorption spectrophotometry**, J. Appl. Poly. Sci. 29 (4) (1984) 1125-1131.

17 S. Collins, S. Davidson, P. H. Greaves, M. Healey and D. M. Lewis, **The natural fluorescence of wool**, J. Soc. Dyer. Color. 104 (9) (1988) 348-352.

18 N. M. Morris and R. J. Berni, **Infrared spectroscopy**, in: J. W. Weaver (Ed.), Analytical Methods for a Textile Laboratory, Am. Assoc. Text. Chem. Color., Research Triangle Park, N. C., 1984, Ch. 11, pp. 265-292.

19 D. O. Hummel and C. Votteler, **Analysis of multicomponent fibers and polymer blends by differential IR**, Angew. Makro. Chem. 117 (1983) 171-194.

20 C. Q. Yang, R. R. Bresee and W. G. Fateley, **Near-surface analysis and depth profiling by FT-IR photoacoustic spectroscopy**, Appl. Spectr. 41 (5) (1987) 889-896.

21 R. S. Davidson, D. King, P. A. Duffield and D. M. Lewis, **Photoacoustic spectro-scopy for the study of adsorption of dyes on wool fabrics**, J. Soc. Dyer. Color. 99 (4) (1983) 123-126.

22 N. Ikuta, T. Sakamoto, T. Kouyama, I. Abe and T. Hirashima, **A nondestructive analysis of silane coupling agents on E-glass cloth by transmission FTIR spectroscopy**, Sen'i Gakkaishi 43 (6) (1987) 313-319.

23 S. Ghosh and G. Brodmann, **At-line measurement of durable press resin on fabrics using the NIR spectroscopy method**, Proc. AATCC Natl. Tech. Conf. 1991, pp. 204-210.

24 F. J. Deblase, M. L. McKelvy, M. Lewin and B. J. Bulkin, **Low-frequency Raman spectra of poly(ethylene terephthalate)**, J. Poly. Sci., Poly. Let. 23 (2) (1985) 109-115.

25 K. Prasad and D. T. Grubb, **Direct observation of taut tie molecules in high-strength polyethylene fibers by Raman spectroscopy**, J. Poly. Sci. Pt. B. Poly. Phys. 27 (1989) 381-403.

340

26 F. Horii, A. Hirai and R. Kitamaru, **Cross polarization/magic angle spinning** ^{13}**C-NMR study: molecular chain conformations of native and regenerated cellulose**, in: J. C. Arthur, Jr. and H. L. Needles (Eds.), Polymers for Fibers and Elastomers, Am. Chem. Soc. Symp. Series 260, Washington, D. C., 1984, pp. 27-42.

27 A. Hirai, F. Horii and R. Kitamaru, **CP/MAS** ^{13}**C NMR study of never-dried cotton fibers**, J. Poly. Sci. Pt. C. Poly. Let. 28 (1990) 357-361.

28 C. M. Carr and W. V. Gerasimowicz, **A carbon-13 CPMAS solid state NMR spectroscopic study of wool: effects of heat and chrome mordanting**, Text. Res. J. 58 (7) (1988) 418-421.

29 B. C. Fong, **19F-NMR spectrometric method for identifying the structure of the fluoroaliphatic group in fluorochemical finishes**, Text. Res. J. 58 (5) (1988) 304-306.

30 D. W. Dwight, J. E. McGrath and J. P. Wightman, **ESCA analysis of polymer structure and bonding**, in: T. K. Wu and J. Mitchell, Jr. (Eds.), Polymer Analysis, Appl. Poly. Symp. 34 (1978) 35-47.

31 M. M. Millard, **Surface analysis of fibers and polymers by x-ray photoelectron spectroscopy: industrial applications**, in: L. A. Casper and C. J. Powell (Eds.), Industrial Applications of Surface Analysis, Am. Chem. Soc. Symp. Series 199, Washington, D. C., 1982, pp. 143-202.

32 C. N. Carr, S. F. Ho, D. M. Lewis, E. D. Owen and M. D. Roberts, **Photoelectron spectroscopy and the surface chemistry of wool**, J. Text. Inst. 76 (6) (1985) 419-424.

33 C. Q. Yang, R. R. Bresee and W. G. Fateley, **Studies of chemically modified poly (ethylene terephthalate) fibers by FT-IR photoacoustic spectroscopy and X-ray photoelectron spectroscopy**, Appl. Spectr. 44 (6) (1990) 1035-1039.

34 T. L. Vigo, A. Muschelewicz and K.-Y. Wei, **Monitored photooxidation of polypropylene nonwovens**, in: R. B. Seymour and T. Cheng (Eds.), Advances in Polyolefins, Plenum Press, New York, 1987, pp. 499-512.

35 P. E. McGovern, J. Lazar and R. H. Michel, **The analysis of indigoid dyes by mass spectrometry**, J. Soc. Dyer. Color. 106 (1)(1990) 22-25.

36 A. Meybeck and J. Meybeck, **Electron spin resonance**, in: F. Happey (Ed.), Applied Fiber Science, Vol. 1, Academic Press, New York, 1978, Ch. 13.

37 M. H. Miles and K. L. DeVries, **Fracture generation of free radicals in Kevlar 49 fibers**, Polym. Mater. Sci. Eng. 52 (1985) 73-75.

38 C. M. Player, Jr. and J. A. Dunn, **Chromatographic methods**, in: J. W. Weaver (Ed.), Analytical Methods for a Textile Laboratory, Am. Assoc. Text. Chem. Color., Research Triangle Park, N. C., 1984, Ch. 13, pp. 315-394.

39 K. R. Beck and D. H. Pasad, **Liquid-chromatographic determination of rate constants for the cellulose-dimethyloldihydroxyethyleneurea reaction**, J. Appl. Poly. Sci. 27 (4) (1982) 1131-1138.

40 A. S. Gozdz and H.-D. Weigmann, **Surface characteristics of intact fibers by inverse gas chromatography**, J. Appl. Poly. Sci. 29 (12) (1984) 3965-3979.

41 T. Shaw and M. A. White, **The chemical technology of wool finishing**, in: M. Lewin and S. Sello (Eds.), Handbook of Fiber Science and Technology. Vol. II. Chemical Processing of Fibers and Fabrics. Functional Finishes. Pt. B., Marcel Dekker, New York, 1984, p. 394.

42 B. B. Wheals, P. C. White and M. D. Paterson, **High-performance liquid chromatographic method utilising single or multi-wavelength detection from the comparison of disperse dyes extracted from polyester fibres**, J. Chromatog. 350 (1) (1985) 205-215.

43 R. M. E. Griffin, T. G. Kee and R. W. Adams, **High-performance liquid chromatographic system for the separation of basic dyes**, J. Chromatog. 445 (2) (1988) 441-448.

44 G. B. Guise and G. C. Smith, **Gel permeation chromatography of a polyamide-epichlorohydrin resin and some other cationic polymers**, J. Chromatog. 235 (2) (1982) 365-376.

45 C. M. Ladisch, Y. Yang, A. Velayudhan and M. R. Ladisch, **A new approach to the study of textile properties with liquid chromatography**, Text. Res. J. 62 (6) (1992) 361-369.

46 G. J. Shugar and J. A. Dean, **The Chemist's Ready Reference Handbook**, McGraw-Hill, New York, 1990, Chapter 21: Thermal Analysis.

47 K. Slater, **The thermal behaviour of textiles**, Text. Prog. 8 (3) (1976), pp. 80-90.

48 S. Ghosh and J. E. Rodgers, **The determination of the heat-set temperature and heat history of carpet yarns by thermal stress analysis**, Mell. Textilber. 67 (1) (1986) 26-29.

49 H. Sakabe, H. Ito, T. Miyamoto and H. Inagaki, **States of water sorbed on wool as studied by differential scanning calorimetry**, Text. Res. J. 57 (2) (1987) 66-72.

50 T. Hatakeyama, Y. Ikeda and H. Hatakeyama, **Effect of bound water on structural change of regenerated cellulose**, Makromol. Chem. 188 (8) (1987) 1875-1884.

51 M. Fukuda, K. Ohtani, M. Iwasaki and H. Kawai, **Fundamental studies on the interaction between moisture and textiles. Pt. VI. Measurement of the heat of moisture sorption by a heat transfer type calorimeter**, Sen'i-Gakkaishi 43 (11) (1987) 567-577.

52 H. A. Hristov and J. M. Schultz, **Thermal response and structure of PET fibers**, J. Poly. Sci., Pt. B., Poly. Phys. 28 (10) (1990) 1647-1663.

53 M. Yatagai, M. Komaki and T. Hashimoto, **Applying differential scanning calorimetry to detergency studies of oily soil**, Text. Res. J. 62 (2) (1992) 100-104.

54 R. J. Berni, V. W. Tripp and J. J. Hebert, **X-ray diffraction and fluorescence**, in: J. W. Weaver (Ed.), Analytical Methods for a Textile Laboratory, Am. Assoc. Text. Chem. Color., Research Triangle Park, N. C., 1984, Ch. 12, pp. 293-312.

55 N. S. Murthy, **Analysis of X-ray diffraction data from polymers by profile fitting**, Am. Lab. 14 (11) (1982) 70-75.

56 R. Hagege, **Electron diffraction and dark field on ultrathin sections of textile fibers**, in: A. French and J. Gardner (Eds.), Fiber Diffraction Methods, Am. Chem. Soc. Symp. Series 199, Am. Chem. Soc., Washington, D. C., 1979, Chapter 17.

57 G. M. Venkatesh, R. E. Fornes and R. D. Gilbert, **Wide angle X-ray diffraction studies of nylon-ionomer fiber blends**, J. Appl. Poly. Sci. 28 (7) (1983) 2247-2260.

58 A. K. Kulshreshtha, M. V. S. Rao and N. E. Dweltz, **Equatorial small-angle X-ray scattering study of porosity and pore size distribution in polyester and their influence on dyeing behavior**, J. Appl. Poly. Sci. 30 (8) (1985) 3423-3443.

59 A. D. French, W. A. Roughead and D. P. Miller, **X-ray diffraction studies of ramie cellulose I** in: R. Atalla (Ed.), Structure of Cellulose, Am. Chem. Soc. Symp. Ser. 340, Washington, D. C., 1987, pp. 15-37.

60 Z. Zheng, S. Nojima, T. Yamane and T. Ashida, **Superstructural model for small-angle X-ray scattering: application to nylon 6 fiber**, Macromol. 22 (11) (1990) 4362-4367.

61 I. S. Polikarpov, I. I. Shiiko, G. I. Kotylar, B .D. Semak and G. F. Pugachevskii, **A polarographic method for the evaluation of the light stability of cotton fabrics**, Izv. Vyssh. Uchebn. Zaved., Tekhnol. Legk. Pron. sti. 25 (2) (1982) 27-31.

62 D. Katović I. Piljac and I. Soljačić, **Determination of iron and copper in textile materials by anodic stripping voltammetry**, Text. Res. J. 55 (1) (1985) 20-23.

63 S. H. Yoon, **Determination of formaldehyde released from resin treated fabrics**, Text. Chem. Color. 17 (4) (1985) 33-37.

64 M. Kudrna and M. Mitterpachová, **Orientation of polypropylene fibre**, Colloid and Poly. Sci. 261 (11) (1983) 903-907.

65 R. D. Ford and P. G. H. Bakker, **The acoustical properties of various carpet and underlay combinations**, J. Text. Inst. 75 (3) (1984) 164-174.

66 S. Osaki, **Dielectric anisotropy of nonwoven fabrics by using the microwave method**, Tappi 72 (5) (1989) 171-175.

67 B. Coté, C. Forget, N. Thérien and A. D. Broadbent, **Measuring the water content of a textile fabric using a radio frequency sensor**, Text. Res. J. 61 (12) (1991) 724-728.

68 K. Nettelnstroth, **Methods for a simple determination of double refraction on synthetic fibers**, Mell. Textilber. 63 (3) (1982) 212-215.

69 H. H. Yang, M. P. Chouinard and W. J. Lingg, **Birefringence of highly oriented fibers**, J. Poly. Sci., Poly. Phys. 20 (6) (1982) 981-987.

70 V. B. Gupta and D. R. Rao, **Birefringence of wool**, Text. Res. J. 61 (9) (1991) 510-516.

71 J. G. Fatou, **Optical microscopy of fibers**, in: F. Happey (Ed.), Applied Fibre Science, Vol. I., Academic Press, New York, 1978, Chapter 3.

72 P. R. Blakey and B. Micklethwaite, **Scanning electron microscopy and X-ray microanalysis**, in: F. Happey (Ed.), Applied Fibre Science, Vol. I, Academic Press, New York, 1978, Chapter 9.

73 R. Hesse, **Microscope methods in textile testing**, Mell. Textilber. 67 (7) (1986) 518-521.

74 G. D. Danilatos, V. N. E. Robinson and R. Postle, **An environmental scanning electron microscope for studies of wet wool fibres**, 6th Quinquinneal Intl. Wool Text. Res. Conf. Proc. Vol. II (1980) 463-471.

75 W. Zhao, N. A. G. Johnson and A. Willard, **Investigating wool fiber damage by image analysis**, Text. Res. J. 56 (7) (1986) 464-466.

76 D. Jose, N. R. S. Hollies and S. M. Spivak, **Instrumental techniques to quantify textural change in carpet. Pt. I: Image analysis**, Text. Res. J. 56 (10) (1986) 591-597.

77 D. Jose, N. R. S. Hollies and S. M. Spivak, **Instrumental techniques to quantify textural change in carpet. Pt. II: Goniophotometry**, Text. Res. J. 58 (4) (1988) 185-190.

78 Y. Wu, B. Pourdeyhimi and S. M. Spivak, **Texture evaluation of carpets using image analysis**, Text. Res. J. 61 (7) (1991) 407-419.

79 C. Luo and R. R. Bresee, **Appearance evaluation by digital image analysis**, Text. Chem. Color. 22 (2) (1990) 17-19.

80 A. Eickmeier, L. Wefers, T. Bahners and E. Schollmeyer, **Speckle photographic analysis of polymer films**, Text. Praxis 44 (10) (1989) 1114-1121.

81 K. Schicktanz, **Practical experience with automatic cloth inspection by the laser-scan system**, Mell. Textilber. 71 (10) (1990) 786-789.

82 F. W. Billmeyer, Jr. and M. Saltzman, **Principles of Color Technology**, 2d Ed., Wiley, New York, 1981, 240 pp.

83 R. S. Hunter and R. W. Harold, **The Measurement of Appearance**, 2d Ed., Wiley-Interscience, New York, 1987, 411 pp.

84 J. H. Nobbs, **Kubelka-Monk theory and prediction of reflectance**, Rev. Prog. Color Rel. Topics 15 (1985) 66-75.

85 S. Moradian and B. Rigg, **The quantification of metamerism**, J. Soc. Dyer. Color. 103 (5/6) (1987) 209-213.

86 T. Braddock, **Accuracy of metameric indices in relation to visual assessments**, J. Soc. Dyer. Color. 108 (1) (1992) 31-41.

87 F. J. J. Clarke, R. McDonald and B. Rigg, **Modification to the JPC79 colour difference formula**, J. Soc. Dyer. Colour. 100 (1984) 128-132 and 180-183.

88 R. McDonald, **Acceptability and perceptibility decisions using the CMC colour difference formula**, Proc. AATCC Natl. Tech. Conf. 1987, pp. 190-195.

89 S. L. Jay, **Practical use of the CMC color difference equation**, Proc. AATCC Natl. Tech. Conf. 1987, pp. 199-202.

90 M. R. Luo and B. Rigg, **BFD (l:c) colour-difference formula. I. Development of formula**, J. Soc. Dyer. Color. 103 (2) (1987) 86-94.

91 M. R. Luo and B. Rigg, **BFD (l:c) colour-difference formula. II. Performance of the formula**, J. Soc. Dyer. Color. 103 (3) (1987) 126-132.

92 F. T. Simon, **Shade sorting by the 555 system**, Am. Dyest. Reptr. 73 (3) (1984) 17-26.

93 J. R. Aspland, C. W. Jarvis and J. P. Jarvis, **A review of shade sorting techniques**, Proc. AATCC Natl. Tech. Conf. 1988, pp. 118-120.

94 K. McLaren, **The development of shade sorting in the United Kingdom**, Proc. AATCC Natl. Tech. Conf. 1987, pp. 196-198.

95 J. R. Aspland, C. W. Jarvis and J. P. Jarvis, **An improved method for numerical shade sorting**, Text. Chem. Color. 19 (5) (1987) 21-25.

96 D. R. Teel, J. R. Aspland and J. P. Jarvis, **Color consistency and practical aspects of shade sorting using CCC**, Proc. AATCC Natl. Tech. Conf. 1991, 139-142.

97 R. Griessner, **Instrumental measurement of fluorescence and determination of whiteness: review and advances**, Rev. Prog. Color Rel. Topics 11 (1981) 25-36.

98 T. F. Chong, **Instrumental measurement and control of colour**, Rev. Prog. Color Rel. Topics 18 (1988) 47-55.

99 R. McDonald, **Colour communication in the 90s**, Proc. AATCC Natl. Tech. Conf. 1991, pp. 148-152.

100 R. E. Vanderhoeven, **Conversion of a visual to an instrumental color matching system: an exploratory approach**. Text. Chem. Color. 24 (5) (1992) 19-25.

101 C. H. Giles, D. G. Duff and R. S. Sinclair, **The relationship between dye structure and fastness properties**, Rev. Prog. Color Rel. Topics 12 (1982) 58-65.

102 M. Langton, **Development of methods of test for colourfastness and reference materials for their use**, Rev. Prog. Color. Rel. Topics 14 (1984) 176-186.

103 W. Schön, **Colorfastness today**, Text. Praxis. Intl. 45 (1) (1990) 30-32.

Chapter 6

TEXTILE PERFORMANCE: END USE AND RELEVANT TESTS

6.1 OVERVIEW

Textile performance is ultimately related to the end use conditions of a material. Numerous laboratory tests have been devised and are continually being refined to simulate end use conditions. Many of the standard test methods may be found in annual or updated technical manuals published by the American Association of Textile Chemists and Colorists (AATCC), American Society for Testing Materials (ASTM), International Standards Organization (ISO) and British Standards Institution (BSI). When these types of tests are discussed in this text, the test method number may or may not be given; however, a particular test or type of test may be readily found in the updated technical manual for each of these organizations. Ultimately, field trials and actual end use studies afford the most reliable estimate of textile performance; however, such studies are expensive and time-consuming and by necessity are less frequently conducted than laboratory simulations.

The physical, chemical and biological influences on textiles affect their end use performance. Although all agents (chemical and biological as well as physical) affect textile performance at the fiber, yarn and fabric levels, emphasis will be given to fabrics since they represent the largest class of textile structures in a variety of applications. Also, many additional interactions and phenomena occur at the fabric level that may not occur at the fiber and yarn levels. Thus, a fabric is usually the most complex and representative form of a textile structure that is subjected to these agents and influences in most end uses. These fundamental influences will be first discussed separately. However, complex interactions of all influences (such as those observed in weathering, storage, comfort, wear and refurbishing) usually have the greatest effect on how textiles perform for their intended use. These will be discussed later in **Section 6.5** under **multiple influences** on performance.

6.2 PHYSICAL AGENTS AND ASSOCIATED TEST METHODS

The first broad class of factors that affect textile performance are physical agents and influences. These may be further subdivided into mechanical deformation and degradation, tactile and associated visual properties of textile products (such as wrinkling, buckling, drape and hand) after their use and manufacture, and their response to heat, liquids and static charge.

6.2.1 Mechanical deformation and failure

Such processes occur because fibers in fabrics are subjected to variable and complex modes of deformation. These include shrinkage, tensile extension, compression, bending or flexing, frictional rubbing, torsion or twisting (shear), entanglement and ultimately fatigue or failure. Moreover, fibers undergo various degrees of delayed and immediate recovery after removal of these deformations. The more complex phenomenon of textile wear is a combination of some or all of these mechanical forces as well as other influences, and is discussed in more detail under **multiple agents** or influences that affect textile performance.

6.2.1a Dimensional stability. The mechanisms by which fabrics are dimensionally stabilized and destabilized and techniques to improve their dimensional stability have been previously discussed in **Chapter 4**. In addition to standard tests that measure the amount of shrinkage in fabrics after laundering or other types of refurbishing, there have been several studies for detecting, predicting and elucidating shrinkage of fabrics. Particular emphasis has been given to wool and cotton fabrics and to knits since these types of fibers and constructions are the most problematic in this regard.

The role of hygral expansion in prototype woven wool fabric continues to be evaluated and investigated. Postle and co-workers determined that hygral expansion in commercial wool fabrics is most strongly influenced by the level of weave crimp in the fabric, but that construction of the weave becomes more important at high moisture regains. They also describe an experimental apparatus that simultaneously measures the changes in wool fabric length and moisture regain with changing relative humidity (ref. 1). Cookson and his colleagues determined that hygral expansion measurements of woven wool fabrics are influenced by the drying method for wet fabrics in their relaxed state. More realistic values are obtained when the fabric is allowed to dry at room temperature before oven-drying (ref. 2). Four steam shrinkage tests were shown to give different results for wool fabrics due to different values for hygral expansion (reversible shrinkage) and relaxation shrinkage (irreversible shrinkage). Simple and rapid tests were recommended and based on wetting the fabric (with or without steaming to remove relaxation shrinkage), then drying it to constant weight at 105°C (ref. 3).

A comprehensive study to predict the weights, shrinkages and relaxed dimensions of 20 different cotton knit fabrics (representative types such as interlock, rib and single jersey) was conducted by Heap and his colleagues (ref. 4, 5). Two methods of relaxation were used: (a) standard tumble washing and drying and (b) washing for only the first cycle, but merely wetting out the fabric before the second and subsequent drying cycles. The latter procedure was adopted as a standard method for this investigation (called Starfish) with five cycles required to reach a fully relaxed state (ref. 5). A measuring frame of fixed dimension has also been used to determine the wash stability of cotton knit fabrics. This approach appeared to give good inter-laboratory correlations but there were still discernible differences between results obtained in wear trials and those in laboratory laundering simulations (ref. 6). An "intelligent" sensor has recently been developed to measure geometric changes (essentially shrinkage) in all types of woven as well as knitted fabrics. The system has the advantage of being a non-destructive, non-contact technique. It is based on a digitized image sequence analyzed with a computer-based pattern recognition program. It is suitable for on-line processing measurements of fabrics and could also be employed in wear and laundering studies if desired (ref. 7).

6.2.1b <u>Fundamental mechanical properties and their relation to failure</u>. The physical properties of textile fibers, particularly their mechanical and related properties, are comprehensively and critically analyzed and discussed in the classic text by Morton and Hearle (ref. 9). For woven fabrics, application of force in the longitudinal and/or transverse directions extends the fibers until they break or fail, and thus produces a load-elongation or stress-strain curve. This curve (Figure 6.1) is essentially a composite average of the stress-strain behavior exhibited by individual fibers and yarns in the fabric. The shape and slope of the curve vary with fiber type and are also influenced by many other factors.

From these stress-strain curves, various fabric mechanical properties that are related to their end use performance may be obtained. These include tensile or breaking strength (tenacity), elongation-at-break, their initial modulus (from the initial slope of the stress-strain curve) and work-of-rupture or energy-to-rupture (area under the curve in Figure 6.1). These specific properties vary widely with fiber type and crimp, yarn twist, fabric geometry and constructions, and fabric finishes and coatings.

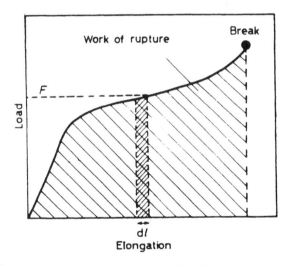

Figure 6.1. Representative stress-strain curve (ref. 8). Courtesy of The Textile Institute.

Many of the synthetic fibers presently produced have high tenacities (e.g., polyester, polyamides, gel-spun polyethylene and aramids), and thus have superior tensile properties and fatigue behavior compared to low or medium tenacity synthetics and to many of the natural fibers. Textile materials with high energy-to-rupture values are usually desirable, and their mechanical performance correlates well in applications where impact loading or repeated deformation occurs. In many other applications however, stresses and strains on the fiber are usually low and cumulative. Under these conditions, fiber failure occurs when the load-elongation characteristics

350

culminate in a substantial loss of elastic recovery of the fiber when it is repeatedly extended and relaxed. Although tearing strength of fabrics gives a quantitative ranking of materials in this regard, the failure characteristics in this mode are not amenable to the analysis of stress-strain curves generated in the extension of fabrics to the break or subjected to cyclic loading.

There are various ways in which fibers in fabrics ultimately undergo failure or fatigue. Microscopy, particularly scanning electron microscopy (SEM), has become an extremely valuable technique in determining how deformation and stresses of textile structures lead to failure and in characterizing the modes and different types of fiber failure that occur. Such fiber fracture may be classified as (a) brittle elastic failure, (b) ductile crack growth, (c) axial splitting, (d) fracture perpendicular to the fiber axis, (e) failure along kink bands, and (f) axial splitting due to tensile fatigue. Many of these phenomena are illustrated in Figure 6.2-6.5 (ref. 9).

In brittle elastic failure, the fiber fails to extend due to its high modulus and rapid failure occurs due to crack initiation at a surface fault. This type of failure is very characteristic of glass fibers. Ductile crack growth occurs in many synthetic fibers (polyamides, polyesters and some acrylics) with slow formation of numerous cracks

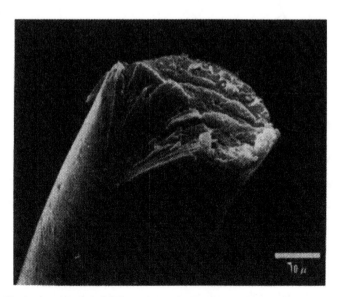

Figure 6.2. Typical example of failure through ductile crack growth (ref. 9). Courtesy of John Wiley and Sons Publishers.

followed by opening to a large V-shape subsequent to catastrophic failure. As observed in Figure 6.2, this is a relatively clean type of break.

Axial splitting or fibrillation is very common in natural fibers such as cotton (Figure 6.3). This type of failure is consistent with the fibrillar structure of the cotton and its

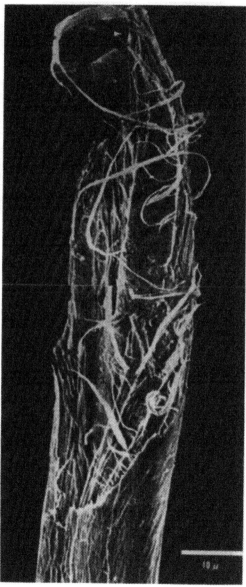

Figure 6.3. Cotton fiber fractured after axial splitting (ref. 9). Courtesy of John Wiley and Sons Publishers.

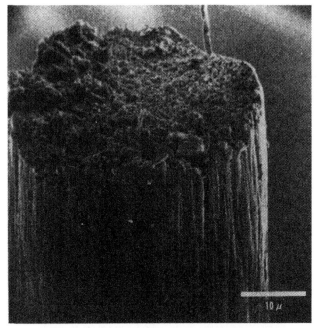

Figure 6.4. Example of acrylic fracture perpendicular to the fiber axis (ref. 9). Courtesy of John Wiley and Sons Publishers.

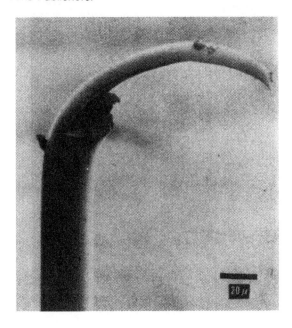

Figure 6.5. Axial splitting due to tensile fatigue in nylon (ref. 9). Courtesy of John Wiley and Sons Publishers.

reversals along the length of the fiber. Certain types of acrylic fibers, rayon, wool and crosslinked cotton fibers undergo fracture perpendicular to their fiber axis (Figure 6.4). Initial cracks occur in a weak region of groups of fibrils and fracture happens across a radial plane. Bending of polyester fibers causes lines to appear at 45° to the fiber axis. These lines are called kink bands and repeated flexing or bending of the fibers cause failure along these bands. Very high tenacity or high performance fibers such as the aramids and certain aliphatic polyamides fail by axial splitting due to tensile fatigue (Figure 6.5). Initial fracture occurs at the surface, then turns to run along the fiber for several fiber diameters until the load-bearing cross section is small enough for tensile failure.

Tensile and particularly tear strength tests are not generally applicable nor particularly meaningful for fabrics in which the yarns or fibers are arranged in random directions (e.g., knits and nonwovens) rather than being perpendicular to each other. Normally, the bursting strength, the amount of force needed to rupture a fabric, is the most reliable measurement of the mechanical integrity of knit and nonwoven fabrics. Biaxial stress and radial force apparatuses have also been developed and employed in an attempt to predict the mechanical performance of these types of fabrics by subjecting them to uniform and reproducible mechanical forces. Computer simulations and models have been devised to predict the failure of nonwoven fabrics and their tensile properties. Britton and his colleagues describe a computer simulation method to predict nonwoven fabric failure based on cluster analysis. In this analysis, the displaced fabric has a realistic degree of randomness or disorder and is strained to the yield point (ref. 10). A two-dimensional computer generated model to predict point-bonded strength of polyester nonwovens agreed with experimental stress-strain curves observed for these fabrics (ref. 11).

Fiber type and other factors (such as photodegradation and biodeterioration) markedly influence the type and mechanism of fiber fatigue or failure. For example, polyamide fibers that have not been degraded by ultraviolet light usually undergo failure by ductile crack growth, but have a fracture morphology characteristic of tensile failure when they are exposed to light. As we will observe later, failure modes are even more complex for fibers in fabrics that have been worn, undergo pilling and been subjected to prolonged and repeated laundering.

Failure of fibers in fabrics rarely occurs due to a simple extension to break. It is due to the complex interaction of stresses and strains that result from repeated bending or flexing, twisting abrasion, compression and stretching. Fibers in fabrics can undergo deformations due to repeated twisting and untwisting that lead to functional failure. This type of failure is called torsional fatigue. Although there are no standard fabric tests available to simulate the twisting and untwisting that fibers undergo in end use, the breaking twist angle of various types of fibers has been measured. Values as low as 2.5-5.0° were observed for glass fibers and as high as 58-63° for polyamide fibers (ref. 8). Torsional fatigue testers in conjunction with SEM demonstrated that this type of fiber failure was similar to axial splitting observed in fibers such as cotton (ref. 12, 13).

Although the majority of mechanical tests for textiles measure their tensile and tear strength or resistance, fabrics in use are also subjected to mechanical forces such as bending or flexing, torsion or twisting, compression and frictional rubbing. Tests that accelerate the mechanical degradation of fabrics to simulate their use or predict their ultimate wear and/or failure, are usually performed under high levels of fiber stress and strain and the fabrics are usually subjected to severe abrasive processes. Thus, these mechanical test results vary markedly in their reliability for predicting the end use performance of textiles. If the objective is to predict the ultimate failure of textile under high stress and strain influences, then tests that measure breaking and tearing strength, elongation-at-break and energy-to-rupture are useful and appropriate. However, tests that attempt to predict more subtle and long-term effects of wear and abrasion are not always indicative of the actual behavior of fabrics during their use. These abrasion tests usually only measure the effect of one type of abrasion (e.g., edge, flat or flex) on the behavior of a fabric to the point of its failure. For example, in a flex abrasion device, the fabric is abraded to the point of failure or for a specified number of cycles by holding it under tension, then bending or flexing it as it is rubbed or abraded unidirectionally by an abrasive bar. Correlation of these results with actual fabric performance is meaningful only when the primary function of the fabric is its ability to withstand repeated flexing and bending under constraint. Similar devices are used to determine the edge and the flat abrasion behavior of fabrics, but only measure one mode of abrasion. This is in contrast to the real mode in which textile materials fail during use. Such failure is usually composed of various combinations of forces applied in a cyclic manner in the fatigue process, i.e., the cumulative effects of deformation and recovery. When fabrics are subjected to severe or multiple abrasion processes by appropriate test methods, the results are more useful in predicting the resistance of textiles to ultimate failure. Abrasion devices that subject fabrics to such complex modes of deformation do so in a unidirectional, multidirectional or uniform rotary manner, then evaluate visually and mechanically the properties of the fabric surface. The complexity of abrasion is typified by the dramatic difference in the failure mode of untreated cotton fibers (typical axial splitting in Figure 6.6) compared to the same fibers coated with a crosslinked polyol that undergo transverse cleavage more typical of a synthetic fiber (Figure 6.7). Both fibers were taken from fabrics abraded with a standard ASTM flex abrader (ref. 14).

Besides the variety of standard and experimental abrasion testers described by Hearle (ref. 9), there have been newer devices constructed to simulate more realistic

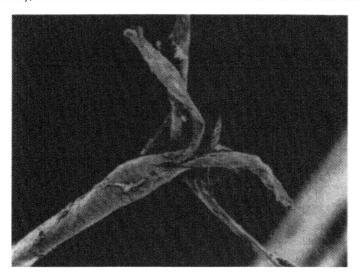

Figure 6.6. Scanning electron micrograph of untreated cotton fibers flexed to failure (ref. 14).

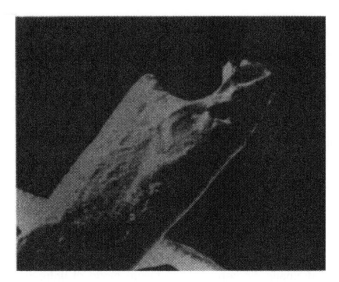

Figure 6.7. Scanning electron micrograph of treated cotton fibers (blended DMDHEU/PEG-1000, cured 3 min/90°C) flexed to failure (ref. 14).

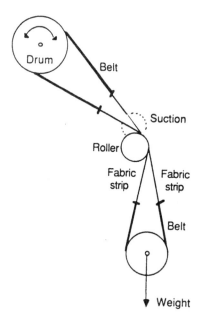

Figure 6.8. Apparatus for producing fabric-to-fabric abrasion. A reversing motor drives the drum, causing fabric strips to rub against each other as they ride over the roller under tension caused by the weight. The dashed circle is the end of a suction tube that carries particles from the space between the fabrics to a particle counter (ref. 15). Courtesy of Textile Research Institute.

abrasion processes in fabrics. Such approaches include a fabric-on-fabric attrition device (ref. 15) and an apparatus that simulates conditions of single fiber transfer from fabrics and garments (ref. 16, 17). The fabric-on fabric attrition apparatus depicted in Figure 6.8 was constructed to simulate more realistic abrasion processes that occur in wear due to subtle changes in fiber, yarn and fabric stresses and deformations. This apparatus was also useful in determining that abrasive wear rate of fabrics depends on contact pressure between fabrics and on the direction of curvature in the abraded zone (ref. 15). Similar considerations were taken into account in the development of an abrasion machine that evaluates structural changes in fabrics in terms of fiber transfer, i.e., the release and relocation of individual fibers in a fabric. Modeling procedures indicated that fiber denier, fiber length and weave type in woven fabrics were the most important factors in influencing the number, length and configuration of the transferred fibers (ref. 16, 17). While neither of these machines have been adopted in standard test procedures, both apparatuses give reasonable profiles and tendencies of various fabrics to undergo abrasive wear.

In a plenary paper and lecture, Hearle critically and comprehensively reviewed our current understanding of fiber structure and its relationship to mechanical performance and failure. He considers our knowledge of structural and mechanical aspects of natural fibers to be very thorough, but does not foresee any immediate improvement in properties and performance. In contrast, he believes there are more opportunities to improve the structure and performance of synthetic fibers such as polyamide and polyester because our current understanding of their morphology and salient features is still rudimentary. Modern mathematical concepts and theories such as fractals, nonlinear systems and chaos and quantum effects offer promise to enhance our understanding of fiber structure and mechanics (ref. 18).

In the same monograph and symposium, Schwartz and Backer and their colleagues offer a model of the tensile failure process in woven fabrics that is in good agreement with actual fabric tensile failure. Three major factors are identified: (a) strength distribution of constituent yarns, (b) recovery length of the yarn in fabric subsequent to a break and (c) local load redistribution subsequent to a break. The model is capable of predicting the load at which the first yarn breaks, the number of isolated yarn breaks before fabric failure, critical crack length at fabric failure and fabric tensile strength (ref. 19).

6.2.1c <u>Effect of heat on mechanical properties</u>. The exposure of fabrics to heat or increasing temperature may be detrimental or beneficial to their mechanical properties and performance depending on the conditions employed. While the thermal oxidation and destruction of fabrics due to combustion are well documented, there have been few detailed studies on more subtle effects of heat on the mechanical and related properties of fabrics.

Richards investigated the effect of prolonged heating (2-120 hr. at 125-180°C) on changes in dimensional stability, tear and tensile strength for untreated and flame-retardant wool, cotton and cotton/polyester blends and for untreated polyester and polyamide fabrics (ref. 20). Tearing strength was reduced more readily for all types of fabrics than tensile strength while dimensional stability was not materially effected. All fabrics except the 100% polyester had undesirable changes in their color. The Arrhenius equation (where k = rate of reaction, A = Arrhenius factor, E = energy of

$$k = A \exp(-E/RT) \tag{6.1}$$

activation, R = the gas constant and T = temperature in °K) was suggested as a method of calculating the useful lifetime of a textile at a given temperature (ref. 20). In contrast to conventional aliphatic polyamide fibers, aramid fibers have excellent thermal stability and retain all their mechanical properties at temperatures (ca. 250°C) where nylon 66 fiber melts.

A series of investigations by Postle and co-workers investigated the beneficial effect of "deageing" wool and synthetic fibers in fabrics by annealing below the glass transition temperature of each fiber type at their equilibrium moisture regain values. Physical ageing of fabrics adversely affects their stiffness and other important properties. A theory was developed to provide a method of predicting the viscoelastic properties of a textile from its fiber properties. When fabrics were annealed under appropriate conditions their bending stress relaxation rate was reduced to a minimum and resulted in maximum fabric recovery from bending and wrinkling. The annealing process was satisfactorily applied to wool/polyester blend fabrics to improve their viscoelastic properties (refs. 21-23).

6.2.1d <u>Surface deformation due to rubbing and friction</u>. Fabrics may also be rubbed in one or more directions across their surface in a manner that does not produce failure, but that does produce unacceptable changes in aesthetics, fabric surface properties and dimensional stability. In the case of wool, felting shrinkage may occur. With other types of fibers, depending on the severity of rubbing and of frictional forces, pills may form on the fabric surface and/or the fabric may acquire a substantial static charge. If the rubbing and friction is severe, fiber failure may occur due to prolonged or extensive surface abrasion.

A variety of test methods and instruments have been developed and devised to measure and predict fabric performance with regard to felting, pilling and development of static charge. Since most felting shrinkage occurs during laundering, test methods associated with this phenomenon will be discussed later (**refurbishing** under **multiple agents** or influences). Development of static charge and test methods will also be discussed later in this chapter under electrostatic influences. Standard test methods that evaluate the tendency of fabrics to pill consist of devices that brush the fabric surface or that randomly tumble them to generate pills. Ratings of from 1 to 5 are normally assigned with 1 indicating the poorest behavior towards pilling and 5 little or no tendency for pilling. However, wear and laundering studies and more sophisticated laboratory simulations and evaluations usually give the most meaningful prediction of the propensity of fabrics to undergo pilling.

Simulation of low level torsional fatigue in cotton/polyester fiber blends on repeated cycling did not always produce fiber fatigue. In mild torsion, it resulted in the formation of pills that resemble those in worn garments composed of fiber blends of identical composition (ref. 24). It was also observed in cotton/polyester blend fabrics that are worn and form pills that the cotton fiber is a passive participant and that pilling is due to the entanglement only of the polyester fibers. This result was consistent with the low tenacity and strength of cotton fibers (relative to synthetic fibers), since the former will be removed from the fiber surface due to breakage while the latter are strong enough to remain and entangle on the surface. When fiber fatigue did occur in synthetic fibers subjected to repeated low level twisting or torsion, this fatigue was the same as that for these fibers in actual end use situations (ref. 24).

360

Pill formation in various fabrics is believed to be initiated by the formation of free fiber ends due to progressive migration of surface fibers, subsequent entanglement of loose fiber ends, and progressive growth of the pill in size and in density. Cooke has comprehensively and critically evaluated the role of fiber fatigue on pilling concerning fuzz fatigue, fiber entanglement and pill growth and ultimate wear-off of pills due to fiber failure and attrition (refs. 25-27). With the aid of SEM, initial fuzz formation was observed to occur with concomitant damage to fibers that are free-standing before subsequent pill formation (ref. 25). The second phase is the entanglement of the fuzz into balls of fibers by a pull-out/roll up mechanism conducive

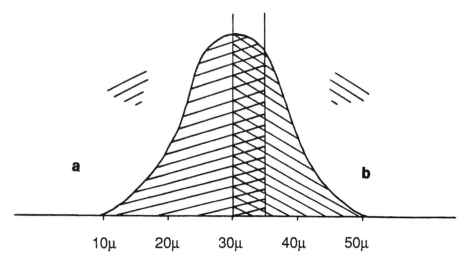

Figure 6.9. Diagram illustrating the pull-out/rollup mechanism of pill growth: (a) fibre removal by selective migration and pilling; (b) fibre removal by multiple fatigue cracking (adapted from ref. 27). Courtesy of The Textile Institute.

to pill growth (Figure 6.9). The process of pilling was also noted to be self-limiting over most parts of garments. However, continued generation of pills in cuffs and elbows caused severe thinning by repeated formation of pills and their subsequent wear-off in that abraded area (ref. 27).

A comparative study of five different laboratory pilling instruments was carried out on interlock knit fabrics composed of various fiber types and blends (100% polyester, 100% wool and blends of wool/polyester, cotton/polyester, rayon/polyester and acrylic/polyester). The investigators considered only one of the five instruments (random tumble pilling device) suitable for correlation to actual wear trial results. Several fiber, yarn, fabric and dyeing and finishing parameters were also varied to determine their importance in pilling propensity. These parameters included: fiber type and length, cross-section, crimp and fineness; yarn twist, linear density, hairiness; fabric construction, weight and stitch length; and processing variables such as washing and scouring, pH, steaming and thermosetting (ref. 28).

A corollary property of fabrics that undergo surface deformation due to friction and rubbing is the generation of lint or particle release. Although the tendency to form lint is not important in many apparel uses, this property is of paramount importance for garments worn in "clean rooms" by health care personnel and in computer manufacturing facilities. In the former application, microparticles may be responsible for causing some postoperative complications. In the latter application, such particles are undesirable in computer chip and manufacture of other computer components. A broad range of linting propensity was observed for eight different fabrics that were rubbed against themselves in a laboratory device. A disposable nonwoven fabric composed of all cellulosic fibers produced over 600 times the lint of a woven, combed and texturized 50/50 cotton polyester fabric (ref. 29). More quantitative results of lint generation were determined by use of a particle counter attached to the fabric attrition device described earlier. These investigators determined that durable press cotton fabrics produce unacceptably high levels of lint relative to most other types (ref. 15). Protective clothing based on high tenacity polyethylene fabrics has been developed that meets the strict criteria for use in clean rooms. However, these fabrics have little moisture transport and content and improvements are needed to provide thermal comfort as well as low particle release.

6.2.1e Bagging, buckling, wrinkling and other deformations. Less severe forms of fabric and fiber deformation do not produce catastrophic failure but may still produce unacceptable surface changes and aesthetics. This type of deformation can be further classified as bagging, buckling and wrinkling. Bagging may be defined as the nonrecoverable stretch of fabrics that increases their volume and thus produces an unacceptable aesthetic appearance. Buckling and wrinkling are more pronounced irreversible deformations in which the fabric has regions that are out of the normal two dimensional plane. Buckling normally denotes a more orderly out-of-plane deformation limited to certain regions while wrinkling usually denotes a random, comprehensive out-of-plane deformation for fabrics. Both phenomena are considered aesthetically unacceptable for fabrics and especially for garments.

Bagging of woven fabrics was satisfactorily predicted by measuring volume changes and statistically correlating such changes with their hysteresis behavior in tensile, bending and shearing deformations and in their dynamic creep behavior. These bagging simulations and correlations were valid for both stretch and conventionally woven fabrics (ref. 30).

Recovery from deformation due to wrinkling is normally determined by both subjective and objective test methods that have evolved with the development of durable press or crease resistant fabrics. In the subjective test method, wrinkling is induced in a fabric by twisting and wrinkling in an apparatus under a predetermined load at standard conditions of temperature and relative humidity. The fabric is subsequently reconditioned to its relaxed state, then visually compared by observers to it to three-dimensional replicas having varying degrees of wrinkling. In the objective test, a strip of fabric is folded for a specified time under a predetermined load. After removal of the load, the crease recovery angle of the fabric is measured (values in warp + fill directions). For durable press performance, fabrics having dry and/or wet crease recovery angles above 280° are considered acceptable.

The lack of agreement of the objective crease recovery method between laborator-ies and its poor correlation with the visual replica test have led to the development of other tests and studies on buckling and wrinkling. A digital image analysis technique (also discussed in **Chapter 5** under **optical methods** of analysis) for appearance evaluation of fabric wrinkling was considered a viable and effective alternative to subjective fabric appearance ratings given by trained observers (ref. 31). The buckling behavior of an aramid fabric was simulated by incorporating effects of fabric weight and nonlinear bending stiffness derived from theoretical models (ref. 32).

Wrinkling of garments was analyzed by deforming a fabric cylinder in axial compression (ref. 33). It was determined that buckling and wrinkling patterns in knitted fabrics (bellows type) were different from those observed in woven fabrics (diamond or lobe type) (Fig. 6.10). A mathematical model was then proposed to represent more accurately the complex buckling and wrinkling behavior of woven fabrics. This model, based on moire topography and experimental observations, may be represented by two groups of contour hyperbolas and is considered different from and more realistic than a Yoshimura buckle pattern shown in Figure 6.10 (b) (ref. 34).

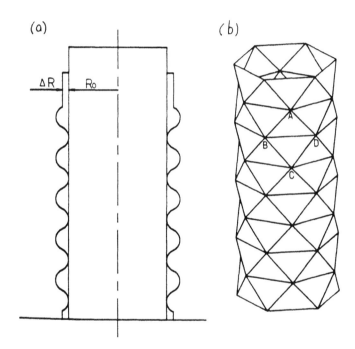

Figure 6.10. Buckling modes: (a) bellows type; (b) Yoshimura type buckle pattern of N = 6. R_o = radius of the core cylinder and ΔR = clearance between core and fabric cylinders and ABCD is a representative lobe (ref. 32). Courtesy of Textile Research Institute.

In addition to changes in aesthetic and mechanical properties caused by extension, surface and edge rubbing, and bending, tests have been devised to predict and measure the snag resistance of fabrics. Most of the tests are based on subjective judgments or evaluations after the fabrics are subjected to mechanical devices that simulate snagging. In one test, a miniature mace with sharp protruding points is bounced against a fabric wrapped around a rotating cylinder. Alternatively, the fabric is loosely sewn around a bean bag, then randomly tumbled in a chamber containing pins against which the fabric can catch and snag. Snagging has been shown to be highly dependent on the direction and severity with which the textile yarns and fibers are pulled from the fabric and on the tension at which the fabric is mounted on a surface. Thus, snagging behavior of fabrics observed in test devices must be carefully correlated with actual snagging of fabrics in use to obtain meaningful and useful results.

6.2.1f   Mechanical tests for special fabrics.   Fabrics with special features or constructions require either additional tests or modification of existing tests for conventional fabrics to characterize adequately their mechanical and related properties. Coated fabrics must be evaluated not only for their mechanical integrity and behavior but also evaluated for their bonding integrity of the coating to the fabric. A compendium of all standard textile test methods for coated fabrics lists numerous abrasion devices, tests for adhesion of coatings to fabrics and for bond strengths of coated and laminated fabrics (ref. 35).

Another type of fabric that requires special consideration is triaxially woven fabric. Schwartz and colleagues have defined a unit cell for this type of fabric and subjected it to biaxial loading in an attempt to compare it to conventionally woven and leno woven fabrics under equivalent conditions. Although their model correctly predicted initial tensile behavior of triaxial fabrics under biaxial loading, no true comparison could be made between the tearing behavior of fabrics with a leno weave and a triaxial weave. The study also concludes that no conventional equivalent exists between triaxial fabrics and conventionally woven fabrics for measurement of their mechanical properties and behavior (ref. 36).

6.2.1g   Bending, stiffness and drape.   The flexural rigidity or stiffness of fabrics relates to their overall mechanical integrity and comfort. Drape, the ability of a fabric to orient itself into folds when influenced by gravity, is markedly dependent on the stiffness and bending length. Thus, test methods and fundamental studies have focused on the mechanics of bending, stiffness and the measurement and prediction of fabric drape. The cantilever method is the preferred test for measuring fabric stiffness. In this test, a rectangular strip of fabric overhangs from a horizontal support; one-half of the overhang length is the bending length, and the flexural rigidity is the bending length of the fabric normalized for its weight per unit area. The heart loop test is used to measure stiffness for fabrics that are extremely soft or supple or that have a tendency to twist or curl. The fabric strip is formed into a heart-shaped loop whose length is measured when the fabric is hung vertically under its own weight; flexural rigidity and bending length are then calculated from the length measured.

An instrument has been designed that effectively transfers the method of overhanging fabric length to the electronic measurement of stiffness (ref. 37). The instrument is based on the principle of cantilever bending and has been used to compare 60 different fabrics by this method and the frequently used Shirley Stiffness meter. Although the electronic device is more objective and rapid than conventional stiffness testers, it offered no advantage in predicting drapeability of fabrics relative to other stiffness testers (ref. 38).

The random nature of fiber bonding in nonwoven fabrics presents a complex structure concerning their bending and stiffness characteristics. Thus, a series of studies was conducted to investigate important characteristics of fibers to determine the mechanics of bending of commercially available nonwoven fabrics made by various processing techniques and offered methods to improve nonwoven fabric flexibility (refs. 39-42). Spunbonded nonwoven fabrics were typified by polyamide fibers (nylon 66) held together by uniformly dispersed junction bonds. It was concluded that the relatively high bending stiffness of these fabrics originates from low degrees of relative fiber motion due to the high density of bonds at fiber contact points. For lightweight fabrics of this type, cross sections are composed of substantial amounts of unconstrained surface layers of fibers while heavier fabrics behave like continuum plates due to fiber constraints (ref. 39). Spunbonded nonwoven fabrics made from continuous thermoplastic polypropylene filaments and thermally spot bonded were also evaluated for fiber characteristics that would impart optimum flexibility. Based on a combined approach of experiments and modelling, optimum bending flexibility for these fabrics was achieved by incorporating some crimp to give them effective axial extensibility and considerable freedom of fiber motion.

The mechanics of bending and design for optimum flexibility of two different types of print-bonded nonwovens were also investigated (ref. 41, 42). A combined approach of analytical modelling and experimentation indicated that staple viscose and polyester fibers with a latex print binder had optimum flexibility when minimum binder was used to retain lamination strength yet retain freedom of fiber motion (ref. 41). The other nonwoven fabric investigated was composed of viscose staple fibers arranged to have elliptical hole patterns print-bonded by acrylic latex in strips. As in earlier studies, this fabric was modelled as a two phase composite of bonded and nonbonded phases. This fabric was very flexible in its transverse direction and reasonably flexible in its machine direction, but was not as drapeable as a woven fabric (ref. 42).

Drape or drapeability of a fabric is usually determined or expressed in terms of its drape coefficient. A representative technique for measuring drapeability of a fabric consists in placing a circular swatch of fabric over a metal disc. The diameter of the fabric is generally twice that of the disc. A light is used to project an image of the draped fabric that is traced on paper. In more recent methods, optical devices scan the contours and folds of the drape fabric to compute its drape coefficient. This coefficient may range from 0 (no drape) to 100 (a fully drapeable fabric).

A psychological scale for fabric stiffness "Stifs" was devised from measured drape coefficients and subjective stiffness evaluations of fabrics by a panel. This psychological scale is based on the formula:

$$\log \psi = \frac{1}{b} \log (a + b + \log D) + C \tag{6.2}$$

where ψ = psychological magnitude, D = drape coefficient, b = a regression coefficient based on fractionation judgments while a and C are constants (ref. 43). In a related study, the subjective evaluation of drape was dynamically simulated by cutting a fabric sample to a tapered geometric shape, folding the cut fabric in half, then gripping it vertically downwards in the clamp of a specially designed device (ref. 44). To duplicate the effect of the fabric bending under its own weight, a glass needle of predetermined weight was fixed to the sample at 0.5 cm from the tip of adhesive tape. The instrument arm is then rotated in 5° increments from 0 to 45°, 45°

to -45° then back to 45° to generate a hysteresis curve for determination of dynamic fabric drapeability.

Multiple regression analysis was used to determine which fabric properties were important in predicting the drape coefficient of knitted fabrics. Stiffness was the most important factor in predicting the drape coefficient of knitted fabrics. However, the drapes of warp knits were more dependent on stiffness, fabric thickness and extension while the drapes of weft knits were dependent primarily on stiffness and shear (ref. 45).

Later studies have focused on mathematical models and analysis to determine and predict fabric drapeability. Drape or more complex buckling deformations may be analyzed by two dimensionless groups J_1 and J_2 that relates bending, membrane and potential energies by the sets of equations:

$$J_1 = U_M/U_B\phi^2 = Yl^2/B \tag{6.3}$$

$$J_2 = U_P\phi/U_B = \gamma gl^3/B \tag{6.4}$$

where U_M = membrane strain energy, U_B = bending strain energy, U_P = potential energy, ϕ = a geometrical measure of the form of deformation, Y = membrane modulus, B = bending stiffness, γg is fabric weight per unit area and l is a characteristic length defining the size of the material (ref. 46). Draping behavior of fabrics were also treated as an orthotropic shell membrane and predicted by applying a geometric non-linear finite element method. Experimental drape coefficients of fabrics were determined on a designed drape tester. The determined values were in excellent agreement with values predicted from the finite-element analysis (ref. 47).

6.2.1h <u>Prediction and measurement of hand</u>. Hand or handle is an elusive quality of fabric that is usually defined as "the subjective assessment of a textile material obtained from the sense of touch" (ref. 48). A critical review surveys techniques for the subjective assessment of hand and relationships between subjective hand assessment and objective measurement from an array of physical properties. For subjective assessments, the terms softness and smoothness are the most frequently used descriptors. Eight physical property parameters correlated positively with subjective hand assessments in earlier investigations. Four were related to bending (elastic flexural rigidity, coercive couple, bending length and loop softness), two were related to shearing (shearing stress and initial shearing modulus) and the other two parameters were coefficient of kinetic friction and drape coefficient (ref. 48).

The extensive and detailed studies of Kawabata and his co-workers (ref. 49, 50) led to the development of instrumentation (Kawabata Evaluation System or KES) and a ranking to determine hand based on fabrics of different construction, weight, weaves and compositions and to fifteen different objective and measurable parameters. These parameters are shown in Table 6.1 and correlate well with primary hand expressions determined for summer and winter suits. The subjective definition of primary hand was arrived at by assigning the per cent of importance to winter and summer suits to the following descriptors: smoothness, crispness, stiffness, fullness and softness, surface appearance and miscellaneous effects. Kawabata describes refinement of his system by differentiating between primary hand value (PH) and total hand value (THV). The latter is an assessment of the overall quality of a fabric in terms of its market value or sale appeal to the consumer (ref. 50).

An extensive evaluation by Mahar and Postle demonstrated that series of ten equations were suitable for translating primary hand of men's suiting fabrics into total hand values by Kawabata's method. These equations were derived for PH values from a panel of experts from Australia, U. S. A., New Zealand and India, but some cultural differences were noted in hand evaluation particularly for summer weight fabrics (ref. 51).

Another study compared hand assessment in the United States and Japan by Kawabata's PH and THV systems and by three other methods: (a) a quantitative nozzle method developed in the United States based on calculation of hand modulus; (b) subjective hand evaluation by an American panel and (c) testing of several mechanical properties (ref. 52). There was good agreement between quantitative approaches used in this comparative study. However, there were also differences in subjective hand evaluation between garments made in the U. S. and Japan due to differences in tailorability of the fabrics made in the different countries (ref. 52).

TABLE 6.1

Fabric mechanical and surface parameters measured under standardized conditions (ref. 50). Courtesy of The Textile Institute.

Parameters	Description	Unit
Tensile LT	Linearity of load/extension curve	none
WT	Tensile energy	gf cm/cm^2
RT	Tensile resilience	%
EM	Extensibility, strain at 500 gf/cm of tensile load	none
Bending B	Bending rigidity	gf cm^2/cm
2HB	Hysteresis of bending moment	gf cm/cm
Shearing G	Shear stiffness	gf cm/deg
2HG	Hysteresis of shear force at 0.5° shear angle	gf/cm
2HG5	Hysteresis of shear force at 5° shear angle	gf/cm
Compression LC	Linearity of compression/ thickness curve	none
WC	Compressional energy	gf cm/cm^2
RC	Compressional resilience	%
Surface MIU	Coefficient of friction	none
MMD	Mean deviation of MIU	none
SMD	Geometric roughness	μm
Thickness T	Fabric thickness	mm
Weight W	Fabric weight	mg/cm^2

Note 1: EM was not included in the original sixteen parameters that were used for prediction of hand values, but it has recently often been used to predict fabric-formability performance in tailoring.

Note 2: Tensile, bending, shearing, and surface properties have both warp and weft directional values for each of their parameters. Suffixes 1 and 2 indicate the warp and weft directional values, respectively, such as LT_1, WT_1, and so on. The absence of a suffix indicates the average value of the two directional values.

The Kawabata system has also been used to evaluate the hand and related properties of a variety of finished fabrics and to evaluate silicone fabric softeners (ref. 53, 54). The investigators were satisfied with the correlations obtained for silicone softeners on woven shirting fabrics, durable press rayon/polyester blend fabrics, sheeting fabrics made from air jet spun yarns, denim and experimental silicone softeners on cotton/polyester blends (ref. 53, 54). An extensive study of the hand of weft-knitted single and double knit fabrics demonstrated that there was good correlation between subjective measurements and those obtained using the Kawabata system (ref. 55).

Despite the widespread acceptance and popularity of the Kawabata system for evaluating hand, alternative approaches to characterize fabric hand by mathematical models have been formulated; however, these approaches have currently gained little acceptance (refs. 56-58). These mathematical approaches are based primarily on the principles of fuzzy cluster analysis. This mathematical treatment is similar to the cluster systems described in **Chapter 5** for **shade sorting** of dyed textiles. In this mathematical analysis, salient properties of fabrics important in hand can be used to sort or cluster fabrics into groups such that the degree of property association is high between members of the same group and low between members of different groups. Marriot's $g^2 \mid W \mid$ criterion (where g = group number and W = value of the determinant of a matrix W containing the sum of squares and products within a group) was used for mathematical modelling in the first paper on this topic (ref. 56). Later studies used fuzzy relation matrices **R** and **R$_s$**, respectively for objective and subjective determination of fabric hand. A weighted vector factor for the matrices (**A** and **A$_s$** respectively for objective and subjective analysis) was also deduced from a survey of a panel of 25 judges for a comprehensive determination of fabric hand (ref. 57, 58). It is likely that intense research activity and methodologies will continue to evolve and be published to accurately define and predict the complex textile properties of both hand and drape.

6.2.2 Thermal properties of textiles

6.2.2a Overview. The thermal behavior of textiles can be essentially characterized and subdivided into the heat and flame resistance of various fibers and their blends and the transmission of heat or heat transfer through textiles under dry and wet conditions. The former is generally of concern in health and safety for protection against burns and asphyxiation by toxic byproducts derived from fires and flammable materials while the latter is usually important in determining thermal comfort. Both areas of thermal behavior have been extensively and periodically reviewed.

6.2.2b Heat and flame resistance. The first noticeable changes in textiles exposed to heat and flame are deterioration of their mechanical properties. This aspect was discussed earlier under mechanical properties (ref. 22). More severe consequences result when textiles are exposed to excess heat and/or conditions where smoldering or spontaneous combustion occur. The heat and flame resistance of various fibers vary considerably. In some instances, heat and flame resistance for a particular fiber type are comparable. In other instances, fibers with good heat resistance have poor flame resistance or conversely fibers with good flame resistance may have poor resistance to heat. Heat resistance is usually poorest for vinyl fibers such as polyvinyl chloride. These types of fibers generally soften and shrink at temperatures as low as 65-70°C and melt at 150°C. Polyurethane fibers have slightly better heat resistance than vinyl fibers but soften and yellow above 100°C. Thermoplastic fibers such as polyethylene, polypropylene and modacrylics generally soften in the range of 130-160°C with accompanying shrinkage and melt at temperatures around 150-170°C. Cotton, rayon, dry wool and silk and polyamides can withstand temperatures up to 150°C for short to moderate periods of time without substantial damage or adverse effects. However, polyamides and rayons may tend to yellow at these temperatures. If wool is wet or is exposed to high relative humidities, its heat resistance is markedly poorer. Wool has been observed to degrade hydrothermally at temperatures above 100°C. Thermoplastic acetate and triacetate fibers and many acrylic and polyester fibers have fairly good heat resistance, since they soften in the range of 190-250°C and melt at temperatures above 250°C. Many inherently flame resistant fibers (e.g., aramids) melt above 300°C while fibers such as polyben-zimidazole (PBI) and glass are able to withstand temperatures of 600°C and 750°C respectively without melting.

The flammability characteristics of various fibers are listed in **Chapter 4** under **flame retardant finishes (Table 4.6)**. The high flammability of cellulosics and the low flammability of wool illustrate that heat and flame resistance of fibers may differ since cellulosics generally have better heat resistance than wool. A variety of test methods have been proposed and adopted for predicting and measuring the flammability of textiles. The impetus for their existence was caused primarily by the passage of the Flammable Fabrics Act in 1953 (USA) and succeeding legislation in several countries mandating minimum flame resistance requirements for general apparel, children's sleepwear, mattresses and upholstery items.

The status of test methods for textile flammability has been extensively and critically reviewed from a variety of perspectives (refs. 59-63). Krasny has written a comprehensive review on flammability evaluation methods (ref. 59) followed by a later review that focuses on apparel flammability and bench-scale simulations (ref. 60). Horrocks compiled an extensive review on the suitability and appropriateness of oxygen-index techniques to predict and measure fabric flammability. He also however lists status of British legislation and standard test methods for all flammable textile materials (ref. 61). A recent review by Damant highlights the flammability of home furnishings but also critically discusses and characterizes fire tests as small scale, bench scale and full or large scale (ref. 62). Brewster and Barker summarized the general requirements and test methods for performance of heat- and flame-resistant protective clothing (ref. 63).

Textile flammability tests may be classified or grouped in several ways. Investigators in this field have grouped them by sample size and how closely the test simulates real conditions, by the parameters measured, by the type of instrumentation and ignition source employed, by standard versus nonstandard methodologies and by the end use of the textile material. The overall methodology approximates a composite of many of the above classifications or in many instances gradients that approach a test of the textiles under actual hazardous conditions. Nevertheless, Table 6.2 attempts to illustrate a condensed and representative list of flammability test methods for textiles that range from small scale, highly instrumented or parameter-specific tests to bench scale tests that normally utilize larger samples but still use flat, regular shaped fabrics rather than more complex textile assemblies such as garments or upholstery. Full scale tests attempt alone to simulate real fire conditions by use of full size and representative materials with real-time test data and interactive simulations of different materials with each other during combustion.

In earlier flammability testing, the most widely used flammability tests for textiles were the 45° test and the vertical flame test or modifications thereof. In these tests, a fabric specimen (preferentially preconditioned in the presence of a dehydrating agent or at very low relative humidity) is mounted in a closed chamber and ignited by a burner whose flame contains known proportions of methane and air. The residual flame time (afterglow in seconds) and char length of the fabrics is then measured to determine its suitability as a flame retardant material. The 45° test is used for rapid and inexpensive screening to determine minimal flame resistance of general apparel. It is not considered to be an acceptable substitute for the more stringent vertical flame test. The vertical flame or 90° test is considered an acceptable test to determine the performance of general apparel and children's sleepwear by regulatory agencies and the textile industry. The focus on later testing for apparel, garments and protective clothing has been on simulation of movement of a person. Enhanced flammability, extinguishability of the clothing and amount of heat transfer required to cause serious burns were parameters considered important in these tests. The MAFT apparatus was developed as a bench scale method for evaluating general flammability of apparel by surface ignition time and heat transfer characteristics. The specimen in the chamber simulates parts of representative garments (ref. 60). The development of a thermal protective performance (TPP) test evolved from the increasing importance of standards and performance required for protective clothing for firefighters and for other personnel. Heat flux data were used to compute a burn protective index based on data on the tolerance of human tissue (ref. 64). The TPP device was used to rate the performance of inherently flame-retardant fabrics (such as aramids and polybenzimidazole fibers) and factors critical for protection of the wearer. The two most important factors in burn protection were the presence of moisture and retention of effective thermal thickness (ref. 65).

TABLE 6.2

Classification of representative textile flammability tests

Small Scale and Instrument-specific Methods	
Name or descriptive term	Salient features and principles
45° and vertical flame tests	Determine length of damaged fabric, burn time and duration of afterglow
Pill test for floor coverings	Flammable solid methenamine placed on material; maximum damaged area or time flames to reach certain radius
Heat release tests	Measures maximum temperature, total heat release rate using single or multiple sensors or calorimeters
Extinguishment methods	Influence of oxygen, extinction or limiting oxygen indices
Smoke analysis	Measure optical density after certain time in specially devised chambers
Toxic gas analysis	Determined by various forms of chromatography/mass spectrometry; kinetic and quantitative gas analysis
Bench Scale and Related Methods	
Mushroom apparel flammability tester (MAFT)	Heat sensor method with specimen suspended from round metal plate with ignition source able to initiate burning sideways, up or down
Apparel flammability modeling apparatus (AFMA)	Semicylinder covered by 54 heat sensors to simulate free hanging or close contact of a garment on a moving body
ASTM flooring radiant panel test	Samples mounted horizontally on floor of test chamber, gas-fired radiant panel at 30° ignites material and critical radiant flux measured
Oxygen-consumption based furniture calorimeter	Specially constructed furniture with one feature varied each time ignited by gas burner simulating wastebasket fire; heat release rates determined

TABLE 6.2 (continued)

Full and Large Scale Fire Tests	
Name or descriptive term	Salient features and principles
California Technical Bulletin 121 for public building mattresses	Full scale testing of mattresses in a specified test room facility ignited by newspaper in garbage can. Weight loss, temperature and CO monitored.
Uniform Building Code Standard 42-2 for textile wall coverings and finish	Materials in small room ignited by propane burner in corner in two steps varying Kw ignition and time. Oxygen depletion measured by heat release rate.

A variety of mannequins with heat sensors have been developed by university, military and industrial laboratories to measure the potential of garments to cause burns. Most of the mannequins are interfaced to computers to provide overall profiles of burn injuries and distribution of burns over the entire human body. Representative examples are the construction and design of Thermo-Man[R] and Thermo-Leg. The former is an adult size male mannequin fitted with heat sensors distributed uniformly over the entire body front and back side so that the mannequin can be exposed to varying heat flux. The latter is a full-size, molded leg with heat sensors that can simulate human leg action in running similar to a victim escaping a flash fire (ref. 66).

Extinguishability of fabrics has increasingly received attention not only because of its relevance in predicting burn injuries but because of the realization that smoke, lack of oxygen and production of toxic byproducts from burning materials are the leading causes of death and injuries in fires. The LOI (limiting oxygen index) was originally developed as a test to determine fabric extinguishability. LOI was originally defined as " the minimal volume fraction of oxygen in a slowly rising gaseous atmosphere that will sustain the 'candle-like' burning of a stick of polymer" (ref. 61). The LOI and later improvements and modifications to this test (such as the EOI or extinction oxygen index) have been extensively and critically reviewed by Horrocks and co-workers (ref. 61). The LOI (limiting oxygen index) test ranks fabrics as flammable on the basis of the amount (%) of oxygen in the air required for them to ignite when they are in a vertical position. LOI values of 30 or greater indicate that the fabric is very flame resistant. The original test was criticized because of its unrealistic sample mounting, ventilating and downward-burning conditions. In the EOI standard devised and discussed by Horrocks, the effective oxygen index is defined as the fraction or percent of oxygen in an oxygen/nitrogen mixture giving a persistence-of-burning time of zero. This allows more careful characterization of the burning characteristics of various untreated and flame-retardant fabrics and more predictable results based on ignition time (ref. 61).

Corollary tests for determining smoke density and toxicity of combustible byproducts have also been developed. These are normally used in conjunction with full scale tests to determine flammability behavior of carpets, mattresses, floor coverings and upholstery. The most common approach to determining smoke density is measurement of optical density rather than loss of weight and time-dependent expression of critical smoke density values. As noted earlier in **Chapter 4** (**Table 4.8**), smoke density varies widely with fabrics derived from different fiber types. A variety of instruments have been developed and modified to measure and predict smoke density from burning textiles.

The varieties of gaseous products from combustion of fibrous materials have also been determined (**Table 4.9**). The most important gas is obviously CO or carbon monoxide because of its biological combination with hemoglobin to cause rapid asphyxiation. However, other gases formed such as HCN (hydrogen cyanide) and sulfur dioxide also pose serious health hazards. Evaluation of toxic gases has focused on type and amount by various instrumental techniques and biological response of laboratory animals in terms of death, loss of reflex or motor responses and sensory irritation.

The development of full or large scale tests most closely simulates real fire conditions. Most of these tests have been developed to determine the individual and interactive flammability behavior of textile interiors in residences, commercial buildings and aircraft. Although these large or full scale tests are the most informative the costs of conducting such tests are much more expensive than laboratory and bench scale flammability tests. Damant has critically reviewed such large scale tests (excluding those for commercial aircraft) with the primary focus on methods for evaluating upholstery and mattresses (ref. 62). A furniture calorimeter has been described by Babrauskas that affords a quantitative assessment of the effects of using different fabrics, padding materials and frames. A wastebasket simulation burner was used as the ignition source and the instrument is based on the critical release of heat required to cause combustion. This device is offered as an inexpensive and more reproducible alternative to full-scale room tests (ref. 67).

Simple modifications of the NBS Flooring Radiant Panel test (Table 6.2) correlated very well with a fire model used for large-scale evaluation of burning of floor coverings and carpets. Ten different commercial carpets differing in total weight, pile weight, pile thickness and fiber composition were evaluated in this manner. Modification of the radiant panel test included measurement of smoke emission and heat output as well as flame spread (ref. 68).

Prototype aircraft interior panels (approximately 25% of full-scale dimensions) were evaluated for their flammability characteristics under conditions comparable to those used for evaluation of full-scale room fires. Flashover conditions were generated by a 52-kilowatt propane burner. Flammability of the prototype materials (such as epoxy/aramid, phenolic/graphite and epoxy/glass facings) usually correlated well with material ignition temperature. Panels faced with phenolic-impregnated fiberglas or graphite did not burn under the test conditions employed (ref. 69).

6.2.2c <u>Thermophysical properties</u>. The thermal transmittance (conductance) or thermal resistance (insulative) properties are the most important and frequently measured thermophysical properties of textiles. Since textile structures are usually porous their thermal transmission (also appropriately called heat transfer) is dependent on air layers trapped within a fabric. Thus, such heat transfer is readily measured by a variety of techniques. However, it is quite difficult to predict thermal conductance or insulative values of fabrics because the amount of air varies within fabric interstices. Nevertheless, thermal transmittance (conductance) or resistance (insulative) values of textiles have been instrumentally simulated by a variety of techniques to assess their role in thermal comfort and/or to determine the insulative value of textiles. These techniques include methods based on the ability of a fabric to maintain an equilibrium temperature in its surroundings, determination of the heating or cooling rate of the fabric, and monitoring heat flow through fabrics with heat flow discs or photovoltaic cells.

Most laboratory instruments are adequate for a relative ranking of the thermal transmittance or resistance of fabrics. However, most earlier methods only considered heat transfer by conduction and either limit or ignore convective and radiant contributions of heat transfer or thermal transmittance. Also, most earlier devices measure the thermal transmittance of textiles under dry conditions and do not have provisions for such an evaluation under conditions of moisture transport. The role of heat transfer of fabrics under dry and moist conditions in thermal comfort and overall clothing comfort will be discussed later in the section on **multiple influences** or agents.

In a series of investigations Yoneda and Kawabata (refs. 70-72) conducted a theoretical analysis of transient heat conduction through textiles to relate it to perceived warmness and coolness of fabrics touching the skin. Fundamental analysis of the heat flow through fabrics is considered by explaining conduction that occurs when a plate with good heat conducting properties is placed on top of a sheet-like solid with poor heat conducting properties (e.g., fabrics). A method for measuring thermal conductivity and thermal diffusivity from the transient phenomenon was presented as an application of this analysis (ref. 70). In the second paper, q_{max} was introduced as a measure of predicting the warm/cool feeling of fabrics. q_{max} is defined as the peak value of heat flux that flows out of a copper plate having a finite amount of heat into the fabric surface after the plate contacts the surface. This measurement system simulated well the heat transmission in the skin layer after contact with an outside object with a time lag of 0.2 seconds (ref. 71). The transient heat conduction across two layers having different thermal properties (human skin and fabric) was considered as a mathematical two-layered model to be solved in terms of heat conduction properties for each layer and boundary conditions. Simulation results demonstrated that the warm/cool feeling depends only on the outer surface material and that the thermal insulation properties of the fabrics are governed by density, specific heat, thickness and thermal conductivity (ref. 72).

The relative contributions and importance of conduction, convection, and radiation to the overall thermal transmission or heat transfer through textiles varies markedly with the end use, environmental conditions (e.g. indoors or outdoors) and fabric construction and permeability. For apparel, the clothing insulative value (denoted as clo) and the evaporative heat transfer from the skin (denoted as i_m) are considered to be the most important parameters that contribute to thermal comfort. These parameters will be discussed later under thermal comfort in the overall context of general comfort.

A variety of techniques and methods have been used to determine the insulative value of textile interiors and of especially porous constructions such as nonwovens. A special window test apparatus had been constructed to determine the insulative values of drapes or interior window coverings (ref. 73). The calorimeter operates on a similar principle to the ASTM Guarded Hot Plate since the heat energy from the window is guarded from going in any direction except out into the room in which the apparatus is located. Insulative values of different drapery materials can then be determined by measuring the heat flow difference between a bare and covered window (ref. 73). Predicting the thermal behavior of drapes and wall coverings is much more complex than measuring their thermal behavior. This complexity occurs because convective and radiant effects as well as conductive effects and fabric openness are operative. In contrast, predicting and measuring the thermal effectiveness of carpets and rugs is relatively straightforward since it has been determined to be directly proportional to their thickness and pile density. The thermal conductivity of prototype polyester nonwovens used as the insulative material was readily measured by a nonsteady state cooling method. The procedure consisted of placing a uniformly heated nonwoven fabric between two metal plates at ambient temperature and observed the decrease in centerline temperature with time. Experimental results also indicated that the thermal effectiveness of nonwovens decreases with their compression (ref. 74).

Mathematical models and devices have been described for measuring and predicting the heat transfer through textiles in the presence of moisture, high wind velocities and other dynamic conditions. An apparatus has been constructed that measures heat transfer or flow through textiles under dry and wet conditions in a natural "mixed flow" convective state. This is accomplished by a sensor interfaced to a microprocessor that dynamically measures and records heat flow through the textiles, and the air temperature and relative humidity of the surrounding areas. With this apparatus as a reference, measurement of the surface temperature of fabrics also allows one to predict the actual heat flow through most conventional textile materials if their thickness is known (ref. 75). The mechanisms of heat flow or transfer through fibrous insulative materials such as synthetic fiber batting and mixtures of down and feathers were determined experimentally and compared to a theoretical model. The theoretical model accounts for all observed heat flow that are due primarily to the relative contributions of conductive and radiative heat flow with little contribute due to convection (ref. 76). The heat loss from the surface of a fabric-covered cylinder in an air stream was measured on a simple laboratory apparatus to simulate the more complex and expensive use of a wind tunnel. The measurements were determined for nonwoven polyester fabrics varying in linear density and thickness (ref. 77).

Besides his detailed work on hand and accompanying instrumentation, Kawabata has constructed a device that measures the heat loss of fabrics and that provides valuable information on the combined heat and moisture transport properties of fabrics. This device is called Thermo-Labo (KESF-TL-2D) and is based on the principles of his earlier investigations of the effective thermal conductivity and warm/cool feeling of fabrics (ref. 78). Figure 6.11 shows four modes by which heat insulation measurements may be conducted. Thermal conductivity of fabrics can also be measured by a steady state method without disturbing the moisture regain properties of fabrics. Additional studies by Fujimoto and Seki were conducted on the Thermo-Labo with fabrics ranging from thin woven construction to thick packing materials for winter garments. These investigators also devised an empirical formula for effective thermal conductivity based on a simple model that included radiative as well as conductive heat transfer. They determined that the thermal conductance is

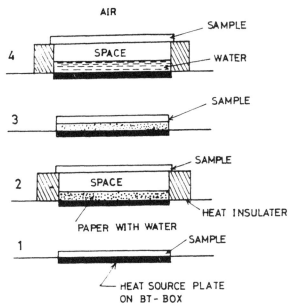

Figure 6.11. Four methods of heat insulation measurement: (1) dry method; (2) skin/space method where a filter paper containing enough water is placed on a heat plate to simulate human skin. A sample is placed on the paper with 7 mm space; (3) skin method. A sample is placed directly on the paper containing the water directly; (4) water/space method (ref. 78). Courtesy of Textile Machinery Society of Japan.

inversely proportional to thickness for relatively thin fabrics but radiative heat transfer was more important than thickness for heat transfer in fabrics thicker than 1 cm (ref. 79).

More current studies have addressed the difficult problem of adequately describing thermal transmittance or heat transfer through moist or wet fabrics (ref. 80, 81). Farnworth has conducted additional experiments and mathematical expressions to model the combined effects of heat and water transport in multi-layered textile structures. Mathematical equations and calculations were constructed in a time-dependent mode and compared favorably to experimental results obtained with a sweating hot plate. The same contributions to heat transfer were observed as those observed in his earlier study without moisture effects (ref. 76). Heat transfer by conduction and radiation were determined to be important as well as vapor transport by diffusion. Observed heat loss under moist or wet fabric conditions is generally explained by the three interactive mechanisms of condensation, absorption and evaporation (ref. 80). A later study used a different approach to elucidate the mechanisms by which heat transfer through moist fabrics occurs. A transient technique similar to one described in an earlier investigation by Yoneda and Kawabata (ref. 70) was used to measure the thermal conductivity of cotton, wool, polypropylene and acrylic fabrics containing varying amounts of water. It was concluded that effective heat transfer in fabrics takes place by conduction, infrared radiation and distillation. The evaporation process is determined primarily by the fiber's moisture sorption process until the fiber reaches its saturation regain. This phenomenon explains the differences in thermal insulative values of very hydrophilic fibers such as wool compared to extremely hydrophobic fibers such as polypropylene (ref. 81).

The heat flow or thermal transmittance through a variety of textiles or related materials can also be dynamically measured by monitoring changes in the surface temperature of the materials. The author and his colleagues satisfactorily predicted the thermal transmittance of very diverse textile materials by measuring the surface temperatures produced by color changes in liquid crystal film strips on the fabric surface. Temperature changes as little as 0.2°C could be detected by this method at different relative humidities in the presence and absence of moisture transport through the textiles (ref. 82). The use of more sophisticated instrumentation such as thermovision cameras (infrared thermography that gives a color-coded surface profile of a material from its surface radiation) offer even greater promise as a technique for dynamically assessing and evaluating the thermal transmission through textiles and related materials. Initial studies of this technique were described by Andrassy and his co-workers for a variety of fabrics and garments (ref. 83). These types of evaluations are also currently being conducted in the author's laboratory to correlate calorimetry values of phase change polymers affixed to fabrics with their thermal performance in various temperature environments.

6.2.3 Gaseous and liquid permeability, sorption and transport

6.2.3a Overview. The interaction and behavior of textile materials, particularly fabrics, with gases and liquids, are important in a variety of end uses. Relevant applications include apparel comfort, diapers, filters, rainwear and all-weather gear and many other end uses. Permeability characteristics of fabrics and layered textile structures to air and to water vapor have been characterized and measured. Liquid sorption, desorption and exsorption (ability to drain off the imbibed liquid) processes in fibrous materials have also been measured and predicted. Extensive studies exist and are being continually refined to elucidate the mechanisms and measurement of the transport of water vapor and liquid water through textile structures under a variety of steady state and dynamic conditions.

6.2.3b Air permeability. The most commonly permeability of fabrics to gases measured is that to air or wind. The ability of a fabric to resist air (low air permeability) or have air freely flowing through it (high air permeability) is dependent primarily on its thickness, porosity, construction, geometry, type and amount of finish and coating and similar factors. Generally, the fabrics are ranked from lowest to highest permeability in amount of cubic feet of air that passes through a square meter of fabric. Two different types of apparatuses are normally employed to force air through fabrics. The most common type of instrument fixes the pressure difference between opposite faces of a fabric then measures the rate of air flow achieved through the fabric. The other type of instrument fixes the air flow rate then measures the pressure difference required to maintain air flow. Other approaches for measuring air permeability include the use of thermistor-type and ionization-type transducers (ref. 84) and the British Standards bubble pressure test that determines the equivalent pore radius and correlates this value to the size of interstices in all types of fabrics that are air-permeable (ref. 85). Air permeability prediction and measurement is more complex for very porous structures such as nonwoven fabrics than it is for less permeable structures. Nevertheless, significant progress has been made in this area. Atwal has measured the air resistance (reciprocal of air permeability) of 140 nonwoven needle-punched fabrics made from 16 different fibers. He has developed an empirical relationship using stepwise multiple regression analysis to relate and predict air resistance or air permeability of nonwovens to the fabric parameters of weight/unit area, thickness, porosity and fineness (ref. 86):

$$r = 15.73 + 141.1m - 0.012 \cdot \frac{h^3}{(1-h)^2} + 29034 \, t/d \qquad (6.5)$$

where r= air resistance, m = weight/unit area, h = porosity and t = thickness and d = fiber fineness.

Air permeability has been related and correlated with numerous other fabric properties. These correlations have usually been made in an attempt to relate them to comfort factors of textiles that will be discussed in detail later in this chapter.

6.2.3c <u>Water vapor transport</u>. The mechanisms of water vapor transport through fabrics are complex and cannot be measured by simple flow techniques used for determining air permeability. Numerous studies have been conducted to measure and predict the transport of water vapor through fabrics. Investigators have concluded that combinations of mechanisms are possible such as capillary wicking and diffusion of gaseous and condensed (liquid) water through fibrous assemblies. A technique has been developed to study moisture transport between fabrics under transient conditions at low moisture contents. Moisture regains of a variety of fabrics (cotton, cotton/polyester and polyester) were varied from less than 3% to over 100%. It was determined that vapor diffusion was the major mechanism of moisture transport between two fabric levels at low moisture content and that wicking did not begin until moisture regain was greater than 30% (ref. 87). Differential scanning calorimetry was used as an analytical technique to measure the water vapor transmission rates through fabrics and relate these values to comfort of protective clothing (ref. 88). A simple apparatus has been devised to measure rapidly water vapor resistance of textiles within an accuracy of at least 1%. The fabric is sandwiched but not compressed between microporous membranes so that it is permeable only to water vapor and impermeable to liquid water and air currents. One side of the sandwich is kept in contact with a pool of liquid water and the other side in contact with a dry air stream (ref. 89). The amount of vaporized water per unit time is measured with a pipette and stopwatch. The vapor resistance R can then be determined from the following equation:

$$m = \frac{O \, \Delta P}{R} \tag{6.6}$$

where m = time, O = area of the sample and ΔP = maximum vapor pressure of water at the test temperature. An improved version of this apparatus has been described by Farnworth and his colleagues. It has the added provision for measuring

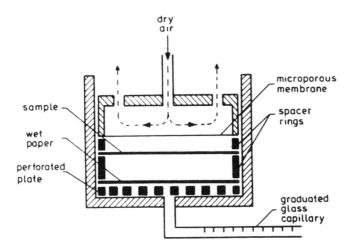

Figure 6.12. Schematic diagram of an apparatus for measuring water vapor resistance under variable relative humidity (ref. 90). Courtesy of Textile Research Institute.

the water vapor resistance of textiles under variable relative humidity (ranges of 35% RH to saturation regain). The wet filter paper supported on the perforated steel plate in Figure 6.12 is the feature of the apparatus that allows variation of relative humidity. Although microporous films showed little difference in their water vapor transport by this method, fabrics and films with hydrophilic coatings had substantial increases in water vapor resistance with decreasing relative humidity (ref. 90).

Dynamic surface wetness for test fabrics was continuously measured to determine the role of dynamic moisture transport in clothing comfort. Sweating skin was simulated by use of a fully wetted chamois heated to skin temperature with a hot plate and variable transformer. A miniature electric hygrometer was used to measure dynamically surface wetness of fabrics comprised of different fiber types and explain comfort differences by observed differences of moisture vapor and rates of changes in both inner and outer fabric surfaces (ref. 91). An apparatus was developed for simultaneously determining moisture sorption and flux of fabrics during a transient period after fabric exposure to a humidity gradient. Figure 6.13 depicts the three possible scenarios of transient events when a fabric is exposed to a relative humidity gradient. In each scenario, moisture flux F occurs through the barrier as a result of a concentration gradient. Simple and complex models of moisture transport fabrics indicate that when diffusion within the fiber is rapid the moisture content is always in

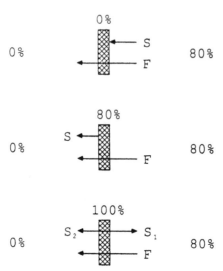

Figure 6.13. Moisture flux is denoted by F and sorption by S. The number above a fabric refers to the atmosphere to which the fabric was initially equilibrated (ref. 92). Courtesy of Textile Research Institute.

sorptive equilibrium with air at the fiber surface. However, when molecular diffusion occurs within the fiber interior, moisture content lags behind changes in moisture content of air at the fiber surface (ref. 92).

In the continuing attempt to relate moisture vapor transport to thermal comfort, the newest apparatus described permits dynamic and simultaneous measurement of both water vapor and heat transport through layered fabrics (ref. 93). The apparatus shown in Figure 6.14 consists of three acrylic frames stacked together in such a manner that two fabrics can be evaluated to simulate layering of garments under conditions of permeability (use of a microporous film barrier) or impermeability (use of aluminum foil for sweat mode). In the permeable mode (water vapor), hydrophobic fabric surfaces showed little difference in their moisture flux compared to hydrophilic fabric surfaces. In the impermeable or sweating mode, the water vapor transfer rate was primarily dependent on the wicking ability of the fabric (ref. 93). Further studies on dynamic heat and water vapor transport for acrylic, cotton, polyester and wool fabrics were conducted on this apparatus. Overall dissipation rate of water vapor depended on rates of water vapor transport and of vapor absorption by fibers. Within a comparable range of fabric structures, water absorption characteristics of fibers were the most important in determining overall water vapor transport rate (ref. 94).

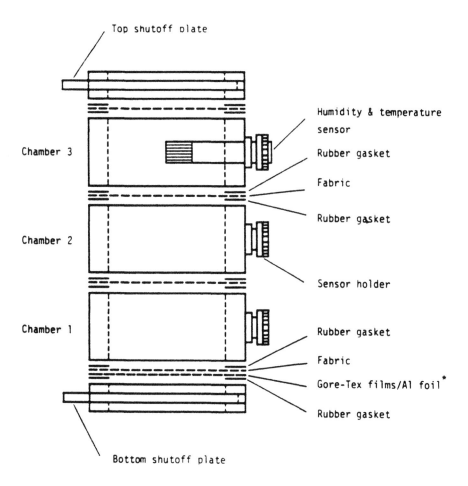

Figure 6.14. Assembly diagram of diffusion column. (*Six layers of Gore-Tex film were used in the water vapor method. Aluminum foil was used in the sweat method) (ref. 93). Courtesy of Textile Research Institute.

6.2.3d <u>Liquid sorption and transport</u>. Mechanisms of liquid flow through fibrous materials have already been discussed in **Chapter 4** under modification of fabrics to increase their **absorbency**. Various types of sorption isotherms characterize the behavior of liquids with fibrous assemblies and are described by several mathematical equations and models to describe the nature of fluid flow. Unsteady states of flow through porous fibers and sorption processes on highly absorbent or superabsorbent fibers lead to more complex interactions and models.

The ability and capacity of fabrics and other textile structures to sorb, desorb and repel water and other liquids have been measured by a variety of laboratory and standardized tests and devices. Absorbency test methods for textiles have been critically reviewed by Chatterjee and Nguyen (ref. 95). Most devices or methods measure the amount of water sorbed as a function of time and/or the capacity of the textile to hold water or other liquids. Both vertical and horizontal wicking tests are used to measure the rate of wetting that occurs due primarily to capillary action. Drop absorbency methods measure the rate of spread of liquid contacting the fabric surface. Demand absorbency tests are considered to give a more reliable measure of fabric wettability. The liquid will enter the absorbent material only when and as long as the sample demands it. The dry fabric is contacted with liquid to allow absorption to occur under a slightly negative hydrostatic pressure (ref. 95). Liquid retention tests of wet or liquid-saturated fabrics essentially measure exsorbency or the capability of the material to retain liquid after some form of applied drainage. Removal of excess liquid is achieved by gravity or centrifugation. The amount of water retained in fibers after centrifugation is usually called water of imbibition and expressed as percentage weight of water retained per weight of fiber. The water repellency of fabrics is usually measured by determining the spray resistance of fabrics to water under conditions simulating rain. Water resistance to fabrics is determined by measuring such resistance with hydrostatic pressure devices. Measurement of this type of resistance has become increasingly important in protective clothing for health care workers to prevent transmission of disease by contact with body fluids.

6.2.4 Electrostatic effects

Static electrical charges may accumulate by contact or rubbing between objects. The relative position of fabrics composed of different fiber types has been determined by many investigators as part of the triboelectric series. The triboelectric series is a list of materials that produce static charges when they are rubbed together in a manner that a material has a positive charge when rubbed with a material below it and a negative charge above it on the list. The general order (starting with the most positive charge) for major fibers types is: wool > polyamide > silk > rayon > cotton > polyester > acrylic > polyolefins.

Various test methods have been developed to simulate the tendency of fabrics to acquire static charges in use. Static charges in fabrics may be generated by direct or indirect contact. Although the development of static charge in textiles may be achieved without frictional contact, it is more easily produced when friction is employed. There are essentially two types of techniques for measuring static charge: (a) those that measure its electrical resistance and (b) those that measure the clinging tendency of the fabric by determining how long it takes to detach itself from a metal plate. Test methods based on resistivity measurements give more reproducible results and measurements than test methods based on charge generation by rubbing and may be conducted at low or high humidities to simulate actual end use conditions (ref. 96).

6.3 CHEMICAL AGENTS AND INFLUENCES

Chemical and/or photochemical exposure of textiles may lead to yellowing or discoloration of undyed fabrics, to fading of dyed fabrics, and/or to degradation of dyed and undyed fabrics. These adverse results are due to depolymerization of the polymer chain in the fiber that may occur by hydrolysis, oxidative processes and/or crosslinking. Textiles have varying degrees of resistance to chemical agents such as water and other solvents, to acids, bases, and bleaches, to air pollutants, and to the photochemical action of ultraviolet light. Resistance to chemical agents is dependent on fiber type, chemical nature of the dyes, additives, impurities, finishes present in the fiber and to a lesser extent on the construction and geometry of the fabric.

6.3.1 Effect of solvents, pH and oxidizing agents

Water and perchloroethylene (or other appropriate dry cleaning fluids) are the two solvents to which most textiles are frequently exposed. This exposure is common because these solvents are used for refurbishing purposes, to remove spots and stains, or accidentally come into contact with fibrous materials. Most synthetic fibers are unaffected by hot or by cold water, but natural and regenerated cellulosic fibers and proteinaceous fibers may undergo dimensional and surface changes and/or the degradation on exposure to water. Cotton fabrics are not adversely affected with regard to their surface aesthetics or degraded on exposure to water or humidity. However, they may undergo shrinkage (reversible hygral contraction) at high relative humidities. Regenerated cellulosic fibers such as rayon are swollen markedly on contact with water. Cellulose acetate is delustered on contact with water while cellulose triacetate is generally unaffected by either hot or cold water. Silk fabrics spot and wool fabrics undergo various types of shrinkage with water with only one type of shrinkage being reversible (hygral expansion with increasing humidity). Wool is also hydrothermally degraded at temperatures above 100°C at high relative humidities.

All natural and most synthetic fibers are unaffected by representative dry cleaning solvents such as perchloroethylene, trichloroethylene and chlorofluorocarbons. Cotton, various rayons, wool, polyester, acrylics and polyurethanes have good resistance to dry cleaning solvents while polyamides generally have excellent resistance to these solvents. However, fibers such as cellulose acetate and polypropylene swell or undergo pronounced shrinkage in trichloroethylene and thus must only be dry cleaned in perchloroethylene.

The resistance of textiles to acids is excellent for wool, polyamide and modacrylic fibers. Hydrocarbon-derived fibers such as polypropylene and polyvinyl chloride as well as glass fibers, are essentially inert to acids and other chemical agents under normal end use conditions. Silk is only slowly attacked by acids while polyester, acrylics and polyurethanes are highly resistant to attack by dilute acids or cold concentrated acids. However, all previously mentioned fibers are either hydrolyzed or severely degraded by hot concentrated acids. In contrast to these fiber types, cellulosic fibers, particularly regenerated fibers, are very susceptible to acid hydrolysis and depolymerization. Although cotton is not affected by dilute acids at ambient temperature, it is readily hydrolyzed by hot dilute acids or cold concentrated acids. Cellulose acetate and triacetate fibers have about the same resistance as cotton to acids, but rayons are most readily attacked by acids relative to other fiber types. The extent of degradation of fabrics exposed to different concentrations of sulfuric acid (5, 10 and 20%) at ambient temperature was determined. Acetate, cotton, rayon, silk, wool and nylon fabrics exposed to the acid were evaluated for their loss of weight and tensile strength. Except for the cellulosic fibers, there was a general similarity for each type of fiber for pattern of acid degradation as a function of exposure time and acid concentration (ref. 97).

The resistance of fibers to alkaline solutions follows similar but not identical trends observed for their resistance to acids. Silk and particularly wool are degraded and damaged by even dilute solutions of weakly basic materials. Dilute alkali saponifies cellulose acetate fibers to unsubstituted cellulose. Cotton is less severely affected by alkali in that it is not chemically degraded but does undergo shrinkage in the slack state (see section in **Chapter 1** on **mercerization**). A variety of synthetic fibers that are usually degraded or hydrolyzed on exposure to hot alkali include: cellulose acetate, acrylic, modacrylic, polyester and polyurethane. Polyamides, polyolefins, polyvinyl chloride, and glass fibers are remarkably resistant to alkali. These fibers are in essence inert to attack by solutions of high pH.

The resistance of textile fabrics in use to various bleaches and oxidizing agents generally parallels their resistance to these agents during purification and preparatory processes (see **Chapter 1** on **bleaching**). However, fabrics and textile products that are in use and/or refurbished usually only come in contact with hypochlorite and peroxide-based bleaches. There is little exposure on end use to other bleaching agents used during processing, e.g., sodium chlorite and peracetic acid. As noted earlier in fabric preparation, the pH at which fibers are bleached is important, since their susceptibility to damage under acid and/or basic conditions is exacerbated by the presence of oxidizing agents.

Although hypochlorite is a very effective bleaching agent for many types of fibers, it is also more sensitive than other types of bleaches to changes in pH in its bleaching effectiveness and in its ability to cause fiber damage. It cannot be used on wool or silk. Wool is irreversibly damaged by OCl⁻, while silk rapidly discolors and dissolves in hypochlorite solutions. Polyurethane and polyamide fibers are not damaged by hypochlorite but do yellow when it is present. Cotton, rayon, cellulose acetate, cellulose triacetate and acrylic fibers may be bleached with hypochlorite at appropriate pH and at low temperatures but may be damaged at high temperatures.

Of all major fiber types, polyester and modacrylic fibers are the least effected by hypochlorite under a variety of conditions.

Peroxide-based bleaches, such as hydrogen peroxide and sodium perborate, are suitable for bleaching wool fabrics without adverse effects. Most other fibers (natural and synthetic) can also be bleached with peroxides without any appreciable fiber damage occurring. Polyolefins, cotton and regenerated cellulosic fibers will be degraded by peroxide bleaches at high temperatures and/or by concentrated solutions of these agents. If available, sodium chlorite may be advantageously employed to bleach a variety of synthetic fibers and particularly to bleach synthetic fiber/natural fiber blends without damage and/or without discoloration.

The development of chemically-resistant protective clothing and textile filtration media have led to test methods relevant to these end uses. A test method useful for selection and performance of textile materials in filtration is based on "critical solution time." This is the time required for a weighted yarn loop to break when immersed in a solvent or solution. The test may also be adapted to determine the suitability of a particular fabric for chemically protective clothing (ref. 98).

6.3.2 Effects of air pollution

As noted earlier in **Chapter 4**, air pollutants affect dyed and undyed textiles by causing them to fade, discolor and undergo chemical degradation. A bibliographic review by Upham and Salvin (ref. 99) is still the most comprehensive and informative treatment of this topic. Sulfur dioxide, both in laboratory and in field exposure studies, appears to be the only air pollutant that significantly degrades undyed fibers, particularly cellulosics and polyamides. The presence of excess moisture or high relative humidity and ultraviolet light usually synergizes and accelerates such degradation. However, most studies have not assessed the relative contributions of degradation caused by air pollutants and photochemical action. It is also not clear whether or not cellulosic fibers are oxidized by SO_2 or undergo depolymerization by acid hydrolysis. The latter could occur by reaction of SO_2 with water to produce H_2SO_3 or H_2SO_4. Other air pollutants such as nitrogen oxides (NO_x) and ozone appear to have little adverse effect on other undyed natural or synthetic fibers (ref. 99).

Air pollutants are usually detrimental to fibers. However, recent studies show that textile furnishings may be beneficially employed to reduce indoor air pollutants by sorbing such pollutants. Wool carpets were shown to be useful in this regard, but other studies indicate that cellulosic fibers are also effective in reducing pollution indoors. This subject has been reviewed and most studies have focused on the use of natural fibers to reduce such indoor pollution (ref. 100).

When fibers are dyed, they undergo different responses to degradation by pollutants than they do when they are undyed. For dyed cellulosic and wool fabrics, only a few dyes undergo fading in the presence of sulfur dioxide. In contrast, nitrogen oxides, particularly NO_2, cause what is historically known as "gas-fume fading." This type of fading has been observed on dyed cotton, rayon, acetate and polyamide fabrics. Various blue dyes are particularly sensitive to fading in the presence of NO_2, but this situation has been alleviated somewhat by incorporating fugitive or substantive inhibitors (various aliphatic and aromatic amines). Ozone also causes cellulosic fibers, cellulose acetate, polyamide, and cotton/polyester permanent press fabrics to fade. Dyes that are susceptible to fading by nitrogen oxides show similar susceptibility to fading by ozone. Moreover, high relative humidities accelerate such fading of dyes in the presence of either nitrogen oxides and/or ozone. Standard test methods exist for evaluating the colorfastness of dye textiles to both nitrogen oxides and ozone. Representative exposure conditions for the resistance of dyed fabrics to sulfur dioxide have also been published (ref. 99).

6.3.3 Photodegradation and photofading

The adverse action of light on textiles has been documented since the Middle Ages. Moreover, lightfastness standards for dyed textiles were proposed as early as 1729 in France by Dufay. However, it is only in the past 30 years that principles have been formulated to explain the photodegradation of light on fibers and on other polymeric materials and substrates. Certain types of fibers, due to their chemical structure and morphology, are more susceptible to photodegradation than other types of fibers. Impurities, additives, dyes, finishing agents and bound polymers may also profoundly influence the photochemical stability of the textile product. Silk, jute and delustered polyamides have particularly poor resistance to ultraviolet light. Several natural and synthetic fibers (cotton, rayon, cellulose acetate, cellulose triacetate and wool) have adequate resistance to ultraviolet light. Polyester, polyvinyl chloride, polyurethane and polyolefin fibers have superior resistance to ultraviolet light. Modacrylic and particularly acrylic fibers have excellent light resistance, while glass fibers are essentially unaffected by ultraviolet light.

The initiation of photochemical degradation, chemical species and reactions involved have been extensively discussed in books and critical reviews (refs. 101-104). As noted in **Chapter 4**, the presence or production of chromophores that serve as sources of photochemical initiation and degradation may be present in or at the end of polymer chains or may be an integral repeating unit in the polymer backbone. A useful approach for describing the interaction of photons with polymers in the solid state has been advanced by Wiles (ref. 101), and is depicted in Figure 6.15.

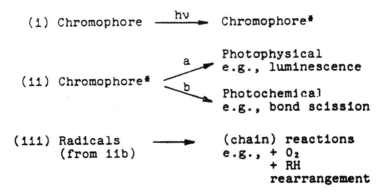

Figure 6.15. Consequences of photon absorption (ref. 101). Courtesy of John Wiley and Sons Publishers.

Initially, certain chromophores in the fiber interact with photons, and undergo photophysical processes such as luminescence or photochemical processes such as bond scission. The latter process produces radicals and reactive species that lead to further degradation of the polymer chain. Energy transfer may occur intermolecularly between an electronically excited molecule ($D^{\cdot}$) and an acceptor molecule (A). This transfer may also occur intramolecularly within the same polymer between a donor in the chain (a chromophoric group in the excited state) and a different segment of the polymer (acceptor) (ref. 102).

A variety of functional groups and chemical species have been identified and detected that initiate and facilitate the photodegradation of fibrous polymers. Chemical species include: molecular and singlet oxygen, peroxy radicals, and organic solvents that generate reactive radicals or lead to products that absorb light in the ultraviolet region. Certain metal oxides and salts, particularly those of zinc, titanium and iron are known to accelerate photodegradation of polymers. Organic impurities in the fibers (such as benzophenone, quinones, polycyclic aromatic hydrocarbons and peroxides) may act as either photosensitizers or photoinitiators or cause fibers to absorb light in the ultraviolet region. Functional groups in polymer chains (particularly carbonyl groups) may facilitate chain scission by Norrish Type I and Norrish Type II processes (ref. 102):

$$—CH_2—CH_2—\underset{O}{C}—CH_2—CH_2— \quad + \; h\nu$$

$$\xrightarrow{\hspace{1cm}} —CH_2—CH_2—\underset{O}{C}. \; + .CH_2—CH_2—$$

(Norrish Type I) $\hspace{3cm}$ (6.7)

$$\xrightarrow{\hspace{1cm}} —CH_2—CH_2—\underset{O}{C}—CH_3 \; + \; CH_2{=}CH—$$

(Norrish Type II)

Standard tests for colorfastness of dyed textiles to light and instrumental methods for determining photodegradation processes in fibers afford useful information to predict fiber performance. Most standard tests involve exposure of fabrics to a simulated noon sunlight source (such as instruments equipped with a xenon or carbon arc lamp and temperature/humidity controls) for either a specified time or until different degrees of perceptible fading or yellowing are observed. Most modern instruments have attachments to simulate the additional detrimental effects of ozone and other air pollutants. In addition to these techniques, more fundamental information on the photophysical and photochemical processes involved in the absorption of light by fibrous materials and other polymers have been critically reviewed (ref. 103). The suitability and appropriate use of light sources (ultraviolet, visible and monochromatic) to which fibers are exposed will markedly influence their performance, and must be related to actual end use exposure conditions. It is possible to determine quantum yield in photosensitized and photodegradative processes by chemical actinometers. The quantum yield is defined as the number of molecules reacting in a particular process divided by the number of quanta absorbed by the system or the number of photons absorbed within a specific time. When fibers are photosensitized, they undergo various types of luminescence such as fluorescence, phosphorescence and chemiluminescence. This luminescence may be detected by changes in the emission and absorption spectra of the fibers and by other techniques such as flash photolysis.

All types of fibers have been studied concerning mechanisms of their photo-stabilization and photodegradation. To indicate the diversity and complexity of the effect that ultraviolet light has on fibers in the presence of dyes, impurities and additives, specific interactions with cellulosic, polypropylene, polyamide and wool fibers will be discussed as well as the interactions between dyes and ultraviolet light and the general phenomenon of yellowing.

The phototendering of cotton fabrics by vat and by other types of dyes composed of anthraquinone and other quinoid structures have been investigated. Two mechanisms have been proposed to explain the photodegradation that occurs. The first mode of photodegradation occurs by oxidation of the cellulosic hydroxyl groups by singlet oxygen. The latter species originates from the reaction of molecular oxygen with the dye in its excited triplet state. The second mode of photodegradation occurs by reaction of activated dye with the cellulosic fiber by a hydrogen-abstraction process. This leads to the production of free radicals that initiate photodegradation. Besides cotton, other types of fibers also undergo phototendering in the presence of dyes that contain quinoid-type structures. Yellow and orange-colored dyes containing a quinoid structure were frequently implicated in phototendering of all fiber types; however, no other structural features of the dyes correlated well with their phototend-ering propensity (ref. 104). These investigators also observed that oxygen is necessary for phototendering to occur, and that high humidity accelerates such degradation. It was also noted that silk and polyamide fibers were the most photodegraded dyed fabrics. Cotton had moderate photodegradation and viscose rayon the least photodegradation.

Much of our understanding of the current processes of photodegradation and photostabilization of polymers and fibrous materials comes from the detailed studies and reviews on polypropylene (ref. 105,106). Although polypropylene is a hydrocarbon chain that consists only of -C-C- and -C-H bonds, it has been demonstrated that charge transfer complexes between polypropylene and oxygen and hydroperoxide oxidation products lead to its photodegradation. Metal, carbonyl, polycylic aromatics and other impurities are photoinitiators that catalyze its decomposition. The formation of free radicals, peroxides, and hydroperoxides, and the occurrence of Norrish type I and II processes have been observed in this polymer. Thus, this demonstrates that certain intermediates lead to depolymerization by chain scission, hydrogen abstraction and other autooxidative pathways.

The ultraviolet behavior and resistance of polyamides are equally well characterized (ref. 107). Fluorescence and phosphorescence emission spectra of nylon polymers vary with the wavelength at which the polymers are excited and provide valuable information on the chemical nature of light-absorbing species of the polyamides. Photooxidation of polyamides occurs through hydroperoxide intermediates that lead to chain scission. Photosensitized degradation of polyamides occurs in the presence of certain pigments and dyes. Accompanying loss in tensile strength in polyamides and in other fibers is particularly pronounced when delustering agents such as TiO_2 are incorporated into the fiber. The mechanism by which TiO_2 accelerates the photodegradation of polyamides and other fiber types has not been completely resolved. However, it has been established that the anatase form of this pigment promotes such degradation while the rutile form is essentially inactive. Some form of active oxygen (either radical ions such as $O2.^-$ or free radicals such as $\cdot OH$) is produced by interactive processes of oxygen or water with photoexcited titanium dioxide. These processes subsequently lead to depolymerization of the polyamides. Many yellow and orange vat dyes of the anthraquinone type were found to catalyze the ultraviolet degradation of most types of fibers including polyamides. Recent evidence favors the hydrogen-abstraction mechanism over the singlet oxygen activation mechanism for this photosensitization discussed earlier for cotton fabrics and vat dyes (ref. 107).

Mechanisms by which dyes fade in solution and on polymers such as fibers have been recently reviewed to update progress made in this area (ref. 108). Important physical factors such as characteristics of the light source, atmospheric composition, humidity, temperature and dye aggregation and concentration influence the rate and tendency of dyes to fade and cause fiber photodegradation. Mechanistic studies on classes of dyes are most extensive for vat anthraquinone dyes and azo dyes but there have also been some studies on triarylmethane and heterocyclic dyes. Azo dyes have the potential for undergoing either photooxidation or photoreduction depending on their structure and the fibrous substrate to which they are bound. There appears to be more direct experimental evidence for reductive rather than oxidative processes in azo dye fading. The preferred mechanism by which most dyed fibers are phototendered is the hydrogen abstraction process rather than the active or singlet oxygen process. A variety of mechanisms have been proposed to explain the detrimental effect that titanium oxide pigments have on fiber photodegradation. These mechanisms include formation of an oxygen radical anion and formation of reactive hydroxyl radicals. However, more careful experimental studies are needed to elucidate fully the mechanisms by which pigments photodegrade fibers (ref. 108).

The effect of light on wool and subsequent photochemical changes that occur in this fiber has historically been an area that has received and continues to receive much attention (ref. 109,110). It has been determined that wool may be either yellowed or photobleached depending on the wavelength and spectral composition of the light to which it is exposed. Yellowing of wool has been related to excitation of various amino acid residues, particularly tryptophan and tyrosine as well as being related to the presence of impurities such as suint (latter contains carbonyl groups). Most mechanisms postulated for the photooxidation of wool ultimately involve reaction of the wool with some form of oxygen to lead to a reactive peroxide species that promotes fiber degradation. More recent investigations suggest that singlet oxygen is involved as a reactive intermediate in the photodegradation of wool and that tryptophan is the principal amino acid responsible for initiating such yellowing and photodegradation. A direct correlation between yellow chromophores and the natural fluorescence of wool fibers has been documented (ref. 111). A decrease in the intrinsic fluorescence was observed after photobleaching of yellow wool. The fluorescence then increased on further irradiation of the photobleached wool with accompanying increases in yellowness.

The general phenomenon of yellowing of all types of fibers has been discussed in a series of papers that address different aspects of this problem (ref. 112). Yellowing of textiles occurs due to the presence of fluorescent brightening agents, phenolic and aromatic amino antioxidants and stabilizers, air pollutants and other chemical species that produces chromophoric substances. Yellowing of optical brighteners on fabrics by exposure to oxides of nitrogen has been documented as well as the formation of quinoid type structures from antioxidants such as BHT (*t*-butylhydroxytoluene) and DTBP (2,6-di-*t*-butylphenol).

6.4 BIOLOGICAL AGENTS AND INFLUENCES

6.4.1 Overview

Textiles and other materials may be adversely affected by various microorganisms and insects. The effect of biological agents on textiles is important for enhancing their end use performance in three areas. Textiles will have desirable aesthetic qualities if they can suppress odor-causing bacteria and other types of odor-causing microorganisms. The hygienic and medical effectiveness of textiles is required to prevent the growth of dermatophytic fungi (those that cause skin disease), pathogenic and potentially lethal microorganisms on fibers and to prevent their infestation by insects. It is also desirable that the textiles be decontaminated from microorganisms and insects if biological attack occurs. Finally, prevention of fiber discoloration and degradation, usually by fungi and insects, prolongs the useful life of the material (ref. 113).

6.4.2 Conducive environments and fiber susceptibility

Suppression or killing of microorganisms that cause odor in apparel is particularly problematic in undergarments and hosiery. In undergarments, the axillae and perineal regions of the body are most susceptible to microbial growth. Although certain synthetic fibers such as polyamides have been observed to retain more odor-causing microorganisms than natural fibers such as cotton, further studies are needed to verify these trends and conclusions. The microflora of the body vary considerably from one area of the body to another in both quantity and type of distribution of microbial species. Therefore, it is not surprising that the types and species of microorganisms vary considerably when they are isolated from worn garments. Similar variability is shown in causative microbial species that have been implicated in body odor. It is generally agreed that the parasitic gram-positive bacteria *Staphylococcus epidermidis* and corneyform bacteria cause body odor in the underarm or axillae area. However, considerably less agreement exists about the causative body-odor producing microorganisms in undergarments (perineal area) and in hosiery (feet). For example, the pathogenic gram-positive bacteria *Staphylococcus aureus* has been implicated in the production of body odor in both the perineal and foot areas. Other microorganisms have also been isolated from worn garments. Gram-negative bacteria and yeasts have been isolated from fabrics that had contact with the perineal area. Various species of dermatophytic fungi have been isolated from hosiery. All of these microorganisms have the ability or potential to cause undesirable body odors (ref. 114).

There are no specific test methods that are widely accepted or practiced for evaluating the resistance of textiles to odor-producing microorganisms. However, gas chromatographs or various olfactometers have been used to detect odoriferous chemical substances in human sweat, e.g., isovaleric acid. Such methods have been adapted to detect such substances in fibrous substrates. Growth of these micro-organisms on textiles is evaluated by test methods that are used for determining the presence and viability of disease-causing and fiber-degrading microorganisms (see Table 6.3 and discussions that follow in the section on test methods).

As in aesthetic applications, no type of fiber exhibits any inherent resistance for preventing the transmission of skin infection and/or pathogenic diseases. Thus, all textiles must be treated with antimicrobial agents and/or disinfected when they are used for hygienic or medical purposes. This is particularly desirable because textiles are good fomites, i.e., inanimate objects that aid in the transmission of disease. This transmission is most prevalent in confined environments such as nursing homes and hospitals. Microorganisms that are frequently observed in the transmission of disease by textiles are the gram-positive bacteria *Staphylococcus aureus* and *Streptococcus pneumoniae*, the gram-negative bacteria *Escherichia coli* and *Pseudomonas aeruginosa*, and viruses such as polio, vaccinia, HIV and hepatitis. The yeast-like fungus *Candida albicans* and other microorganisms have been implicated as the cause of diaper rash. Dermatophytic fungi such as *Trichophyton interdigitale* and *Trichophyton rubrum* are frequently isolated from socks and hosiery of persons that have athletes' feet.

Persistence of various pathogenic bacteria on fibers appears to be influenced more by relative humidity and method of bacterial contamination rather than by differences in fiber type. Low relative humidities and contamination by bacterial aerosols or dust resulted in much longer persistence time than high relative humidities and direct microbial contact with fabrics. Similar trends with regard to persistence times of polio and vaccinia viruses on natural fibers were observed. Although all fiber types had sufficient persistence with viruses to be epidemiologically important, wool had the longest persistence times and cotton the shortest persistence times at comparable relative humidities (ref. 114).

Laundering studies on the persistence and survival of vaccinia and polio viruses were conducted before the current high incidence of infection by various forms of hepatitis and the discovery of the insidious HIV virus. Hot water wash with detergents markedly reduced the amount of detectable polio virus on different types of fabrics. It is generally agreed that HIV viruses are readily deactivated by laundering and also agreed that decontamination of surfaces with hepatitis is much more difficult. The large number of health care and other personnel that are at risk by contact with body fluids containing either hepatitis or HIV viruses has led to the recent OSHA standard that leak and liquid proof protective clothing be worn in these situations (ref. 115).

In contrast to the biological resistance of textiles for aesthetic and medical applications, there is considerable variation among fiber types in their ability to withstand damage and/or discoloration caused by microorganisms and insects. Cellulosic fabrics are the most susceptible to degradation by rot- and mildew-producing fungi and algae. However, impure wool, polyamide and poly(vinyl alcohol) fabrics and polyurethane-coated fabrics have also been reported to be damaged and discolored by bacteria, fungi and algae. There is an extensive amount of literature on cellulolytic and non-cellulolytic fungi, bacteria, yeasts and algae that were isolated from textiles exposed to outdoor environments. The most comprehensive references on this topic are the classic book by Siu on the microbial decomposition of cellulose (ref. 116) and a later review article on this topic (ref. 117). The U.S. Army Quartermaster Corps in the 1940's collected and compiled data on fungi, bacteria, yeasts, and algae isolated from textiles in several different tropical and subtropical locations throughout the world. The most frequently isolated and tested fungi were the noncellulolytic *Aspergillus niger* and the cellulolytic *Chaetomium globosum*. Other species of fungi usually isolated from fabrics were *Memnoniella echinata*, *Penicillium luteum*, and *Myrothecium verrucaria* (ref. 116,117). The degradation of fabrics and other textile items outdoors have been related to certain species of bacteria such as *Actinomycetes*, the blue-green algae *Tolypothrix byssoidea* and species of myxomycetes (ref. 117).

It has been pointed out by Siu (ref. 116) that the number of isolations of a given species is often a reflection of the particular method used. Moreover, factors such as pH, temperature, nutrients, nitrogen source and geographical location of fabric exposure exert considerable influence on the number and type of species of fungi, bacteria, and other microorganisms on textiles outdoors. The population of fungi and other microorganisms on textiles outdoors, in the opinion of Siu, resembles the population in soil.

Both environmental factors (temperature, relative humidity, pH, oxygen, nutrients and light) and structural features in cellulosic fibers and fabrics and in other types of fibers influence the rate and extent of attack by fiber-degrading microorganisms. Fungal growth on fibers is most rapid at relative humidities greater than 80%. However, some species of fungi will grow on fabrics at relative humidities in a lower range (50-65%). Most cellulolytic microorganisms are mesophilic and undergo rapid growth in the range of 25-45°C. This accounts for most of the fiber damage occurring in tropical and subtropical climates. In contrast, microorganisms that undergo growth at other temperature ranges (45-70°C----thermophilic or <10°C---psychrophilic) do not usually cause fiber degradation.

Cellulolytic bacteria are usually active at a pH of 7.0-8.0 and cellulolytic fungi active at a pH of 4.0-6.5. A variety of nutrients are required to promote microbial growth. These include: various trace elements, carbon, nitrogen, phosphorus and vitamins. Thus, periodic cleaning of textile surfaces will minimize microbial growth because nutrients will not be readily available. Ultraviolet and visible light are germicidal to most microorganisms. However, further study is needed to determine its biostatic and biocidal effectiveness against specific microorganisms for a given period of light exposure.

Less accessibility of fabric constructions and individual fibers at the morphological level to physical and chemical agents leads to less attack by microorganisms. If highly oriented and crystalline fibers have not been degraded by mechanical, chemical, or thermal influences, they are more resistant to microbial attack than textiles that have undergone some form of physical or chemical degradation. The former are also more impervious to microorganisms than textiles that have an amorphous morphological structure or that are relative more accessible to chemical agents and reactions. Thus, tightly woven fabrics are less susceptible to attack by microorganisms than are loosely woven textile structures. Survival of linen fabrics on Egyptian mummies tightly wrapped and preserved in an arid environment dramatically illustrate this point.

Textiles vary considerably in their susceptibility to insect damage and infestation (ref. 113). The determining factors are fiber type, species of insect, and the environment in which the textile is stored or used. Wool is very susceptible to attack by various species of moths and beetles. These include: the webbing clothes moth (*Tineola bisselliella*), the varied carpet beetle [(*Anthreus verbasci* (L.)] and the furniture carpet beetle (*Anthreus flavipes LeConte*). Other types of fibers are generally unaffected by these pests. Storage of wool fabrics in low humidities at low temperature will minimize their infestation by moths and beetles. However, these pests can normally withstand much greater climatic extremes and variations than most microorganisms. Thus, insectproof finishes (see **Chapter 4**) or spraying the areas surrounding the wool textiles with insecticides have been employed as control and protection techniques. Also, woolen materials are inserted in sealed containers with volatile substances such as *p*-dichlorobenzene and naphthalene.

Insects such as silverfish and cockroaches attack cellulosic fibers more readily than other types of fibers. This is particularly true if the fibers containing sizing material composed of starch and dextrin or if they are contaminated with organic matter that will serve as a food source. Regenerated cellulosic fibers (such as rayon) that have a low degree of polymerization and/or high accessibility are also more prone to attack by silverfish and cockroaches. Termites may destroy or damage most textile fibers. For example, synthetic fibers and rubber used in underground cables have been destroyed by these pests. Textiles used to store food and feed products may be damaged by different pests. These insects include: the confused flour beetle (*Tribolium confusum*), the red flour beetle (*Tribolium castaneum*) and cadelle larve (*Tenebroides mauritanicus*). Although cellulosic fibers may be prone to attack by these insects, damage to any fiber type for storage of edible products is likely because the insects will ultimately destroy textile structures to gain access to the food.

6.4.3 Laboratory and field tests

Laboratory tests that evaluate the susceptibility and resistance to textiles to microorganisms and insects do not always correlate well with actual behavior in end use. However, these tests are useful for rapid screening of various modified and unmodified fabrics for their ability to withstand biological attack. Table 6.3 lists representative laboratory tests for evaluating antibacterial and antifungal activity of textiles for aesthetic uses, medical uses and for fiber preservation. These tests can be essentially subdivided into those that are qualitative and quantitative in scope.

Most qualitative antimicrobial tests are useful for screening large numbers of fabric samples but provide little reliable information on the durability of antimicrobial finishes to repeated laundering or to prolonged outdoor exposure. For example, the Warburg respiration test for bacteria and the manometric technique for fungi only evaluate one microbial function, an increase or decrease in oxygen consumption. Similar deficiencies exist for the agar plate and parallel streak tests. Both methods are based on the diffusion of active agents off the fiber into the biological medium and

TABLE 6.3

Laboratory tests for antibacterial and antifungal activity (adapted from ref. 113).

Test	Analytical capability	Scope and limitations
Agar plate	Qualitative; bacteriostatic	Activity measured indirectly by diffusing off fabric into culture medium; can be run rapidly for screening samples.
Parallel streak	Qualitative; bacteriostatic	Same as agar plate, except fabric swatches placed perpendicular to streaks of test organisms.
Majors	Semiquantitative; bacteriostatic	Based on amount of titratable acid or alkali produced by organism in presence of treated fiber; reproducibility fair.
Quinn	Quantitative; bacteriostatic or bactericidal; fungistatic or fungicidal	Fabric dried, sterilized, inoculated with test organism, and bacterial or fungal colony count made under high-powered microscope; time-consuming; some fabrics require special sterilization procedures.
AATCC-100	Quantitative; bacteriostatic	Similar to Quinn test, except that fabric not dried before sterilization and % reduction of bacteria in fabric measured by serial dilution techniques; time-consuming.
Lashen	Quantitative; bacteriostatic	Modification of Quinn test, fabric not dried under special conditions or sterilized; not yet accepted as a general method.
Warburg respiration	Qualitative; bacteriostatic	Based on increase of oxygen consumption of bacteria (inhibition of respiration) in presence of treated fibers, rapid screening possible, other bacterial functions not measured.
Manometric	Qualitative; fungistatic	Same principle as Warburg respiration technique applied to fungal growth.
Agar plate	Qualitative; fungistatic	Fabric placed on agar surface, inoculated with test fungus; inspected visually for fungal growth.
Humidity jar	Qualitative; fungistatic	Fabric inoculated with test fungus is sprayed, suspended in water, incubated, inspected visually for fungal growth weekly (up to a maximum of 28 days).

TABLE 6.3 (continued)

Test	Analytical capability	Scope and limitations
Perfusion	Qualitative; fungistatic	Various fabric treatments (as strips) fastened to perfusion bed by adhesive, inoculated with agar discs, incubated by immersion in nutrient, inspected for fungal growth, rapid screening.
Soil burial	Semi-quantitative; fungistatic	Strips of treated fabric buried in soils in trays or beds for a specified period of time, fabric samples then washed and conditioned at standard temperature and humidity. Loss of breaking strength of fabric then determined.

and thus do not measure the actual biological activity of the fiber. However, the qualitative 28 day humidity jar test (visual inspection of fabrics for fungal growth) is considered by the EPA (U. S. Environmental Protection Agency) to be a reliable indicator of the actual mildew resistance of textiles.

Most tests that evaluate the decontamination of textiles by disinfectants, sanitizers and laundry additives are usually variations of the agar plate and parallel streak methods, i.e., they measure the zone of inhibition of textiles against representative microorganisms. Since long term durability is not expected nor required in assessing the decontamination of textiles, these tests are effective for their intended scope. Soil burial tests are useful for predicting rot and mildew resistance of textiles provided the relative humidity, temperature, pH, and soil nutrients closely resemble those encountered in actual outdoor exposure.

The Quinn, AATCC-100, and Lashen tests all measure the actual antimicrobial activity of untreated and treated fabrics, and differ only in certain aspects of their experimental procedure. The differences are primarily those that set the degree and extent of fabric sterilization before inoculation with microorganisms. Although these quantitative tests are tedious and time-consuming, they afford useful and fairly reliable methods for predicting the actual performance of fabrics to repeated laundering and to prolonged outdoor exposure.

In addition to the test methods described in Table 6.3, visual observations (particularly scanning electron microscopy) and various physical, chemical and biological tests have been used to determine initial attack and different stages of damage in fibers due to the deleterious attack and presence of microorganisms and insects. When fiber degradation is visually observable to the human eye, it indicates that the damage is usually very extensive. Discoloration of fabrics by fungi and by algae and the presence of insect excrement are also reliable indicators of substantial fiber deterioration. It was noted earlier that scanning electron microscopy has been used extensively to determine the mode and extent of fiber damage due to wear and abrasion. Although it has not been used to the same extent to determine fiber biodeterioration, it has nevertheless proven quite useful in this regard. With SEM, the initial attack of fibers by fungal hyphae can be detected and it is possible to differentiate between chemical and biological attack of various fibers. Moreover, SEM can determine the species of moth or beetle that degrades wool by identification of the shape of the insect's mandible imprints.

Physical tests that have been used to determine the strength, toughness and wear profiles of fabrics and textile structures can also be used to determine the extent of fiber and structural damage caused by biological attack. These include losses in fabric weight, tensile strength, extensibility and tear strength.

A chemical method that has some utility is based on relating the color of pigments that fungi produce to the particular species present on the fiber. For example, poly(vinyl chloride) floor coverings were determined to be exposed to the following fungal species: *Aspergillus versicolor* (red pigment) and *Penicillium janthinellium* (yellow, partly green pigment) (ref. 113).

Biological and biochemical methods have also been used to detect fiber deterioration. This topic has been critically reviewed by McCarthy who proposed five different rapid screening tests for early detection and measurement of microbial growth on fibers (ref. 118). These methods include (a) bioluminescent assay of ATP with lucifern-*luciferase* mixtures; (b) detection of ergosterol produced by certain classes of fungi; (c) production of ammonia by urealytic bacteria and (d) determination of *esterase* activity by hydrolysis of fluorescein diacetate or with *p*-nitrophenyl acetate.

Fluorescence microscopy was also used to detect many common fiber-degrading fungi (ref. 119). When these textile biodeteriogens were exposed to blue light (350-400 nm), most of them produced some form of auto-fluorescence varying in intensity and characteristic of each fungal species.

Test methods for evaluating the insect resistance of textiles may be conveniently classified by their use or exposure to one of these three environments and pests: (a) crawling insects in museums, commercial buildings and residences, (b) pests on food and feed products stored in fabrics or sacks, and (c) mosquitoes and other outdoor pests (ref. 113).

Most test methods for determining the ability of fabrics to withstand damage by crawling insects are variations of standard tests devised for determining their resistance of wool to various species of moths and beetles (ref. 120). Essentially, known species of insects reared under carefully controlled temperatures and humidities are exposed to the fabric for a specified period (usually four or more weeks). The textile item is then subsequently evaluated by inspecting it for visual damage and/or loss of weight. Similar methodology is used to evaluate the effectiveness of textiles against food pests, but the exposure times are usually much longer (six to eight months) with intermediate periodic inspections to assess insect damage. These latter tests, because of their long exposure periods, are more appropriately called field trials (ref. 113). For determining the effectiveness of fabrics against outdoor pests (e.g., mosquitoes and chiggers), the textile is impregnated with a specific amount of repellent per unit area. It is then placed on a person's arm (with the treated fabric wrapped around it) into a confined environment to determine how long the treated material protects the person from the first insect bite.

The ultimate criteria for biological performance of textiles are actual wear trials and prolonged fabric exposure to representative climatic conditions. Because of public health considerations, such trials are not usually conducted with pathogenic microorganisms or with insects, but are limited to odor-causing and fiber-degrading microorganisms and pests. Because such actual wear trials and exposures are time-consuming and expensive, field trials have been conducted. These are an intermediate testing stage between laboratory methods and prolonged wear or exposure. An excellent example is the use of tropical chambers containing mixed fungal cultures and exposure of fabrics in them to simulate long term tropical outdoor exposure.

6.5 MULTIPLE AGENTS AND FACTORS THAT AFFECT TEXTILE PERFORMANCE

6.5.1 Overview

The serviceability and performance of textiles in their end use are ultimately determined by a complex interrelationship of factors. Thus, it is quite difficult to predict their overall behavior on prolonged use. However, the end use of the textile does allow one to narrow the focus on factors that contribute to its effective lifetime and performance. As shown in Figure 6.16, textiles are affected by their environment, functional/aesthetic factors and the degree and type of refurbishing required. Environmental factors include their exposure to varying degrees of sunlight, humidity, biological agents and air pollutants. Functional and aesthetic influences contribute to their wear and soiling, overall comfort and safety. Refurbishing (laundering and/or

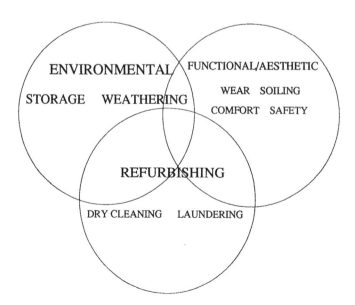

Figure 6.16. Interrelationship of factors affecting overall textile performance.

dry cleaning) exposes the fabrics to mechanical deformations, heat, detergents, bleaches, solvents and other textile auxiliaries. Apparel, carpets, and upholstery are more frequently subjected to mechanical stresses and deformations, to soiling and to refurbishing than textiles for other end uses. Conversely, non-contact items such as drapes, awnings and other textile products are more likely to be exposed primarily to adverse environmental factors (e.g., sunlight, heat, humidity, air to pollutants, and mildew) for prolonged periods of time than are apparel. Thus, each of these three major influences and their interrelationship affects overall textile performance.

6.5.2 Storage

The factors that influence the performance of textiles used and stored indoors vary with fiber type and the amount of exposure that the fabrics receive with regard to sunlight, humidity, temperature changes, and air pollutants. Most systematic studies on the storage of textiles indoors have been made for preserving historic textiles stored and displayed in museums (ref. 121). In the museum environment, all fabrics are susceptible to varying degrees of fading and degradation by both artificial and by natural light. Temperatures under 10°C and relative humidities less than 50% were effective in preventing most types of textile biodeterioration in museums, but woolen materials required mothproofing at all temperatures to prevent insect damage. These recommendations and observations can be used as reasonable guidelines to prolong the useful life of apparel for modern day use. Approaches taken in more recent studies of deterioration of textiles in storage are primarily concerned with deterioration at the fiber level and only focus on microbiological effects (ref. 122).

Although there are no standard or widely accepted test methods for evaluating the ability of textiles to be stored or used indoors for prolonged periods of time, there are a few studies that have made specific recommendations in this area. Screening tests have been devised for determining the tendency of white fabrics to yellow during storage, particularly if they contain fluorescent brightening agents and are exposed to pollutants such as nitrogen oxides (ref. 114). There has been a study that correlated the in-service color fading of carpets with their exposure to atmospheric contaminants. Particular emphasis was given to ozone exposure, but there was also some data relating fading to exposure to nitrogen oxides (ref. 123).

6.5.3 Weathering

Weathering of fabrics or other materials outdoors is a complex process. Such environmental deterioration is influenced by the amount of sunlight, temperature, humidity, air pollutants and particulate matter, rain, wind, snow and the microflora present in the region or climate where the textiles are exposed. It is generally agreed that ultraviolet light, rot and mildew fungi, algae and air pollution are the major factors that contribute to the discoloration and degradation of textiles exposed outdoors (ref. 121, 124). Table 6.4 provides a brief and informative list of the principal causes of textile weathering.

TABLE 6.4

Principal weathering agencies (ref. 124). Courtesy of The Textile Institute.

Light	Acid rain
Atmospheric oxygen	Microorganisms
Oxides of nitrogen and sulfur	Extreme heat and cold
Ozone	Wind
Water	Dust

Accelerated weathering tests have been employed with varying degrees of success for predicting the ability of textiles to withstand outdoor environments. Apparatuses such as the Weatherometer™ provide accelerated exposure simulations for fabrics. In such an apparatus, the fabrics are exposure to simulated noon sunlight (xenon arc light source) at varying degrees of temperature and relative humidity. Soil burial tests and tropical chambers containing representative microflora are also frequently used to assess the ability of treated fabrics to withstand prolonged outdoor exposure.

The overriding consideration for all these simulations is that the fabric be exposed to the most representative conditions and agents present in the climate or location where the textile will be used. For example, in dry, desertlike climates with little rainfall, ability of textiles to withstand degradation by ultraviolet light should be the primary concern. In highly urban areas and certain industrial locations, detrimental effects of air pollutants on materials (at appropriate temperatures/RH) should be emphasized.

Ultimately, long term (usually six months to three years) exposure of fabrics outdoors provides the most accurate information about their ability to withstand weathering processes. However, this involves considerable time and expense. Thus, there are few comprehensive or comparative studies for evaluating the long term weathering behavior of a variety of diverse fiber types and fabric constructions. There have been however, detailed comparative studies conducted by Lünenschloss and co-workers (refs. 125-129) on the weather resistance of yarns composed of representative natural and synthetic fibers. Evaluations of various physical and mechanical properties of these different yarns were made before exposure and after exposure for six months and a year. Properties such as tenacity, energy-to-rupture and flex abrasion were measured. The observation was made that acrylic and poly(vinyl chloride) yarns had the best overall weather resistance. Polyamide and polyester yarns were considerably effected but still superior to cotton yarns. Rayon, other regenerated cellulosics and wool had less than average weathering resistance while jute and silk yarns had extremely poor resistance to weathering. These results must be interpreted with some caution because the effect of air pollution may change the ability of different types of fibers to withstand weathering. For example, it was observed that cotton fabrics had better weather resistance than synthetic fabrics (polyamide and polyester) in areas of low air pollution, but had poorer resistance than the synthetic fabrics in areas of high air pollution (ref. 130). A recent study by Barnett and Slater relates serviceability of six different types of fabrics (cotton, wool, polyester, nylon and two 50/50 cotton/nylon blends of different construction) outdoors to critical values that were retained in their tensile and tear strength and flex abrasion resistance. With the exception of cotton fabrics, all other fabrics undergo rapid degradation in early stages of weathering. Mathematical analysis of this pattern indicates that most fabrics would be functionally inferior at much shorter times than predicted or recommended in earlier studies (ref. 131).

A variety of chemical and instrumental techniques have been employed for assessing the extent and type of damage caused to fibers on exposure outdoors. In some instances, it is difficult to differentiate photodegradation of fibers from degradation of fibers by air pollutants or even from degradation caused by rot and mildew fungi. As in other evaluations, scanning electron microscopy has proved advantageous in visually differentiating specific causes of fiber degradation that occurred outdoors. Photoelectron spectroscopy (ESCA) has also been used to detect surface changes in various films made from synthetic polymers that were exposed outdoors. It offers promise as an instrumental technique to assess and differentiate the various modes of fiber degradation that occur outdoors (ref. 132).

The numerous studies over the years on the measurement and particularly the prediction of the useful lifetime of fabrics exposed to weathering influences clearly indicate certain testing priorities. The actual environmental conditions must be simulated as closely as possible in laboratory chambers and accelerated field tests to effectively predict long-term functional behavior of outdoor textiles. Thus, the proper weighting of weathering influences in Table 6.4 must be simulated as closely as possible in place of extensive long-term exposure outdoors.

6.5.4 Wear

Wear of textiles is also a complex phenomenon and is the overall result of various mechanical stresses and deformations that fibers, yarns, and fabric experience in use. Unfortunately, the term **abrasion** has been frivolously and erroneously used to describe the more complex processes of **wear**. This has resulted in a bewildering array of abrasion instruments that purport to predict the wear life and performance of textile materials. Morton and Hearle (ref. 8) aptly illustrate this point by comparing the effective **abrasion/wear** resistance of representative natural and synthetic yarns and fabrics for a variety of laboratory-type instruments relative to selected wear trials. They observed a marked variation in results from one test method to another even for the same fiber type. The use of scanning electron microscopy by Dweltz and Sparrow (ref. 133) verifies these discrepancies. These investigators observed that untreated and resin-treated cotton fabrics (both degraded from wear in use) bore little resemblance to those obtained from the same fabrics abraded in laboratory instruments. As shown in Figure 6.17 (a) and (b), actual wear produces fibrillation without major cracks or faults in the untreated fabric, but fibrillation and cracks along the spiral angle occurred due to actual wear in the resin-treated fabric. In contrast, when

414

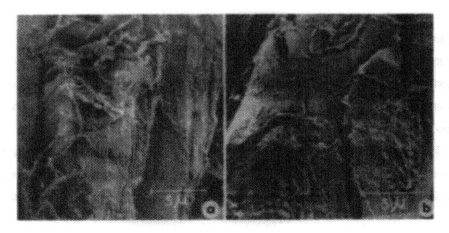

Figure 6.17. Untreated cotton fiber (a) and resin-treated cotton (b) from fabric subjected to actual wear life (ref. 133). Courtesy of Textile Research Institute.

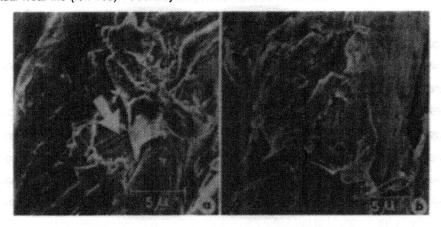

Figure 6.18. Untreated cotton fiber (a) and resin-treated cotton (b) from fabric subjected to conditioned abrasion/Accelorotor with emery liner (ref. 133). Courtesy of Textile Research Institute.

both fabrics were abraded by a random tumble method with an emery liner, untreated and treated fabrics exhibited cuts and separated into small fragments [Figure 6.18 (a) and (b)].

Garment or apparel wear is the first of the two major applications or areas that have received considerable attention in attempts to correlate laboratory simulations and field trials with actual wear performance. The other major area is carpet wear that will be discussed after garment wear.

There have been some investigations in which good correlations were obtained between simulated and predicted garment wear. Barella and co-workers (ref. 134, 135) claim that wear of undergarments and jeans may be predicted and realistically simulated by measuring the force that a garment exerts on the body by fitting it onto a manikin to specified sizes. The fitted garment is then subjected to a series of fatigue cycles, followed by a subsequent measurement of the reduction of force arising from the fatigue and the accompanying stretch that occurs. The garment is then washed under standard or representative conditions. Additional measurements are subsequently made on force and stretch to indicate mechanical and dimensional changes caused by washing and drying the garment.

A very brief but relatively current and comprehensive literature review is available on wear studies (ref. 136). The authors identify the various physical parameters and characterization tests which numerous investigators consider important for predicting apparel wear life from laboratory simulations. Important parameters are resistance to various forms of abrasion, performance after extensive laundering, weight loss of fabrics, scanning electron microscopy and comfort and appearance characteristics. As in other studies, the Martindale abrasion tester is considered to simulate most closely the actual wear of apparel relative to other readily available abrasion and related devices.

An AATCC committee has published realistic guidelines for better laboratory and interlaboratory correlation of simulations to actual wear performance of apparel. Some considerations were control of testing variables such as standardization of testing instruments, correct use of subjective rating scales for appearance, refurbishing conditions and sampling of test fabrics. Equally important emphasis was placed on the closest possible and facile simulation of test methods to actual service conditions of worn garments. Terminologies such as performance property, wear-refurbishing cycle and wear service conditions were defined and recommended as qualifying conditions for development and effective use of laboratory test methods (ref. 137). Newer laboratory abrasion instruments described earlier (such as the fabric-on-fabric and the single fiber transfer devices) may provide better correlation for long term performance of apparel during actual wear than most standard or experimental wear simulation devices (ref. 15, 17). Design features of both of these instruments appear to simulate the subtle influences on fabrics that result in wear and varying degrees of undesirable aesthetic and functional attributes. However, more extensive studies are needed to verify whether these instruments do indeed provide reliable simulations to apparel wear life.

Substantial progress has been made in simulating and predicting carpet wear and overall performance. Carnaby has critically evaluated the mechanics of carpet wear in terms of the behavior of both the pile and the backing. He considers three factors or agents important for predicting the effective wear life of carpets. The first factor is the frictional slippage effect within and between yarns in the pile and in the backing. The second factor is the viscoelastic/plastic behavior of fibers in the pile and the backing. The third factor is the loss of the pile by abrasion due to fiber cutting and to fatigue, and the breaking off of individual fibers and the shedding of loose fibers (ref. 138). A novel instrument with a gauge has been developed to measure rapidly carpet pile thickness without having to shear it off as recommended in other tests. Such direct and non-destructive measurement allows more objective thickness measurements to be rapidly made for manufacturing specifications and during various stages of wear trial and actual use (ref. 139).

A carpet walker has been constructed to simulate the systematic wear that carpets undergo by contract walkers (Figure 6.19). Essential features of this walker are two cylinders with one mounted above the other. The lower cylinder has sneaker sole material fixed to its curved surface driven by an electric motor while the upper cylinder has the capability of fixing carpet to its curved surface. Shearing is simulated by appropriate mechanical action and wear evaluation is assessed by measuring changes in reflectance (L, a, b values) with a Hunter color difference meter. This walker produced similar wear patterns to those obtained by contract walkers but at a much shorter evaluation time (ref. 140).

The mechanisms and variables affecting carpet performance were critically reviewed in a recent paper by Southern and co-workers (ref. 141). There are three mechanisms that have been identified for loss of carpet appearance: fiber loss due to breakage, poor recovery or resiliency of pile fibers and fiber surface matting. The

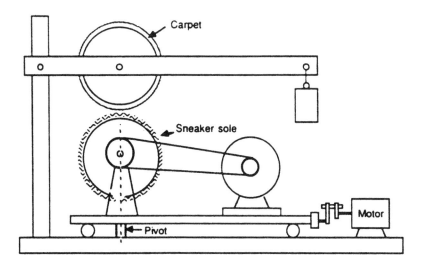

Figure 6.19. The TRI carpet walker (ref. 140). Courtesy of Textile Research Institute.

newer concept of carpet body and its importance to carpet performance is also discussed. Carpet body is essentially the resistance to compression due to a complex combination of intratuft rigidity and tuft-to-tuft interactions. Improvements in carpet performance can be achieved by optimization of carpet body through increased plytwist and reduced bulk and resistance to matting by addition of high shrinkage filaments.

Although upholstery has not received nearly the attention that garments and carpets have with regard to wear, there is a relatively recent study identifying important attributes in upholstery wear. Warfield (ref. 142) reported on a consumer wear study to identify performance differences and attributes in four cotton print upholstery fabrics used on sixty chairs of two slightly different styles (ref. 142). Chairs were actually used for two years in residences then evaluated by the investigator. Soiling was the primary result of extensive wear and use and was very difficult to remove by traditional refurbishing methods. Other types of wear were color changes in the fabric caused by holes, tears, seam breakage and abrasion. Soiling and mechanical wear characteristics varied widely from one chair to another due to different degrees of use and care. This information should be useful for manufacturers and consumers for the design and care of upholstery fabrics.

418

6.5.5 Soiling and staining

Because of the complex and variable composition of materials that soil and stain textiles, simulating actual soiling of textile surfaces has proven to be quite difficult. However, a fibrous structure soiling model has been proposed by Patterson and Grindstaff (ref. 143) that considers particulate and oily soil attachment to fibers (Figure 6.20). The basic problem is the removal of both particulate and oily soils from

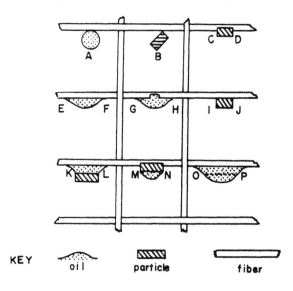

Figure 6.20. Fibrous structure soiling model (ref. 143). Courtesy of Marcel Dekker Publishers.

relatively complex fabric surfaces in such a manner that the soils are not reattached to the surfaces. Oil droplets (A) that have a minimum contact angle with a fiber or fabric surface and particulate matter (B) that only contacts the surface at a single point are readily removed by gentle mechanical action or excess liquid. However, if the particulate matter is embedded well in the fiber (C-D) or adheres, the oil drop makes good contact with the fabric surface (E-F or I-J) or is trapped within a fiber cavity (G-H). These three types of attachment require much more extensive mechanical agitation or liquid detergency to remove soils. There are also more complex types of attachment (K-L or M-N) that involve interfaces between two different types of soils or only partial removal of an oil droplet (O-P) due to cohesional separation. These situations require even more complex techniques and considerations for effective removal of soil from fabrics.

To simulate and predict the soiling behavior of fabrics, synthetic soil compositions have been applied to surfaces by direct contact, as soil solutions or dispersions, by tumbling the fabrics with particulate soils and by transfer to the fabrics from other soiled surfaces. Uniformity of application, particularly by transfer methods, is generally poor. However, there is a method that claims to simulate rapidly and realistically transfer soiling. It consists of inserting clean fabrics with soiled polyurethane cubes into an Accelorotor™ and tumbling them for a short time. This method appears to be applicable for evaluating the transfer of oily, particulate, and mixed soils on untreated fabrics and even on fabrics that contain water and oil repellent finishes (ref. 144).

There are also methods for testing the resistance to fabrics to wet soiling or the redeposition of soils onto fabrics. Unfortunately, most of these methods measure the deposition of soils added to the wash cycle rather than the extent of redeposition of soil desorbed from soiled to unsoiled garments. Soil transfer methods are considered a more accurate method provided the amount of soil desorbed and deposited are measured:

$$\% \text{ soil transfer} = \frac{100 \, s_d}{s_s - s_w} \tag{6.8}$$

To measure the amount of soil deposited the denominator in Eq. 6.8 is kept constant by washing the fabric to be evaluated with a soiled fabric that readily releases soil.

Test methods for evaluating the stain and water repellency of fabrics are more straightforward than soil removal tests in predicting fabric performance. In the case of stains, this essentially involves the application of representative stains to the fabrics with or without controlled pressure. For water repellency testing, fabrics are exposed to water in spray form or by sprinkling devices or by hydrostatic heads. The latter simulate moderate to heavy rainfall on the textile surface (ref. 145, 146). For oil repellency, eight different hydrocarbon liquids with decreasing surface tensions are applied to the fabric surface with an eyedropper. If the fabric is not wet within 30 seconds by the most viscous liquid, the liquid with the next lowest surface tension is applied until the fabric is wet within 30 seconds. This test provides a relatively ranking of oil repellency with effective fluorocarbon finishes providing the highest ranking of eight with the liquid *n*-heptane (ref. 145, 146).

The soiling and staining of carpets has been critically reviewed from the perspective of factors that promote and minimize soiling, that predict such performance and that list requirements for carpets based on the newer stain-blocker technology (ref. 147). Fiber and pile attributes have been improved to both hide soiling and impart good soil and stain resistance. Carpet soiling is primarily a surface phenomenon. Thus, fibers with the least surface area (lowest filament size and modification of cross section) and lowest surface energy (fluorochemical barriers) give the best soil resistance. New stain-blocker technology provides a chemical barrier to acid dyes and to food and beverage spills. However, stain blockers do not prevent stains from greases or oils and must be used in conjunction with fluorochemicals for the best overall carpet performance. Semi-empirical and empirical equations are given for prediction of relative soil deposition on fibers with different properties and for correlation of basic fiber properties to predict the soil rating R for carpets with identical shade and soiling exposure. Parameters in the equations include the level of dirt, surface area and surface energy of test fibers, delustrant level, void ratio and cross-section shape factor of fibers (ref. 147).

A new carpet soiling test has been devised based on the modification of the carpet test walker shown in Figure 6.19. Soil is slowly and uniformly fed to the top cylinder (one on which the carpet is mounted) to simulate soiling by a shoe. Color differences were measured (L, a, b) before and after simulations to determine the extent of carpet soiling. This method is preferable to standard methods that either control only wear patterns or soiling but not both variables at the same time in a realistic simulation (ref. 148). Removal of soil from apparel and from carpets is more appropriately discussed in the later section on **refurbishing**.

6.5.6 Safety

The concept of safety for textile end uses has primarily evolved from health and safety regulations promulgated in the United States and most other industrialized countries. The first safety considerations were addressed by the passage of the

Flammable Fabrics Act for apparel in the U.S.A. in 1953. Standards and regulations currently exist for flame-resistance upholstery and mattresses in residences, lodging and institutions and for occupational protective clothing effective against burns and heat stress. The most recent standard by OSHA (ref. 115) requires that health-care workers and many others exposed to blood-borne pathogens (particularly hepatitis and HIV viruses) wear protective clothing that is impermeable to body fluids and other liquids. Thus, protective clothing against burns, heat stress, chemical agents and biohazards has proliferated over the last several years. Flame-retardant textiles that protect the general public against fires have also received more attention because of injury and death to firefighters and occupants of hotels, institutions and residences.

Performance of protective clothing and relevant test methods are discussed in recent books and monographs partially or totally focused on this topic (refs. 149-152). Fabrics made of gel-spun polyethylene fibers have superior cut and ballistic resistance and are recommended for law enforcement personnel and various occupations where this type of safety is desirable. Many of the inherently flame retardant fibers are now used in protective clothing from heat stress for firefighters and for protection from natural gas fires and explosions by sports car drivers. The TPP test (thermal protective index) described earlier in the section on flame-retardant tests is increasingly being used as a standard to evaluate heat stress performance of garments in these situations. Most of the bench scale and full or large scale flammability tests are used to simulate real fires for evaluation of apparel, upholstery and carpets (ref. 149). Requirements for chemically protective clothing and biologically protective clothing are also described in a text dedicated solely to this topic (ref. 150). Chemically protective clothing must be able to withstand pesticides and hazardous chemicals in civilian applications and toxic warfare agents in military applications. Appropriate simulations have been devised to match the application requirements. Safety of health care workers against biohazards is primarily concerned with leakproof protective clothing effective against contamination by blood-borne pathogens such as the HIV and hepatitis viruses. Thus, tests described earlier for hydrostatic pressure against liquids are the most relevant and appropriate. More regulations and situations will demand and mandate the safety of personnel by appropriate protective clothing. Thus, it is likely that newer and more sophisticated garments and corresponding test methods will evolve to meet these requirements.

6.5.7 Comfort

The concept of comfort related to textile performance has been extensively evaluated and discussed in numerous specific investigations and in many reviews and books. Comfort is comprehensively discussed in two books by Hollies (ref. 153, 154) and in a critical review by Slater (ref. 155). Comfort not only has many contributing factors but also is defined and emphasized differently by various investigators. Thus, the present brief discussion and selection of comfort simulations and studies are inherently biased by the author's perception of comfort. Some studies have focused and continue to focus on the purely thermal and physical aspects of comfort while other studies focus on nonthermal aspects and physiological or psychological aspects. Although there is some disagreement on what the most important factors are that contribute to textile comfort, it is generally agreed that comfort with regard to performance is influenced by a combination of psychological, physiological and physical factors. The "Gestalt" or holistic aspects or interaction of factors (Figure 6.21) most closely approximates "real-world" comfort.

Psychological scales that are used to assess clothing comfort have been investigated and reviewed (ref. 154). A subjective scale containing thirteen descriptive terms for thermal comfort has been formulated. The descriptors range from extremely cold to thermally neutral to hot to the point of illness. Subjective psychological warmth has been mathematically characterized by the equation:

$$S_w = k \, (T_k - 305.7)^{1.6} \qquad (6.9)$$

where T_k is the temperature in °K above thermal neutrality for the skin of the forearm and k is a constant. Contact clothing comfort and various descriptive terms such as damp, cold, scratchy, stiff and nonabsorbent have been used to devise a subjective comfort rating scale for fabrics. Although there has been some progress using a psychological assessment of clothing comfort, additional studies are needed to relate subjective comfort factors to the physiological factors and the physical attributes of fabrics to assess properly overall comfort.

Figure 6.21. Comfort's Gestalt (ref. 154). Courtesy of Dr. Gene J. Pontrelli.

Most of the physiological studies on clothing comfort have evolved from research conducted at the U. S. Army Quartermaster Corps by Goldman (ref. 153) and from similar research conducted by Mecheels in West Germany (ref. 154). Goldman describes clothing in physiological terms by six sets of parameters. The first is that of thermal insulation provided by clothing (R_T or clo, units of thermal resistance). The second is resistance to evaporation (R_E or i_m, permeability index or evaporative resistance). There is also a parameter that relates to wetting out of light clothing in which the fraction of the wetted area and the time required to attain complete wetness at certain activity levels is important. There are effects due to body motion that result in a decrease in insulation with an increase in metabolic activity and the effect of wind velocity (K_V, coefficient of wind velocity). The final two parameters are radiant effects (emissivity for body heat and reflectivity due to solar radiation) and the effect of deliberate change of arrangement of fabric layers on physiology. As noted, most physiological studies measure only one or a few of these parameters, but usually not all of them (ref. 153).

Mecheels has adopted a somewhat different approach for defining clothing comfort in terms of thermophysiological and psychometric comfort ranges. The difference between $T_a$max and $T_a$min defines the useful psychrometric range of clothing. This difference is that between an optimum ambient temperature at which the clothed body will have an uncomfortable, but bearable core temperature and a minimum ambient temperature at which the clothed body with specific heat production will not quite feel cold. $T_a$max is influenced by the total thermal and moisture resistance of clothing (R_c and R_e) and the relative humidity, while $T_a$min is dependent only on the thermal resistance of the clothing. A broader psychrometric range is achieved by greater moisture transport [$i_m = f(R_c/R_e)$] and high clo values of the clothing. The useful range becomes smaller with increased activity of the wearer. Mecheels has determined the resistance values of thermal and moisture transport of fabrics in layers by construction of a simulated skin apparatus. His manikin has flexible limbs, is capable of assuming various positions and simulating metabolic activities (prone position, sitting, standing, walking) and can determine these resistance values under dynamic conditions, thus approximating the behavior of human subjects (ref. 156). Earlier simulation devices such as sweating hot plates and stationary manikins are considered less useful than this versatile manikin.

A review of the effective use of textiles for energy conservation focused on comfort performance of all textile materials indoors (home furnishings and carpets and apparel) rather than only on clothing comfort (ref. 156). Again, this illustrates the relativistic evaluation of comfort by stressing certain factors and parameters rather than others. Emphasis was given in this review to Fanger's approach. Fanger has developed multivariate equations to predict thermal comfort or thermal neutrality indoors (ref. 156, 157). These equations interrelate four environmental factors indoors with clothing and metabolic activity in that environment. The development of such an equation by Fanger (ref. 157) is based on a statistical analysis of over 1,000 human subject responses. Six parameters constitute the equations: metabolic activity of room occupants, clothing insulative or clo value, and the relative humidity, air velocity and air and mean radiant temperature of the room. He considers the mean radiant temperature of the indoor environment to be the most important factor in predicting thermal comfort of room occupants. An instrument based on these varying these parameters from Fanger's thermal comfort equations has also been developed to assist in predicting the thermal comfort of room occupants (ref. 158).

Earlier studies focused primarily on dry heat transfer as the most important physical parameter in clothing comfort. Later studies have either focused on the role of moisture transport or the combined effect of heat and moisture transport in providing comfort. Moisture transport is usually differentiated into moisture-vapor transmission and liquid-moisture transmission. The variety of laboratory instruments discussed earlier to simultaneously measure heat and moisture transport were primarily designed to predict and measure some aspect of thermal or overall clothing comfort (refs. 75, 78, 90-93).

There are numerous other physical properties that have been correlated with clothing comfort. Air permeability and microclimate air exchange have also been determined to play a role in clothing comfort. The porosity of the fabric, its rate of vapor and liquid moisture sorption and loss of moisture through evaporative processes, and other related factors, influence the comfort of the wearer. Similarly, passage of air through fabrics and the exchange of air through the fabrics under various climatic conditions affects comfort. The generation of static charge, the contact of fibers on the skin, and other miscellaneous factors (e.g., hand and contact of the fabric with the skin) also have been determined to influence and change thermal comfort (ref. 153-156).

Two series of recent studies have focused on sensorial comfort affected by changes in moisture in clothing (ref. 159, 160) and interaction between skin changes and comfort perception under a variety of climatic conditions (ref. 161-164). In the sensorial comfort studies, human subjects were asked to used the method of magnitude estimation to determine the intensity of moisture stimuli applied to their backs. Good correlations were obtained for appropriate responses at different moisture levels. This psychophysical method is offered as a means of obtaining closer correlations between objective and subjective measurements than current psychological descriptors (ref. 159, 160).

In the second series of papers, specially constructed jersey knit fabrics were initially evaluated by a modified Kawabata Thermolabo apparatus (one that evaluates warm/cool sensations of fabrics). Heat transfer was shown to be closely related to fabric thickness, bulk density and air volume fraction. The experimental fabrics were further characterized for their hand and related surface properties by the Kawabata KES system. Ten female subjects wore these garments while exercising in a hot, humid environment. Skin temperature, stratum corneum water content and evaporative water loss were significantly related to perceived thermal comfort while capillary blood flow was significant related to overall comfort. Further experiments were recommended to differentiate the relative physiological contributions to thermal comfort (refs. 161-164).

The emergence of high performance garments comprised of hollow fibers or microporous membranes that are liquid impermeable but vapor permeable have led to updated assessments of comfort (ref. 165, 166). Pontrelli has updated his "comfort Gestalt" to include the design of garments to enhance comfort by use of hydrophobic polyester fibers that have high capillary action and that provide better thermal insulation because of air inside a hollow fiber (ref. 165). Bucheck illustrates the importance of the moisture vapor transmission rate in comfort of various fabrics containing microporous membranes. He also demonstrates that it is possible to have a relatively leakproof fabric with good comfort properties by use of the appropriate microporous membrane (ref. 166).

Slater has updated his critical assessment of comfort and stresses the interactive or holistic view of this concept. He discusses the objective and subjective measurement of comfort by physical, physiological and psychological methods. Physical parameters considered to be important are heat transfer, fabric or garment thickness, moisture-vapor and liquid-moisture transmission and wind chill (ref. 167). In a later paper, he proposes fuzzy cluster analysis as a mathematical approach to relate comfort to textile performance (ref. 168). As noted earlier, this type of mathematical approach has been successfully used for color shade sorting of dyed fabrics and for evaluating hand.

Ultimately, the integration of macroclimate environmental, microclimate physiological, psychological and physical factors in wear trials provides the most reliable basis for assessing textile comfort. However, attempts to predict such complex interactions are difficult when a limited number of human subjects or laboratory apparatuses are employed. Successful application of the cluster analysis for specific macroclimate and microclimate environments may reduce the number of important parameters needed to predict accurately the complex phenomenon of comfort.

6.5.8 Refurbishing

The refurbishing of textiles is another aspect of textile performance that is relatively complex because it involves mechanical deformation, exposure to various detergents, bleaching agents, solvents and textile auxiliaries. Heat and mechanical deformation are also required to dry the fabric or textile product. Laundering and drycleaning, the two principal types of refurbishing methods, have been the subject of numerous specific studies and reviews. There are also several publications that give guidelines for the removal of stains prior to refurbishing.

Removal of soil is affected by the characteristics of the textile substrate (hardness, roughness and cross-sectional shape of fibers, fabric construction, yarn twist and structure, swelling during cleaning); properties of the soil (amount present, particle size and geometry of solid soils, affinity for fibers, solubility and/or dispersibility in cleaning media, interactions between oily and particulate soils); and method of refurbishing. Method of cleaning or refurbishing may vary with regard to wet or drycleaning, surfactant type and concentration in the detergent, various detergent additives, amount of agitation, cleaning time and temperature (ref. 143).

Removal of stains from fabrics before refurbishing is best accomplished by treating the material as soon as possible and by using certain techniques for certain groups or classes of stains. The USDA stain removal bulletin lists nine major groups of stains and recommends safe and readily available cleaning agents to address this problem. Ammonia, hydrogen peroxide, dry cleaning solvent, bleach, amyl acetate, acetic acid and other substances are used as stain removal agents. Sponging with drycleaning solvent in as small a wet area as possible is used as the first step in removing many stains. This procedure is usually followed by application of the stain removing agent, then flushing to remove the stain and the stain removal agent. Fabrics are allowed to dry and the procedure is repeated again if required (ref. 169).

Several current and concise publications review the salient features of home laundering and detergency (refs. 170, 171). Laundering of fabrics and of other textile items is usually carried out to remove stains, soils, and unacceptable body odors and to rejuvenate desirable aesthetic properties. Thus, most synthetic laundry detergents contain surfactants. These are compositions that have hydrophobic and hydrophilic properties to remove respectively oily and water-soluble soils. Surfactants may be either soaps made from long chain fatty acids, anionic compounds such as organic sulfonates, nonionics such as polyethylene glycols, cationics such as quaternary ammonium salts or amphoteric substances whose detergency varies with the pH of the detergent solution. While the surfactant comprises 10-30% of a synthetic laundry detergent, there are numerous other materials present. These include: moisture, perfumes, bacteriostats, fluorescent brighteners, oxygen bleaches, foam stabilizers and antifoaming agents, corrosion inhibitors, anti-soil-redeposition agents, builders and chelating agents, and electrolytes and fillers. The percentages of all of these ingredients differ not only with the brand of detergent, but also with the environmental regulations pertaining to effluents in certain regions and specific geographic areas. Designs of washing machines vary from one continent to another. In North America, most washing machines are of a top load design while in Europe the machines load from the front (ref. 170).

The present and future status of detergents used for textiles is lucidly discussed in an article with numerous and informative color photographs and charts (ref. 171). As Figure 6.22 illustrates, the washing process has mostly positive benefits (removal of soils, odor and improvement of aesthetic appearance) but there are also negative consequences such as loss of functional finish and possible mechanical and chemical damage. A variety of detergent formulations have been developed that are suitable for laundering specific fiber and fiber blend combinations, textile constructions and special dyed and printed materials. Further advances and considerations in this area are likely due to the greater increased in use of dyed fabrics and of washable functional and leisure outerwear and sportswear (ref. 171).

REFINISHING OF SOILED TEXTILES BY

Washing	**Removal**
Bleaching	**of dirt**
Softening	
Antistatic treatment	**Replacing**
Dressing/starching	**finish**
Aromatics to	**Additional**
improve odor	**benefits**

Figure 6.22. Textiles-demands of washing process (adapted from ref. 171). Courtesy of Melliand Textilberichte.

Drycleaning and commercial laundering recommendations are frequently made by trade organizations such as the International Fabricare Institute in the United States. This organization was largely responsible for the content of the USDA stain removal bulletin that is frequently requested by the public. Removal of a diversity of stains and drycleaning of just about every type of garment or material are frequently discussed in their publications. An excellent example of such information is the

430

publication of international care label symbols for garments shown in Figure 6.23. This removes language barriers and allows garments to be professionally drycleaned

Figure 6.23. International care labeling system (ref. 172). Courtesy of International Fabricare Institute.

and refurbished by all commercial establishments throughout the world. The successful refurbishment of textiles is the last but perhaps most complex event in the amazing journey from their production to their performance.

GENERAL REFERENCES FOR TEXTILE PERFORMANCE: END USE AND RELEVANT TESTS

AATCC Technical Manual, Vol. 67, Am. Assoc. Text. Chem. Color, Raleigh, N.C., 1992, 396 pp.

1991 Annual Book of ASTM Standards, Section 7: Textiles. Vol. 07.01. Textiles-Yarns, Fabrics and General Test Methods, Am. Soc. Test. and Mater., Philadelphia, PA., 1991, 952 pp.

1991 Annual Book of ASTM Standards, Section 7: Textiles. Vol. 07.02. Textiles-Fibers, Zippers, Am. Soc. Test. and Mater., Philadelphia, PA., 1991, 844 pp.

L. Fourt and N. R. S. Hollies, **Clothing: Comfort and Function**, Marcel Dekker, New York, 1970, 254 pp.

N. R. S. Hollies and R. F. Goldman (Eds.), **Clothing Comfort: Interaction of Thermal, Ventilation, Construction and Assessment Factors**, Ann Arbor Science Publishers, Ann Arbor, Michigan, 1977, 189 pp.

D. S. Lyle, **Performance of Textiles**, Wiley, New York, 1977, 592 pp.

W. E. Morton and J. W. S. Hearle, **Physical Properties of Textile Fibres**, Heineman, London, 1975, 660 pp.

B. Rånby and J. F. Rabek, **Photodegradation, Photo-oxidation and Photostabilization of Polymers**, Wiley, New York, 1975, 573 pp.

R. G. H. Siu, **Microbial Decomposition of Cellulose**, Reinhold, New York, 1951, 531 pp.

M. J. Schick (Ed.), **Surface Characteristics of Fibers and Textiles, Pts. I and II**, 1975 and 1977, Marcel Dekker, New York, 669 pp.

K. Slater, **Textile degradation**, Text. Prog. 21 (1/2) (1991) 1-158.

T. L. Vigo, **Preservation of natural textile fibers–historical perspectives**, in: J. C. Williams (Ed.), Preservation of Paper and Textiles of Historic and Artistic Value, Adv. Chem. Ser. 164, Am. Chem. Soc., Washington, D. C., 1977, Chapter 14.

T. L. Vigo, **Protection of textiles from biological attack**, in: M. Lewin and S. B. Sello (Eds.), Handbook of Fiber Science and Technology. Vol. II. Chemical Processing of Fibers and Fabrics. Functional Finishes, Pt. A., Marcel Dekker, New York, 1983, Chapter 4.

S. H. Zeronian, **Conservation of textiles manufactured from man-made fibers**, in J. C. Williams (Ed.), Preservation of Paper and Textiles of Historic and Artistic Value, Adv. Chem. Ser. 164, Am. Chem. Soc., Washington, D. C., 1977, Chapter 15.

TEXT REFERENCES FOR TEXTILE PERFORMANCE: END USE AND RELEVANT TESTS

1 R. C. Dhingra, R. Postle and T. J. Mahar, **Hygral expansion of woven wool fabrics**, Text. Res. J. 55 (1) (1985) 28-40.

2 P. G. Cookson, A. G. De Boos, A. F. Roczniok and N. G. Ly, **Measuring hygral expansion in woven wool fabrics**, Text. Res. J. 61 (6) (1991) 319-327.

3 A. M. Wemyss and M. A. White, **Dimensional stability testing of wool in finishing and tailoring**, Mell. Textilber. 68 (7) (1987) 509-513.

4 S. A. Heap, P. F. Greenwood, R. D. Leah, J. T. Eaton, J. C. Stevens and P. Kehrer, **Prediction of finished weight and shrinkage of cotton knits--the Starfish project. Pt. I. Introduction and general overview**, Text. Res. J. 53 (2) (1983) 109-119.

5 S. A. Heap, P. F. Greenwood, R. D. Leah, J. T. Eaton, J. C. Stevens and P. Kehrer, **Prediction of finished weight and shrinkage of cotton knits--the Starfish project. Pt. II. Shrinkage and the reference state**, Text. Res. J. 55 (4) (1985) 211-222.

6 N. Greenblau, **The assessment of the dimensional stability of knitted cotton fabrics**, Proc. 1985 AATCC Natl. Tech. Conf., 153-163.

7 R. Massen and U. Winkler, **"Intelligent" sensor for non-contact and non-destructive measurement of shrinkage in knitted and woven fabrics**, Mell. Textilber. (6) (1985) 441-442.

8 W. E. Morton and J. W. S. Hearle, **Physical Properties of Textile Fibers**, Heineman, London, 1975, Chapter 17.

9 J. W. S. Hearle, B. Lomas and A. R. Bunsell, **The study of fiber fracture**, in L. H. Princen (Ed.), Scanning Electron Microscopy of Polymers and Coatings, Appl. Poly. Symp. 23, Wiley-Interscience, New York, 1974, pp. 147-156.

10 P. N. Britton and A. J. Sampson, **Computer simulation of the mechanical properties of nonwoven fabrics. Pt. III. Fabric failure**, Text. Res. J. 54 (7) (1984) 425-428.

11 T. H. Grindstaff and S. M. Hansen, **Computer model for predicting point-bonding nonwoven strength, Part I**, Text. Res. J. 56 (6) (1986) 383-388.

12 I. E. Clark and J. W. S. Hearle, **Influence of temperature and humidity on the biaxial-rotation fatigue of fibres**, J. Text. Inst. 73 (6) (1982) 273-280.

13 K. E. Duckett and B. C. Goswami, **A multi-stage apparatus for characterizing the cyclic torsional fatigue behavior of single fibers**, Text. Res. J. 54 (1) (1984) 43-46.

14 T. L. Vigo, J. S. Bruno and W. R. Goynes, **Enhanced wear and surface characteristics of polyol-modified fibers**, in: L. Rebenfeld (Ed.), Science and Technology of Fibers and Related Materials, Appl. Poly. Symp. 47 (1991) 1-32.

15 G. E. R. Lamb, S. Kepka and B. Miller, **Studies of fabric wear. Pt. I: Attrition of cotton fabrics**, Text. Res. J. 59 (2) (1989) 61-65.

16 P. A. Annis and R. R. Bresee, **An abrasion machine for evaluating single fiber transfer**, Text. Res. J. 60 (10) (1990) 541-548.

17 P. A. Annis, R. R. Bresee and T. R. Cooper, **Influence of textile structure on single fiber transfer from woven fabrics**, Text. Res. J. 62 (5) (1992) 293-301.

18 J. W. S. Hearle, **Understanding and control of textile fiber structure**, in: L. Rebenfeld (Ed.), Science and Technology of Fibers and Related Materials, Appl. Poly. Symp. 47 (1991) 1-32.

19 M. C. Boyce, M. L. Palmer, M. H. Seo, P. Schwartz and S. Backer, **A model of the tensile failure process in woven fabrics**, in: L. Rebenfeld (Ed.), Science and Technology of Fibers and Related Materials, Appl. Poly. Symp. 47 (1991) 383-402.

20 H. R. Richards, **Thermal degradation of fabrics and yarns. Pt. I. Fabrics**, J. Text. Inst. 75 (1) (1984) 28-36.

21 X. Tao and R. Postle, **Physical ageing and annealing in fibers and textile materials. Pt. I. Physical ageing in single wool fibers and textile assemblies**, Text. Res. J. 57 (7) (1987) 387-395.

22 S. Şukigara, R. C. Dhingra and R. Postle, **Physical ageing and annealing in fibers and textile materials. Pt. II. Annealing behavior of textile materials produced from wool and manmade fibers**, Text. Res. J. 57 (7) (1987) 479-489.

23 X. Tao, S. Sukigara, R. Postle and R. C. Dhingra, **Physical ageing and annealing in fibers and textile materials. Pt. III. Physical ageing and annealing in blended textile assemblies**, Text. Res. J. 57 (7) (1987) 601-610.

24 B. C. Goswami, K. E. Duckett and T. L. Vigo, **Torsional fatigue and the initiation mechanism of pilling**, Text. Res. J. 50 (8) (1980) 481-485.

25 W. D. Cooke, **The influence of fibre fatigue on the pilling cycle. Pt. I. Fuzz fatigue**, J. Text. Inst. 73 (1) (1982) 13-19.

26 W. D. Cooke, **The influence of fibre fatigue on the pilling cycle. Pt. II. Fibre entanglement and pill growth**, J. Text. Inst. 74 (3) (1983) 101-108.

27 W. D. Cooke, **The influence of fibre fatigue on the pilling cycle. Pt. III. Pill wear-off and fabric attrition**, J. Text. Inst. 75 (3) (1984) 201-211.

28 A. Naik and F. López-Amo, **Pilling propensity of blended textiles**, Mell. Textilber. 63 (6) (1982) 416-423.

29 E. M. Buras and R. H. Harris, Jr., **Generation of lint from fabrics**, Text. Chem. Color. 15 (1) (1983) 1-3.

30 H. Yokura, S. Nagae and M. Niwa, **Prediction of fabric bagging from mechanical properties**, Text. Res. J. 56 (12) (1986) 748-754.

31 C. Luo and R. R. Bresee, **Appearance evaluation by digital image analysis**, Text. Chem. Color. 22 (2) (1990) 17-19.

32 A. Shinohara, Q.-Q. Ni and M. Takatera, **Geometry and mechanics of the buckling wrinkle in fabrics. Pt. I. Characteristic of the buckling wrinkle**, Text. Res. J. 61 (2) (1991) 94-100.

33 A. Shinohara, Q.-Q. Ni and M. Takatera, **Geometry and mechanics of the buckling wrinkle in fabrics. Pt. II. Buckling model of a woven fabric cylinder in axial compression**, Text. Res. J. 61 (2) (1991) 100-105.

34 T. G. Clapp and H. Peng, **Buckling of woven fabrics. Pt. III. Experimental validation of theoretical models**, Text. Res. J. 60 (11) (1990) 641-645.

35 Anon., **Physical testing of coated fabrics: an alphabetical listing of test methods**, J. Coated Fabrics 17 (3) 1987) 14-21.

36 P. Schwartz, R. Fornes and M. Mohamed, **An analysis of the mechanical behavior of triaxial fabrics and the equivalency of conventional fabrics**, Text. Res. J. 52 (6) (1982) 388-394.

37 S. M. Lee and A. S. Argon, **The mechanics of the bending of non-woven fabrics. Pt. I. Spunbonded fabric (Cerex)**, J. Text. Inst. 74 (1) (1983) 1-11.

434

38 S. M. Lee and A. S. Argon, **The mechanics of the bending of non-woven fabrics. Pt. II. Spunbonded fabric with spot bonds (Fibretex)**, J. Text. Inst. 74 (1) (1983) 12-18.

39 S. M. Lee and A. S. Argon, **The mechanics of the bending of non-woven fabrics. Pt. III. Print-bonded fabric (Masslinn)**, J. Text. Inst. 74 (1) (1983) 19-30.

40 S. M. Lee and A. S. Argon, **The mechanics of the bending of non-woven fabrics. Pt. IV. Print-bonded fabric with a pattern of elliptical holes (Keybak)**, J. Text. Inst. 74 (1) (1983) 31-37.

41 A. R. Kalyanaraman and A. Sivaramakarishnan, **An electronic instrument to measure stiffness of fabrics**, Text. Res. J. 53 (9) (1983) 573-575.

42 A. R. Kalyanaraman and A. Sivaramakarishnan, **An electronic fabric stiffness meter--performance evaluation with the known instruments**, Text. Res. J. 54 (7) (1984) 497-484.

43 H. M. Elder, S. Fisher, G. Hutchinson and S. Beattie, **A psychological scale for fabric stiffness**, J. Text. Inst. 76 (6) (1985) 442-449.

44 T. S. Ranganathan, D. Ramaswamy, K. S. Jayaraman, R. Sanjeevi, V. Arumugan, S. Das and A. Vaidyanathan, **A new approach to drape**, J. Text. Inst. 77 (3) (1986) 226-228.

45 M. L. Gaucher, M. W. King and B. Johnston, **Predicting the drape coefficient of knitted fabrics**, Text. Res J. 53 (5) (1983) 297-303.

46 J. W. S. Hearle and J. Amirbayat, **Analysis of drape by means of dimensionless groups**, Text. Res. J. 56 (12) (1986) 727-733.

47 J. R. Collier, B. J. Collier, G. O'Toole and S. M. Sargand, **Drape prediction by means of finite-element analysis**, J. Text. Inst. 82 (1) (1991) 96-107.

48 B. C. Ellis and R. K. Garnsworthy, **A review of techniques for the assessment of hand**, Text. Res. J., 50 (4) (1980) 231-238.

49 S. Kawabata. **Standardization of Hand Evaluation**, HESC, The Textile Machinery Society of Japan, Osaka, Japan, 2nd Ed., 1980.

50 S. Kawabata and M. Niwa, **Fabric performance in clothing and clothing manufacture**, J. Text. Inst. 80 (1) (1989) 19-43.

51 T. J. Mahar and R. Postle, **Fabric handle equations for Australia, New Zealand, India and U.S.A.**, J. Text. Mach. Soc. Japan 37 (9) (1984) T143-T147.

52 H. M. Behery, **Comparison of fabric hand assessment in the United States and Japan**, Text. Res. J. 56 (4) (1986) 227-240.

53 R. L. Barker and K. M. Vohs, **Assessment of the handle of finished fabrics**, Proc. Am. Assoc. Tex. Chem. Color. 1986 Natl. Tech. Conf., pp. 23-34.

54 A. J. Sabia and A. M. Pagliughi, **Use of Kawabata instrumentation to evaluate silicone fabric softeners**, Proc. Am. Assoc. Tex. Chem. Color. 1986 Natl. Tech. Conf., pp. 102-106.

55 P.-L. Chen, R. L. Barker, G. W. Smith and B. Scruggs, **Handle of weft knit fabrics**, Text. Res. J. 62 (4) (1992) 200-211.

56 N. Pan, K. C. Yen, S. J. Zhao and S. R. Yang, **A new approach to the objective evaluation of fabric handle from mechanical properties. Pt. III. Fuzzy cluster analysis for fabric handle sorting**, Text. Res. J. 58 (10) (1988) 565-571.

57 M. Raheel and J. Liu, **An empirical model for fabric hand. Pt. I. Objective assessment of light weight fabrics**, Text. Res. J. 61 (1) (1991) 31-38.

58 M. Raheel and J. Liu, **An empirical model for fabric hand. Pt. II. Subjective assessment**, Text. Res. J. 61 (2) (1991) 79-82.

59 J. F. Krasny, **Flammability evaluation methods for textiles**, in: M. Lewin and E. Pearce (Eds.), Flame-retardant Polymeric Materials, Vol. 3, Plenum Press, 1982, pp. 155-200.

60 J. F. Krasny, **Apparel flammability: accident simulations and bench-scale tests**, Text. Res. J. 56 (5) (1986) 287-303.

61 A. R. Horrocks, M. Tunc and D. Price, **The burning behaviour of textiles and its assessment by oxygen-index methods**, Text. Prog. 18 (1/2/3) (1989) 205 pp.

62 G. H. Damant, **Burning issues: the role of furnishings**, J. Fire Sci. 9 (1991) 3-43.

63 E. P. Brewster and R. L. Barker, **A summary of research on heat resistant fabrics for protective clothing**, Am. Ind. Hyg. Assoc. J. 44 (2) (1983) 123-130.

64 I. Shalev and R. L. Barker, **Analysis of heat transfer characteristics of fabrics in an open flame exposure**, Text. Res. J. 53 (8) (1983) 475-482.

65 R. L. Barker and Y. M. Lee, **Analyzing the transient thermophysical properties of heat-resistant fabrics in TPP exposures**, Text. Res. J. 57 (6) (1987) 331-338.

66 W. P. Behnke, A. J. Geshury and R. L. Barker, **Thermo-man[R] and thermo-leg: large scale test methods for evaluating thermal protective performance**, in: J. P. McBriarity and N. W. Henry (Eds.), Performance of Protective Clothing, ASTM STP 1133, ASTM, Philadelphia, PA., 1992, pp. 266-280.

67 V. Babrauskas, **Upholstered furniture heat release rates: measurements and estimation**, J. Fire Sci. 1 (1/2) (1983) 9-32.

68 L. Benisek, M. J. Palin and R. Woollin, **Fair and realistic flammability tests for carpets**, J. Fire Sci. 6 (1/2) (1988) 25-41.

436

69 A. Abramowitz and T. I. Eklund, **Model tests of aircraft interior panel flammability**, J. Fire Sci. 3 (2) (1985) 129-140.

70 M. Yoneda and S. Kawabata, **Analysis of transient heat conduction and its application. Pt. 1. The fundamental analysis and application to thermal conductivity and thermal diffusivity measurements**, J. Text. Mach. Soc. Japan Trans. Eng. Ed. 29 (4) (1983) 73-83.

71 M. Yoneda and S. Kawabata, **Analysis of transient heat conduction and its application. Pt. 2. A theoretical analysis of the relationship between warm/cool feeling and transient heat conduction in skin**, J. Text. Mach. Soc. Japan Trans. Eng. Ed. 31 (4) (1985) 79-85.

72 M. Yoneda and S. Kawabata, **Analysis of transient heat conduction and its application. Pt. 3. An analysis of two-layered body problem**, J. Text. Mach. Soc. Japan Trans. Eng. Ed. 34 (1) (1988) 1-6.

73 J. J. Anderson, **Guarded window test apparatus for determining the insulation value of interior window coverings**, Text. Res. J. 52 (10) (1982) 643-651.

74 J. R. Martin and G. E. R. Lamb, **Measurement of thermal conductivity of nonwovens using a dynamic method**, Text. Res. J. 57 (12) (1987) 721-727.

75 C. B. Hassenboehler, Jr. and T. L. Vigo, **A mixed flow thermal transmittance tester for textiles**, Text. Res. J. 52 (8) (1982) 510-517.

76 B. Farnworth, **Mechanisms of heat flow through clothing insulation**, Text. Res. J. 53 (12) (1983) 717-725.

77 G. E. R. Lamb and M. Yoneda, **Heat loss from a ventilated clothed body**, Text. Res. J. 60 (7) (1990) 378-383.

78 S. Kawabata, **Development of a device for measuring heat-moisture transfer properties of apparel fabrics**, J. Text. Mach. Soc. Japan Trans. 37 (8) (1984) T130-141.

79 T. Fujimoto and N. Seki, **Effective thermal conductivity and role of radiative transfer in many kinds of clothing materials**, J. Text. Mach. Soc. Japan Trans. 40 (2) (1987) T13-22.

80 B. Farnworth, **A numerical model of the combined diffusion of heat and water vapor through clothing**, Text. Res. J. 56 (11) (1986) 653-665.

81 A. M. Schneider, B. N. Hoschke and H. J. Goldsmid, **Heat transfer through moist fabrics**, Text. Res. J. 62 (2) (1992) 61-66.

82 T. L. Vigo, C. B. Hassenboehler, Jr. and N. Wyatt, **Surface temperature measurements for estimating heat flow through textiles**, Text. Res. J. 52 (7) (1982) 451-456.

83 M. Andrassy, V. Hranilovic and D. Raffaelli, **Examination of the thermal conductivity of textiles by thermography**, Texstil 31 (4) (1982) 209-214.

84 K. Gniotek, **Possibility of using transducers for measuring the air permeability of fabrics. II. Thermistor and ionization methods**, Prezglad Wlokienniczy 36 (6) (1982) 369-371.

85 British Standards Institution, **Method for measurement of the equivalent pore size of fabrics (bubble pressure test)**, British Standard BS 3321, British Standards Institution, London, 1986, iii + 5 pp.

86 M. S. Atwal, **Factors affecting the air resistance of nonwoven needle-punched fabrics**, Text. Res. J. 57 (10) (1987) 574-579.

87 M. M. Adler and W. K. Walsh, **Mechanisms of transient moisture transport between fabrics**, Text. Res. J. 54 (5) (1984) 334-343.

88 M. Day and P. Z. Sturgeon, **Water vapor transmission rates through textile materials as measured by differential scanning calorimetry**, Text. Res. J. 56 (3) (1986) 157-161.

89 C. A. van Beest and P. P. M. M. Wittgen, **A simple apparatus to measure water vapor resistance of textiles**, Text. Res. J. 56 (9) (1986) 566-568.

90 B. Farnworth, W. A. Lotens and P. P. M. M. Wittgen, **Variation of water vapor resistance of microporous and hydrophilic films with relative humidity**, Text. Res. J. 60 (1) (1990) 50-53.

91 K. Hong, N. R. S. Hollies and S. M. Spivak, **Dynamic moisture vapor transfer through textiles. Pt. I. Clothing hygrometry and the influence of fiber type**, Text. Res. J. 58 (12) (1988) 697-706.

92 J. A. Wehner, B. Miller and L. Rebenfeld, **Dynamics of water vapor transmission through fabric barriers**, Text. Res. J. 58 (10) (1988) 581-592.

93 J.-H. Wang and H. Yasuda, **Dynamic water vapor and heat transport through layered fabrics. Pt. I. Effect of surface modification**, Text. Res. J. 61 (1) (1991) 10-20.

94 J.-H. Wang and H. Yasuda, **Dynamic water vapor and heat transport through layered fabrics. Pt. II. Effect of the chemical nature of fibers**, Text. Res. J. 62 (4) (1992) 227-235.

95 P. K. Chatterjee and H. V. Nguyen, **Mechanism of liquid flow and structure property relationships**, in: P. K. Chatterjee (Ed.), Absorbency, Elsevier Publishers, Amsterdam, 1985, pp. 29-84.

96 S. B. Sello and C. V. Stevens, **Antistatic treatments**, in: M. Lewin and S. B. Sello (Eds.), Handbook of Fiber Science and Technology. Vol. II. Chemical Processing of Fibers and Fabrics. Part B. Functional Finishes, Marcel Dekker, New York, 1984, Chapter 4.

97 B. Guoping and K. Slater, **The progressive deterioration of textile materials. Part V. The effect of acid treatment on fabric tensile strength**, J. Text. Inst. 81 (1) (1990) 59-68.

438

98 V. Rossbach and N. Karunaratna, **Testing the stability of technical textiles by the "critical solution time" method**, Mell. Textilber. 66 (3) (1985) 223-226.

99 J. B. Upham and V. S. Salvin, **Effect of Air Pollutants on Textile Fibers and Dyes**, EPA-650/3-74-008, Research Triangle Park, N. C., 1975, 85 pp.

100 B. Walters, B. C. Goswami and T. L. Vigo, **Sorption of air pollutants onto textiles**, Text. Res. J. 53 (6) (1983) 354-360.

101 D. M. Wiles, **Photo-oxidative reactions of polymers**, in: B. Rånby and J. F. Rabek (Eds.), Long-Term Properties of Polymers and Polymeric Materials, Appl. Poly. Symp. 35, Wiley, New York, 1979, pp. 235-241.

102 B. Rånby and J. F. Rabek, **Accelerated reactions in photo-degradation of polymers**, in: B. Rånby and J. F. Rabek (Eds.), Long-Term Properties of Polymers and Polymeric Materials, Appl. Poly. Symp. 35, Wiley, New York, 1979, pp. 243-263.

103 B. Rånby and J. F. Rabek, **Photodegradation, Photo-oxidation and Photostabilization of Polymers**, Wiley, 1975, New York.

104 J. F. McKellar and N. S. Allen, **Photochemistry of Man-Made Polymers**, Applied Polymer Science Publishers, London, 1979.

105 D. J. Carlsson and D. M. Wiles, **The photooxidative degradation of polypropylene. Pt. 1. Photooxidation and photoinitiation processes**, J. Macromol. Sci., Revs. Macromol. Chem. C14 (1) (1976) 65-106.

106 N. S. Allen, **Recent advances in the photo-oxidation and stabilization of polymers**, Chem. Soc. Rev. 15 (1986) 373-404.

107 N. S. Allen and J. F. McKellar, **The photochemistry of commercial polyamides**, Macromol. Revs. 13 (1978) 241-281.

108 N. S. Allen, **Photofading mechanisms of dyes in solution and polymer media**, Rev. Prog. Color. Rel. Topics 17 (1987) 61-71.

109 P. A. Duffield and D. M. Lewis, **The yellowing and bleaching of wool**, Rev. Prog. Color. Rel. Topics 15 (1985) 38-51.

110 K. Schäfer and B. Groger, **Photostabilization of wool**, Mell. Textilber. 72 (6) (1991) 206-211.

111 K. Schäfer, **The natural fluorescence of wool**, J. Soc. Dyer. Color. (5/6) (1991) 345-346.

112 P. W. Harrison (Ed.), **Update on yellowing**, Text. Prog. 15 (4), The Textile Institute, Manchester, 47 pp.

113 T. L. Vigo, **Protection of textiles from biological attack**, in: M. Lewin and S. B. Sello (Eds.), Handbook of Fiber Science and Technology. Vol. II. Chemical Processing of Fibers and Fabrics. Functional Finishes, Pt. A., Marcel Dekker, New York, 1983, Chapter 4.

114 T. L. Vigo and M. A. Benjaminson, **Antibacterial fiber treatments and disinfection**, Text. Res. J. 51 (7) (1981) 454-465.

115 T. L. Vigo, **Protective clothing against biohazards**, in: M. Raheel (Ed.), Protective Clothing Systems and Materials, Marcel Dekker, New York, 1994, Chapter 9.

116 R. G. H. Siu, **Microbial Decomposition of Cellulose**, Reinhold, New York, 1951, 531 pp.

117 A. J. Desai and S. N. Pandey, **Microbial degradation of cellulosic textiles**, J. Sci. Ind. Res. 30 (1971) 598-606.

118 B. J. McCarthy, **Rapid methods for the detection of biodeterioration of textiles**, Intl. Biodeter. 23 (1987) 357-364.

119 B. J. McCarthy and P. H. Greaves, **Auto-fluorescence from standard textile biodeteriogens**, Mell. Textilber. 71 (11) (1990) 913-914.

120 D. M. Lewis and T. Shaw, **Insectproofing of wool**, Rev. Prog. Color. Rel. Topics 17 (1987) 86-94.

121 T. L. Vigo, **Preservation of natural textile fibers-historical perspectives**, in: J. C. Williams (Ed.), Preservation of Paper and Textiles of Historic and Artistic Value, Adv. Chem. Ser. 164, Am. Chem. Soc., Washington, D. C., 1977, Chapter 14.

122 K. Slater, **Textile degradation**, Text. Prog. 21 (1/2), The Textile Institute, Manchester, 1991, 158 pp.

123 M. R. Nipe, **Atmospheric contaminant fading**, Text. Chem. Color. 13 (6) (1981) 136-146.

124 J. R. Holker, **Weathering of textiles**, Textiles 17 (3) (1988) 64-71.

125 J. Lünenschloss and H. Stegherr, **The effect of weathering on the properties of textile fibers. I.**, Textil-Praxis 15 (9) (1960) 931-939.

126 J. Lünenschloss and H. Stegherr, **The effect of weathering on the properties of textile fibers. II.**, Textil-Praxis 15 (10) (1960) 1011-1017.

127 J. Lünenschloss and H. Kurth, **The effect of weathering on the properties of textile fibers. III.**, Textil-Praxis 15 (11) (1960) 1146-1150.

128 J. Lünenschloss and H. Kurth, **The effect of weathering on the properties of textile fibers. IV.**, Textil-Praxis 15 (12) (1960) 1283-1289.

129 J. Lünenschloss and H. Kurth, **The effect of weathering on the properties of textile fibers. V.**, Textil-Praxis 16 (1) (1961) 51-56.

130 A. H. Little and H. L. Parsons, **The weathering of cotton, nylon and terylene fabrics in the United Kingdom**, J. Text. Inst. 58 (10) (1967) 449-462.

131 R. D. Barnett and K. Slater, **The progressive deterioration of textile materials. Pt. V. The effects of weathering on fabric durability**, J. Text. Inst. 82 (4) (1991) 417-425.

132 A. Dilks and D. T. Clark, **ESCA studies of natural weathering phenomena at selected polymer surfaces**, J. Poly. Sci., Poly. Chem. 19 (11) (1981) 2847-2860.

133 N. E. Dweltz and J. T. Sparrow, **A SEM study of abrasion damage to cotton fibers**, Text. Res. J. 48 (11) (1978) 633-636.

134 A. Barella, R. M. Sauri, C. Polo, J. M. Etayo, J. A. Salmurri and L. Viertel, **Simulation of wearing and assessment of garments. Correlation with a practical wear trial and the cost advantage of the method**, Bull. Sci. Inst. Textile de France 8 (30) (1979) 135-149.

135 A. Barella, R. M. Sauri, C. Polo, J. M. Etayo, J. A. Salmurri and L. Viertel, **Simulation and diagnosis: extension to other garments**, Bull. Sci. Inst. Textile de France 8 (32) (1979) 315-332.

136 M. Jacob and V. Subramaniam, **A review of literature on wear studies**, Colourage 35 (7) (1988) 17-18.

137 R. S. Merkel and I. B. Miller, **Recommendations for investigating the degree of correlations between lab tests and wear tests**, Text. Chem. Color. 19 (4) (1987) 23-25.

138 G. Carnaby, **The mechanics of carpet wear**, Text. Res. J. 51 (8) (1981) 514-519.

139 J. Lappage, J. Bedford and D. Crook, **The WRONZ carpet pile thickness gauge**, J. Text. Inst. 75 (3) (1984) 229-234.

140 G. E. R. Lamb, S. Kepka and B. Miller, **Studies of appearance retention in carpets**, Text. Res. J. 60 (2) (1990) 103-107.

141 J. Southern, J.-P. Yu, W. Baggett and R. Miller, **Fundamental physics of carpet performance**, in: L. Rebenfeld (Ed.), Science and Technology of Fibers and Related Materials, Appl. Poly. Symp. 47 (1991) 355-371.

142 C. L. Warfield, **Upholstered furniture: results of a consumer wear study**, Text. Res. J. 57 (4) (1987) 192-199.

143 H. T. Patterson and T. H. Grindstaff, **Soil release by textile surfaces**, in: M. J. Schick (Ed.), Surface Characteristics of Fibers and Textiles. Pt. II., Marcel Dekker, New York, 1977, Chapter 12.

144 E. Kissa, **Soil-release finishes**, in: M. Lewin and S. B. Sello (Eds.), Handbook of Fiber Science and Technology. Vol. II. Chemical Processing of Fibers and Fabrics. Functional Finishes, Pt. B., Marcel Dekker, New York, 1984, Chapter 3.

145 B. M. Lichstein, **Stain and water repellency of textiles**, in: M. J. Schick (Ed.), Surface Characteristics of Fibers and Textiles. Pt. II., Marcel Dekker, New York, 1977, Chapter 13.

146 E. Kissa, **Repellent finishes**, in: M. Lewin and S. B. Sello (Eds.), Handbook of Fiber Science and Technology. Vol. II. Chemical Processing of Fibers and Fabrics. Functional Finishes, Pt. B., Marcel Dekker, New York, 1984, Chapter 2.

147 D. Jose, P. Hauck and P. Singh, **Advances in carpet-fibre technology**, in: L. Cegielka (Ed.), Carpets: Back to Front, The Textile Institute, Manchester, Text. Prog. 19 (3) (1988) 60-70.

148 G. E. R. Lamb, **A new carpet soiling test**, Text. Res. J. 62 (6) (1992) 325-328.

149 T. L. Vigo and A. F. Turbak (Eds.), **High-Tech Fibrous Materials: Composites, Biomedical Materials, Protective Clothing, and Geotextiles**, Am. Chem. Soc. Symp. Ser. 457, Am. Chem. Soc., Washington, D. C., 1991, 398 pp.

150 M. Raheel (Ed.), **Protective Clothing Systems and Materials**, Marcel Dekker, New York, 1994, 262 pp.

151 J. P. McBriarity and N. W. Henry (Eds.), **Performance of Protective Clothing**, ASTM STP 1133, ASTM, Philadelphia, PA., 1992, 1023 pp.

152 P. Bajaj and A. K. Sengupta, **Protective clothing**, Text. Prog. 22 (2/3/4) (1992) 1-117.

153 L. Fourt and N. R. S. Hollies, **Clothing: Comfort and Function**, Marcel Dekker, New York, 1970, 254 pp.

154 N. R. S. Hollies and R. F. Goldman, **Clothing Comfort: Interaction of Thermal, Ventilation, Construction and Assessment Factors**, Ann Arbor Science, Ann Arbor, Mich., 1977, 189 pp.

155 K. Slater, **Comfort properties of textiles**, The Textile Institute, Manchester, Text. Prog. 9 (4) (1977) 1-91.

156 T. L. Vigo and C. B. Hassenboehler, Jr., **Effective use of textiles for energy conservation**, in: T. L. Vigo and L. J. Nowacki (Eds.), Energy Conservation in Textile and Polymer Processing, ACS Symp. Series No. 107, American Chemical Society, Washington, D. C., 1979, Chapter 18.

157 P. O. Fanger, **Thermal Comfort: Analysis and Applications in Environmental Engineering**, McGraw-Hill, New York, 1970.

158 B. W. Olesen, **Thermal comfort**, Bruel and Kjaer Techn. Rev. 2 (1982) 3-41.

159 M. M. Sweeney and D. H. Branson, **Sensorial comfort. Pt. I. A psychophysical method for assessing moisture sensation in clothing**, Text. Res. J. 60 (7) (1990) 371-377.

160 M. M. Sweeney and D. H. Branson, **Sensorial comfort. Pt. II. A magnitude estimation approach for assessing moisture sensation**, Text Res. J. 60 (8) (1990) 447-452.

161 K. L. Hatch, S. S. Woo, R. L. Barker, P. Radhakrishnaiah, N. L. Markee and H. I. Maibach, *In vivo* cutaneous and perceived comfort response to fabric. Pt. I. **Thermophysiological comfort determinations for three experimental knit fabrics,** Text. Res. J. 60 (8) (1990) 405-412.

162 R. L. Barker, P. Radhakrishnaiah, S. S. Woo, K. L. Hatch, N. L. Markee and H. I. Maibach, *In vivo* cutaneous and perceived comfort response to fabric. Pt. II. **Mechanical and surface related comfort property determinations for three experimental knit fabrics,** Text. Res. J. 60 (8) (1990) 490-494.

163 K. L. Hatch, N. L. Markee, H. I. Maibach, R. L. Barker, S. S. Woo and P. Radhakrishnaiah, *In vivo* cutaneous and perceived comfort response to fabric. Pt. III. **Water content and blood flow in human skin under garments worn by exercising subjects in a hot and humid environment,** Text. Res. J. 60 (9) (1990) 510-519.

164 N. L. Markee, K. L. Hatch, H. I. Maibach, R. L. Barker, S. S. Woo and P. Radhakrishnaiah, *In vivo* cutaneous and perceived comfort response to fabric. Pt. IV. **Perceived sensations to three experimental garments worn by subjects exercising in a hot, humid environment,** Text. Res. J. 60 (10) (1990) 561-568.

165 G. J. Pontrelli, **Comfort by design,** Mell. Textilber. 70 (12) (1989) 906-910.

166 D. J. Bucheck, **Comfortable clothes through chemistry,** Chem. Tech. 21 (1991) 142-147.

167 K. Slater, **The assessment of comfort,** J. Text. Inst. 77 (3) (1986) 157-171.

168 G. H. Rong and K. Slater, **A new approach to the assessment of textile performance,** J. Text. Inst. 83 (2) (1992) 197-208.

169 Home and Garden Bulletin No. 62, **Removing Stains from Fabrics,** USDA, Agricultural Research Service, 1977, 26 pp.

170 J. Lloyd and C. Adams, **Domestic laundering of textiles,** Textiles 18 (3) (1989) 72-80.

171 R. Weber, P. Krings and J. Hoffmeiester, **Detergents for textiles of today and tomorrow,** Mell. Textilber. 71 (6) (1990) 471-479.

172 M. Scalco and B. Yu, **Care symbols,** Intl. Fabricare Inst. Bulletin 21 (1) (1992) 9-10.

AUTHOR INDEX

Pages listed for author (including editor) references that are cited in the body of the text are **boldface.** All other pages listed are for references at the end of chapters.

SUBJECT INDEX

472

Printed and bound by CPI Group (UK) Ltd, Croydon, CR0 4YY

03/10/2024

01040428-0012